42.50

ALUMINIUM IN BIOLOGY AND MEDICINE

The Ciba Foundation is an international scientific and educational charity. It was established in 1947 by the Swiss chemical and pharmaceutical company of CIBA Limited—now CIBA-GEIGY Limited. The Foundation operates independently in London under English trust law.

The Ciba Foundation exists to promote international cooperation in biological, medical and chemical research. It organizes about eight international multidisciplinary symposia each year on topics that seem ready for discussion by a small group of research workers. The papers and discussions are published in the Ciba Foundation symposium series. The Foundation also holds many shorter meetings (not published), organized by the Foundation itself or by outside scientific organizations. The staff always welcome suggestions for future meetings.

The Foundation's house at 41 Portland Place, London W1N 4BN, provides facilities for meetings of all kinds. Its Media Resource Service supplies information to journalists on all scientific and technological topics. The library, open five days a week to any graduate in science or medicine, also provides information on scientific meetings throughout the world and answers general enquiries on biomedical and chemical subjects. Scientists from any part of the world may stay in the house during working visits to London.

ALUMINIUM IN BIOLOGY AND MEDICINE

A Wiley–Interscience Publication

1992

JOHN WILEY & SONS

Chichester · New York · Brisbane · Toronto · Singapore

Published in 1992 by John Wiley & Sons Ltd
Baffins Lane, Chichester
West Sussex PO19 1UD, England

Other Wiley Editorial Offices

John Wiley & Sons, Inc., 605 Third Avenue,
New York, NY 10158-0012, USA

Jacaranda Wiley Ltd, G.P.O. Box 859, Brisbane,
Queensland 4001, Australia

John Wiley & Sons (Canada) Ltd, 22 Worcester Road,
Rexdale, Ontario M9W 1L1, Canada

John Wiley & Sons (SEA) Pte Ltd, 37 Jalan Pemimpin #05-04,
Block B, Union Industrial Building, Singapore 2057

Suggested series entry for library catalogues:
Ciba Foundation Symposia

Ciba Foundation Symposium 169
x + 316 pages, 56 figures, 22 tables

Library of Congress Cataloging-in-Publication Data
Aluminium in biology and medicine.
      p. cm.—(Ciba Foundation symposium; 169)
    Editors, Derek J. Chadwick (organizer) and Julie Whelan.
    "A Wiley–Interscience publication."
    Includes bibliographical references and indexes.
    ISBN 0-471-93413-5
    1. Aluminum—Physiological effect—Congresses. 2. Aluminum—
Pathophysiology—Congresses. 3. Alzheimer's disease—
Pathophysiology—Congresses. I. Chadwick, Derek. II. Whelan,
Julie. III. Series.
    [DNLM: 1. Aluminum—congresses. W3 C161F v. 169]
QP535.A4A485 1992
612'.01524—dc20
DNLM/DLC
for Library of Congress   92-14002
                                     CIP

British Library Cataloguing in Publication Data
A catalogue record for this book is
available from the British Library

ISBN 0 471 93413 5

Phototypeset by Dobbie Typesetting Limited, Tavistock, Devon.
Printed and bound in Great Britain by Biddles Ltd, Guildford.

Contents

Participants

J. D. Birchall ICI Advanced Materials, PO Box 11, The Heath, Runcorn, Cheshire WA7 4QE, UK

J. A. Blair Pro-Vice Chancellor's Office, Aston University, Aston Triangle, Birmingham B4 7ET, UK

J. M. Candy MRC Neurochemical Pathology Unit, Newcastle General Hospital, Westgate Road, Newcastle upon Tyne NE4 6BE, UK

J. P. Day Department of Chemistry, University of Manchester, Oxford Road, Manchester M13 9PL, UK

J. A. Edwardson MRC Neurochemical Pathology Unit, Newcastle General Hospital, Westgate Road, Newcastle upon Tyne NE4 6BE, UK

G. Farrar Pharmaceutical Sciences Institute, Aston Triangle, Birmingham B4 7ET, UK

J. K. Fawell Water Research Centre, Henley Road, Medmenham, PO Box 16, Marlow, Bucks SL7 2HD, UK

T. P. Flaten* National Institute of Neurological Disorders & Stroke, NCI-FCRDC Building 376, PO Box B, Frederick, MD 21702-1201, USA

R. M. Garruto Laboratory of Central Nervous System Studies, National Institute of Neurological Disorders and Stroke, National Institutes of Health, Building 36, Room 5B-21, Bethesda, MD 20892, USA

J. L. Greger Department of Nutritional Sciences, University of Wisconsin, 1415 Linden Drive, Madison WI 53706, USA

P. M. Harrison Department of Molecular Biology & Biotechnology, University of Sheffield, Sheffield S10 2TN, UK

Present address: Wessels Gate 17, N-7043 Trondheim, Norway.

G. V. W. Johnson   Department of Behavioral Neurobiology, Spark
Center Rm 1011, University of Alabama at Birmingham, UAB Station,
Birmingham, AL 35294-0017, USA

R. S. Jope   Department of Behavioral Neurobiology, University of
Alabama at Birmingham, School of Medicine, UAB Station,
Birmingham, AL 35294-0017, USA

D. N. S. Kerr   Royal Postgraduate Medical School, Hammersmith
Hospital, Du Cane Road, London W12 0NN, UK

J. Klinowski   Department of Chemistry, University of Cambridge,
Lensfield Road, Cambridge CB2 1EW, UK

D. R. McLachlan   Centre for Research in Neurodegenerative Diseases,
University of Toronto, Tanz Neuroscience Building, 6 Queen's Park
Crescent W, Toronto, Ontario, Canada M5S 1A8

R. B. Martin   Department of Chemistry, University of Virginia,
McCormick Road, Charlottesville, VA 22903, USA

C. N. Martyn   MRC Environmental Epidemiology Unit, University of
Southampton, Southampton General Hospital, Southampton SO9 4XY,
UK

D. P. Perl   Department of Pathology, Division of Neuropathology,
Mount Sinai School of Medicine, One Gustave L. Levy Place,
New York, NY 10029, USA

O. H. Petersen   The Physiological Laboratory, University of Liverpool,
Brownlow Hill, PO Box 147, Liverpool L69 3BX, UK

Ch. Schlatter   Institute of Toxicology, ETH and University of Zürich,
CH-8603 Schwerzenbach, Switzerland

M. J. Strong   Department of Clinical Neurological Sciences, University
Hospital, Room 10 LY1, 339 Windermere Road, PO Box 5339, London,
Ontario, Canada N6A 5A5

G. B. van der Voet   Toxicology Laboratory, University Hospital, PO Box
9600, NL-2300 RC Leiden, The Netherlands

M. K. Ward Department of Medicine, Royal Victoria Infirmary, Queen Victoria Road, Newcastle upon Tyne NE1 4LP, UK

R. J. P. Williams Inorganic Chemistry Laboratory, University of Oxford, South Parks Road, Oxford OX1 3QR, UK

C. M. Wischik MRC Laboratory of Molecular Biology, Hills Road, Cambridge CB2 2QH, UK

H. M. Wisniewski Institute for Basic Research in Developmental Disabilities, 1050 Forest Hill Road, Staten Island, NY 10314-6399, USA

P. F. Zatta Department of Biology, University of Padova, Via Trieste 75, I-3513 Padova, Italy

Aluminium in biology: an introduction

R. J. P. Williams

Inorganic Chemistry Laboratory, University of Oxford, South Parks Road, Oxford OX1 3QR, UK

The easiest way to an understanding of the roles played by an element in the chemistry of living systems is to start from the periodic table:

IA	IIA	IIIA	IVA	V	Transition metals	IIIB
Li	Be	B	C	......		
Na	Mg	**Al**	Si	P	....	
K	Ca	*M*	....		Cr Mn Fe .... Ga	

where M stands for a great number of elements including scandium (Sc), yttrium (Y) and the fourteen lanthanides (Da Silva & Williams 1991). Of this group of elements, Li, Be and B are rare elements in the cosmos, because of the mode of nuclear synthesis and the stability of the nucleus of carbon. Even so, boron is essential for plant life. All the other elements shown above are more abundant and, with the exception of Al, M, Cr and gallium (Ga), are used extensively by biology and are in fact essential for all life. In chemical terms this means that biological systems have created a strong artificial divide between cationic elements, metals, and anionic, non-metal elements by ignoring the elements which exist only as M^{3+} ions in water (Fig. 1). Iron and manganese escape exclusion through the switch to Fe^{2+} and Mn^{2+}. A similar restriction of acceptability applies in Group IV to elements such as germanium (Ge) and titanium (Ti), which widens the gap between biologically useful metals and useful non-metals. What is wrong with M^{3+}, and especially Al^{3+}, as far as biology is concerned?

Many points will be highlighted in the following chapters. Some of them will be introduced here so that the flow of the symposium will be apparent.

The first point to make is that aluminium is not easily available, and this is generally true of all M^{3+} ions. Hydrolysis at $pH = 7$ ensures that while virtually all Groups IA, IIA and divalent ions of transition metals from Mn to Zn are available and soluble, and the elements from Group V onwards are hydrolysed to anions and so become soluble and available (Fig. 1), the elements

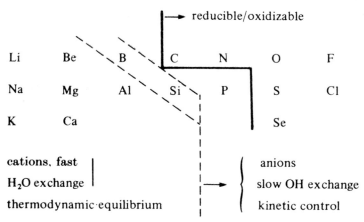

FIG. 1. Some important properties which divide the lighter elements and are used in biology. Na^+, K^+, Mg^{2+} and Cl^- are simple ions used as fast diffusing messengers or in structures. Much silicon and some boron is used structurally. C, N, O and S, with a little selenium, are used in many forms as central structural organic elements. Phosphorus as phosphate is also extensively employed. This leaves on one side the rarely used and relatively rare Li^+, Be^{2+} and F^-, together with the common element, aluminium.

from Groups III and IV are lost as oxide precipitates. In addition, iron, as Fe^{3+}, is heavily hydrolysed and is not readily available, so that biology has had to devise siderophores and proteins (transferrins) to obtain it. Now, the fact that biology could learn to handle the uptake of Fe^{3+} means that it has always had the capability to devise means of obtaining Al^{3+} and M^{3+} generally. However, it did not do so and does not do so. Why?

The parallel chemistries of Fe^{3+} and Al^{3+} (M^{3+}) raise other problems, because Al^{3+} could perhaps utilize Fe^{3+} systems of uptake and transport. Is this a problem? Similarly, methods for removing Al^{3+} in biology, if they exist, must run into the problem of protecting Fe^{3+}, unless there are selectivity factors. Notice, for example, that most of the insoluble minerals of the earth undergo easy isomorphous substitution of Al and Fe—for example, in asbestos. We have to look for subtle chemical differences to see where iron and aluminium differ in order to understand how nature has avoided aluminium and can manage to do this in the presence of iron.

A suggestion may not be out of order. Al^{3+} is exceedingly small, with a radius of 0.54 Å, while Fe^{3+} has a radius of 0.65 Å. There are always likely to be heavy steric constraints around Al^{3+}. However, we also know that Fe^{3+} has a preference for nitrogen-donor ligands which exceeds that of Al^{3+}. Taking these two factors together, we should not be too surprised that the difference between the hydrolysis constants giving hydroxides and the binding constants for ligands such as $EDTA^{4-}$ favours Fe^{3+} binding to $EDTA^{4-}$, while

TABLE 1   Some thermodynamic binding data for aluminium and closely related elements

Property	Al^{3+}	Ga^{3+}	Fe^{3+}
Solubility product (OH^-)	32.0	38.0	38.8
log K ($EDTA^{4-}$)	16.3	21.7	25.8
log K (F^-)	6.4	5.5	5.3
log K (OH^-)	9.0	11.1	11.4

For more extensive data on Al^{3+} (and on Ca^{2+}) see Tam & Williams (1985).

aluminium is more likely to be found as the hydroxide (Table 1). Even Ga^{3+}, which has a somewhat similar affinity for $EDTA^{4-}$, is weaker in its binding than Fe^{3+}, although Ga^{3+} is intermediate between Fe^{3+} and Al^{3+} in size. We find that transferrin binds to Ga^{3+} some 100-fold less tightly than does Fe^{3+}. We may wonder if Ga^{3+} is a better marker for Fe^{3+} or for Al^{3+}.

The above considerations concern thermodynamic binding properties, but we should also consider rates of reaction of M^{3+}. It has been known for long enough that the on/off reactions of M^{3+} ions, even removal of water from them, are slow. Their reactions are very sticky so that (unlike M^+ and M^{2+} ions) they are of little value in message transmission which demands exchange (compare K^+, Na^+ and Ca^{2+} and the lesser value of the slower Mg^{2+}). Even in acid/base catalysis (compare the relatively fast acting Mg^{2+}, Mn^{2+} and Zn^{2+} but the lesser value of the slower Ni^{2+}) and gene control (as for example by the faster Fe^{2+}, Mn^{2+}, Zn^{2+}) the trivalent ions may be unsuitable because of slowness of exchange. All such factors influence the way in which evolution 'assesses' the value of an element. In a comparative sense, does Al^{3+} have no positive advantage and therefore can it only inhibit useful functions of other elements?

Only when we have this chemistry in mind can we turn to the potentially hazardous influence of M^{3+} and especially Al^{3+} in biology—once it is admitted. Here, we divide the effects of Al^{3+} in the environment (problems generated by acid rain) from those in biological fluids. Clearly, in acidic water Al^{3+} is damaging to aquatic life, especially fish gills and probably plant root systems. We must ask why this is so. We may suspect that Al^{3+} is displacing Ca^{2+} and Mg^{2+} at external sites, such as the root cell wall and the extracellular structures of fish gills (Tam & Williams 1986). Does the same logic apply in the digestive system? We know that Al^{3+} salts are used to remove excess phosphate by precipitation in patients on dialysis, but this leads to high Al^{3+} concentrations elsewhere and consequent hazards.

When we come to the inside of the cell, we approach a new milieu. There is virtually no calcium, but much Mg^{2+} and a variety of phosphate compounds. Al^{3+} (M^{3+}) has a high affinity for phosphate. This consideration (and a similar one arises through the affinity of Al^{3+} for fluoride) leads to obvious thoughts about the possible problems of Al^{3+} poisoning, but (and there are many buts) how could Al^{3+} ever come to be in the cell?

One of the possible modes is through the carriers of iron. In bacteria these are siderophores, in plants maybe citrate, and in animals, transferrins. We shall need to know if Al^{3+} binding to these molecules is strong enough and if the shapes of its complexes are sufficiently similar to those of Fe^{3+} for it to be carried into the cell by these carriers. Of course, this is not enough. It must also be released from the internalized carrier. Assuming that a step in the uptake of Fe^{3+} by these carriers is reduction to Fe^{2+}, this path is not open to Al^{3+}, but what if the step is just acidification?

There are, then, many chemical and biochemical issues before we come to the problems which could be related to Alzheimer's disease. It is here we must be more cautious, because while aluminium may be a problem, it is clear that there may well be genetic factors and (resultant) proteins or glycoproteins generated through 'mistaken' syntheses. These aspects will be discussed in several chapters and they introduce new points for consideration. How good are the analytical methods for measuring aluminium levels inside cells? How well characterized is aluminium's association with target proteins, and with early signs of Alzheimer's disease? To what degree is Alzheimer's disease inherited?

Let us go so far as to say that maybe aluminium is not yet known to be the *cause* of any known disease of those who live on a normal diet (Sherrard 1991). This must not stop us asking whether it exacerbates any diseases. At the same time, it must not hide from us the fact that Al^{3+} is often found associated with plaques outside cells within damaged nervous tissue of brain. We know, in fact, that this association is diagnostic in many cases of the diseased state in post mortem analysis. Why has Al^{3+} become selectively bound in plaques? This must tell us the nature of the specific lesion. Other trivalent ions are purposefully introduced to be used in diagnostics—Ga^{3+}, Gd^{3+}, In^{3+} and so on. What do we understand about their relative chemistries? Will a combination of the use of high valent ions lead us to a set of diagnostic tools for disclosing what Al^{3+} is indicating to us?

I close leaving the problems as they were. Let us see what light can be shed upon them in the next few days.

References

Da Silva JJRF, Williams RJP 1991 The biological chemistry of the elements. Oxford University Press, Oxford

Sherrard DJ 1991 Aluminum—much ado about something. N Engl J Med 324:558–559

Tam SC, Williams RJP 1985 Electrostatics and biological systems. Structure and Bonding 63:103–151

Tam SC, Williams RJP 1986 One problem of acid rain: aluminium. J Inorg Biochem 26:35–44

Aluminium speciation in biology

R. Bruce Martin

Chemistry Department, University of Virginia, Charlottesville, VA 22903, USA

Abstract. Before we can understand the role of Al^{3+} in living organisms we need to learn how it interacts with molecules found in biological systems. The only aluminium oxidation state in biology is $3+$. In aqueous solutions there are only two main Al(III) species: the hexahydrate Al^{3+} at pH < 5.5 and the tetrahedral aluminate at pH > 6.2. In the blood plasma, citrate is the main small molecule carrier and transferrin the main protein carrier of Al^{3+}. In fluids where the concentrations of these two ligands are low, nucleoside di- and triphosphates become Al^{3+} binders. Under these conditions Al^{3+} easily displaces Mg^{2+} from nucleotides. When all three classes of ligands are at low concentrations, catecholamines become likely Al^{3+} binders. Double-helical DNA binds Al^{3+} weakly and under no conditions should it compete with other ligands. Al(III) in the cell nucleus probably binds to nucleotides or phosphorylated proteins. Al^{3+} undergoes ligand exchange much more slowly than most metal ions: 10^5 times slower than Mg^{2+}.

1992 Aluminium in biology and medicine. Wiley, Chichester (Ciba Foundation Symposium 169) p 5–25

To understand the roles of an element in an organism we need to know not only the gross amount present, but also the locale and species—valence state and complexes or compounds—into which the element enters. In the case of Al^{3+}, strong complexing by citrate or transferrin prevents Al^{3+} from acting as a surrogate for Mg^{2+} in reactions with nucleotide. The chemistry of aluminium is relatively simple. It reacts 10^7 times faster than Cr^{3+}, its hydroxide is much more soluble than that of Fe^{3+}, and it exhibits only one oxidation state in biological systems, Al^{3+}. Metallic aluminium is too reactive to be found free in Nature, and the metal is won from its ores only with difficulty. Thus there is no oxidation-reduction chemistry to Al^{3+} in biology.

We begin our evaluation of Al^{3+} by considering its ionic radius. In both mineralogy and biology, comparable ionic radii are frequently more important than charge in determining behaviour. The effective ionic radius of Al^{3+} in six-fold coordination is 54 pm. By way of comparison, other values are Ga^{3+}, 62; Fe^{3+}, 65; Mg^{2+}, 72; Zn^{2+}, 74; Fe^{2+}, 78; and Ca^{2+}, 100 pm (Martin 1986a). On the basis of the radii, though quite small, Al^{3+} is closest in size to Fe^{3+} and Mg^{2+}, and it is to these ions that we compare Al^{3+}. Ca^{2+} is much larger, and in its favoured eight-fold coordination exhibits a radius of 112 pm, yielding a volume nine times greater than Al^{3+}. For this and other reasons we have argued that in biological systems Al^{3+} will be competitive with Mg^{2+}, rather than Ca^{2+} (Macdonald & Martin 1988, Martin 1988a). Both Al^{3+} and Mg^{2+} favour oxygen donor ligands, especially phosphate groups (Martin 1990b). Al^{3+} is 10^7 times more effective than Mg^{2+} in promoting the polymerization of tubulin to microtubules (Macdonald et al 1987). *Wherever there is a process involving Mg^{2+}, seek there an opportunity for interference by Al^{3+}.*

Al^{3+} hydrolysis

Whatever ligands may be present, understanding the state of Al(III) in any aqueous system demands awareness of the species that Al(III) forms with the components of water at different pH values. (We use Al(III) as a generic term for the 3 + ion when a specific form is not indicated.) In solutions more acid than pH 5, Al(III) exists as an octahedral hexahydrate, $Al(H_2O)_6^{3+}$, usually abbreviated as Al^{3+}. As a solution becomes less acidic, $Al(H_2O)_6^{3+}$ undergoes successive deprotonations to yield $Al(OH)^{2+}$, $Al(OH)_2^{+}$ and soluble $Al(OH)_3$, with a decreasing and variable number of water molecules (Martin 1988a, 1991b). Neutral solutions give an $Al(OH)_3$ precipitate that redissolves, owing to the formation of tetrahedral aluminate, $Al(OH)_4^{-}$, the primary soluble Al(III) species at pH > 6.2.

The four successive deprotonations from $Al(H_2O)_6^{3+}$ to yield $Al(OH)_4^{-}$ squeeze into an unusually narrow pH range of less than one log unit with pK_a values of 5.5, 5.8, 6.0 and 6.2 (Ohman 1988). In contrast, the corresponding four normal deprotonations from $Fe(H_2O)_6^{3+}$ span 6.6 log units with pK_a values of 2.7, 3.8, 6.6 and 9.3. The narrow span for Al^{3+} is explained by the cooperative nature of the successive deprotonations resulting from a concomitant decrease in coordination number (Martin 1991b).

The upper half of Fig. 1 shows the distribution of free metal ion and mononuclear hydrolysed species based on the four successive pK_a values given above. Thus only two species dominate over the entire pH range, the octahedral hexahydrate $Al(H_2O)_6^{3+}$ at pH < 5.5, and the tetrahedral $Al(OH)_4^{-}$ at pH > 6.2, while there is a mixture of hydrolysed species and coordination numbers between 5.5 < pH < 6.2 (Martin 1988a, 1991b). These equilibria must be considered in all solutions containing Al(III). If in addition other ligands are incapable of holding Al(III) in solution, it becomes necessary to include the solubility equilibrium.

The lower half of Fig. 1 applies to solutions saturated with amorphous $Al(OH)_3$. The dashed straight line of slope 3 in the lower half of the figure gives the molar concentration of the free metal ion, $[Al^{3+}]$. The solid curve represents the total concentration of the free metal ion and all mononuclear hydrolysed forms, T_{Al}, with the distribution shown in the upper half of the figure. From the lower part of Fig. 1 we learn that large amounts of $Al(OH)_3$ dissolve in acidic stomachs.

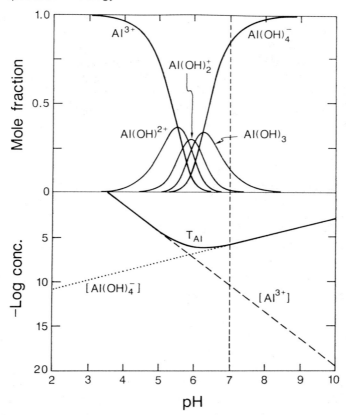

FIG. 1. Al^{3+} hydrolysis. *Upper half*: Mole fraction of soluble species as a function of pH. *Lower half*: For saturated solutions of amorphous $Al(OH)_3$, the negative logarithm of molar concentration of the free ion, $[Al^{3+}]$, is shown as a straight dashed line; and the sum over all species present, T_{Al}, as a curved solid line. The straight dotted line represents the concentration of tetrahedral aluminate ion, $[Al(OH)_4^-]$. (From Martin 1991b by permission of the publisher. © 1991 Elsevier Science Publishing Co. Inc.)

Figure 1 also shows as a dotted line the molar concentration of aluminate, $[Al(OH)_4^-]$, the main species at pH > 6.2. At any pH the ordinate distance between the straight lines for $[Al(OH)_4^-]$ and $[Al^{3+}]$ gives the logarithm of their concentration ratio. Thus at pH 7.0 the molar ratio of $[Al(OH)_4^-]/[Al^{3+}]$ is given by $10^{4.5} = 3 \times 10^4$. Therefore, $Al(OH)_4^-$ (aluminate) should be the starting point for thinking about aluminium in biological systems.

What happens when an $AlCl_3$ solution is administered at a local concentration of 0.01 M Al(III) to a tissue at pH 7? Ascent of the dashed pH 7 line in Fig. 1 indicates that the permissible free Al^{3+} is only $10^{-10.3}$ M and that of all soluble forms is $10^{-5.7}\,M = 2\,\mu M$. Unless the remainder of the added Al(III) has been complexed by other ligands, it will form insoluble $Al(OH)_3$. From the upper part of Fig. 1 we see that of the soluble forms, 85% is $Al(OH)_4^-$ and 14% $Al(OH)_3$. Of the Al(III) administered at pH 7, only $2\,\mu M$ appears in soluble forms, most of which is $Al(OH)_4^-$, since most of the added Al(III) precipitates or coordinates to nearby ligands. To keep Al(III) in solution, we can administer it as a complex.

pAl $= -\log\ [Al^{3+}]$

In addition to the hydrolysis features already considered, the amount of free, aqueous Al^{3+} in solution depends upon several variables: ligands present, their stability constants with Al^{3+}, and the mole ratio of total Al(III) to total ligand. For ligands with protons competing with metal ion for binding sites in the pH range of interest, the pH is also a variable. Thus, instead of simple association of metal ion and basic ligand, $Al^{3+} + L \rightarrow AlL$, the relevant reaction may become displacement of a proton from the acidic ligand by the metal ion, $Al^{3+} + HL \rightarrow AlL + H^+$. For ligands containing amino, phenolate and catecholate groups, the amount of free, aqueous Al^{3+} in neutral solutions becomes pH dependent. Thus, for these ligands, the listed stability constants overstate effective binding strengths and need to be lowered to reflect competition of the proton with the metal ion for basic binding sites. The most practical method to allow for proton–metal ion competition at a ligand is to calculate *conditional* stability constants applicable to a single pH (Martin 1986c, 1988a, 1991a). Conditional stability constants may also allow for the deprotonation of metal ion-coordinated water that yields more stable complexes with increasing pH in some Al^{3+} complexes, such as with citrate, nitrilotriacetate and EDTA.

Results from the quantitative evaluation of conditional stability constants are revealingly expressed as the negative logarithm of the free Al^{3+} concentration, $-\log\ [Al^{3+}] = pAl$. By analogy with pH, higher pAl values represent smaller amounts of free Al^{3+}. Table 1 and Figs. 2 and 3 show the conclusions. Table 1 lists conditional stability constants and pAl values at pH 6.6 and 7.4 for several systems with $1\,\mu M$ total Al(III) under the conditions indicated in the table (Martin 1991a). Weak Al^{3+} binders appear at the top and strong binders at the bottom of Table 1. The increasing pAl values as one goes down the table indicate decreasing free Al^{3+} concentrations. Thus, since 0.1 mM citrate lies lower in Table 1 then does 1 mM ATP^{4-}, we predict that citrate will withdraw Al^{3+} from ATP^{4-}, and experimentally citrate has been used for this purpose (Womack & Colowick 1979). Despite high normal stability constants as indicated in the second column, salicylate (Ohman & Sjoberg 1983) and catecholamines (Kiss et al 1989) bind Al^{3+} relatively weakly because competition from the proton in the strongly basic ligands leads to the low conditional stability

TABLE 1 Negative logarithm of free Al^{3+} concentration (pAl) [a]

		pH 6.6		pH 7.4	
Complex or ligand[b]	Log K_s	Log $K_{6.6}$	pAl	Log $K_{7.4}$	pAl
DNA	<5.6	<5.6	<7.3	<5.6	<7.3
Salicylate, 0.2 mM	12.9, 10.6	6.3, 4.0	9.1	7.1, 4.8	10.7
Amorphous Al(OH)$_3$	Insoluble		9.1		11.5
Al^{3+} to Al(OH)$_4{}^-$			9.1		12.1
Catecholamines	15.6, 13.0	7.4, 4.8	9.7	9.0, 6.4	12.8
Kaolinite[e]	Insoluble		10.2		12.6
AlPO$_4$	Insoluble		11.7[c]		12.1[d]
(HO)$_2$AlO$_3$POH$^-$	See text		11.7[c]		12.9[d]
Nitrilotriacetate (NTA)	11.1	10.0	11.7	11.6	13.3
2,3-DPG, 3 mM	12.5	11.6	12.2	12.2	13.1
ATP, 1 mM	7.9, 4.6	8.9	12.3	9.8	13.0
Citrate, 0.1 mM	8.1	11.3	13.3	12.7	14.7
Transferrin				13.6, 12.8	15.3
F$^-$, 5 mM, with OH$^-$			14.9		15.1
EDTA	16.2	13.1	14.8	14.7	16.4
Desferrioxamine	24.1	16.8	18.4	19.2	20.8

[a] 1 µM total Al(III) except for insoluble salts. Equilibria in addition to those related to listed log K$_c$ values often required to calculate pAl.
[b] 50 µM ligand unless otherwise noted.
[c] 10 mM total phosphate.
[d] 2 mM total phosphate.
[e] Al$_2$(OH)$_4$Si$_2$O$_5$ with 5 µM Si(OH)$_4$, typical of plasma.

constants listed in the third and fifth columns. Kaolinite is the least soluble aluminium silicate (Martin 1990a).

Figures 2 and 3 show for several ligands pAl as a function of pH. A 1 µM total Al(III) is assumed, as indicated by the left pAl = 6 intercept of the curves on the ordinate at low pH, where in all cases the aluminium appears unbound as Al^{3+}. As the pH increases, strong differentiation in binding strength occurs, with maltol the weakest and desferrioxamine the strongest Al^{3+} binder. The rectangle at pH 7.4 labelled Tf represents a range of reported values for transferrin under blood plasma conditions (Harris & Sheldon 1990, Martin et al 1987).

The dashed curve labelled Al(OH)$_4{}^-$ near pH 9 in Figs. 2 and 3 represents the pAl of 1 µM total Al(III) allowed by the hydrolysis depicted in the upper

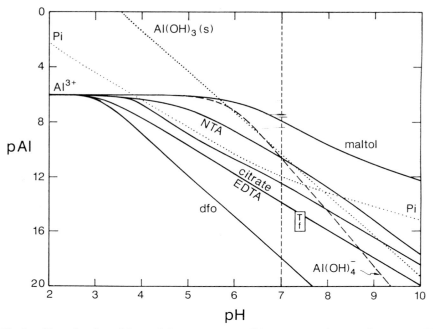

FIG. 2. Negative logarithm of free aqueous Al^{3+} concentration (pAl) versus pH allowed by several introduced Al(III) complexes and solids. The dotted straight line of slope 3 represents amorphous $Al(OH)_3(s)$ and the dotted curve, 4.5 mM inorganic phosphate. Dashed and all solid curves refer to 1 μM total Al(III), beginning as Al^{3+} at pAl = 6 at low pH on the left-hand ordinate. The dashed curve shows the decrease of free Al^{3+} due to metal ion hydrolysis leading to $Al(OH)_4^-$ at bottom right. Solid curves (with mole ratios) represent maltol (3:1), nitrilotriacetate (NTA) (2:1), citrate (4:1), EDTA (2:1) and desferrioxamine, dfo (2:1). The rectangle labelled Tf refers to 50 μM transferrin under blood plasma conditions. (From Martin 1991a by permission of the publisher.)

half of Fig. 1. The left-hand ordinate intercept of the dashed curve is also at pAl = 6 where the predominant species is Al^{3+}. The dashed curve and the dotted line labelled $Al(OH)_3(s)$ apply to all aqueous solutions containing Al(III). The dotted line of slope 3 represents the pAl allowed by the solubility of an amorphous $Al(OH)_3$.

Phosphate binding

In the human body, extracellular fluids contain about 2 mM total phosphate at pH 7.4 and intracellular fluids about 10 mM total phosphate at pH 6.6. Al^{3+} forms an insoluble salt with phosphate, often designated as $AlPO_4$ or sometimes as $AlPO_4 \cdot 2H_2O$, corresponding to the composition of the mineral

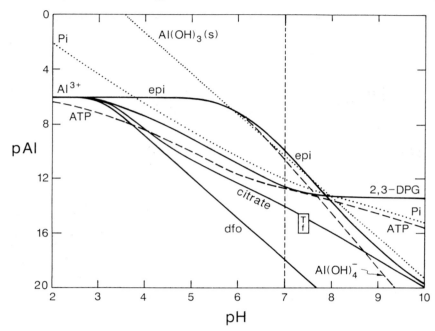

FIG. 3. Negative logarithm of free aqueous Al^{3+} concentration (pAl) versus pH allowed by several natural Al(III) complexes and solids. The dotted straight line of slope 3 represents amorphous $Al(OH)_3(s)$, and the dotted curve, 4.5 mM inorganic phosphate. Dashed and all solid curves refer to 1 μM total Al(III) beginning as Al^{3+} at pAl = 6 at low pH on the left-hand ordinate. The dashed curve shows the decrease of free Al^{3+} due to metal ion hydrolysis leading to $Al(OH)_4^-$ at bottom right. Solid curves (with mole ratios) represent adrenaline (epinephrine), epi (10:1), 2,3-diphosphoglycerate, DPG (3000:1), ATP (dashed curve, 1000:1), citrate (100:1) and desferrioxamine, dfo (2:1). The rectangle labelled Tf refers to 50 μM transferrin under blood plasma conditions.

variscite. For intracellular fluids at pH 6.6 containing 10 mM total phosphate we find $-\log [Al^{3+}] = pAl = 11.7$, while for extracellular fluids at pH 7.4 containing 2 mM total phosphate, pAl = 12.1 (Martin 1991a). This pair of values appears in the seventh row of Table 1 and represents extremely low maximum free Al^{3+} concentrations.

The dotted curve labelled P_i in Figs. 2 and 3 shows the pAl permitted by the solubility of $AlPO_4$ in the presence of an inorganic phosphate concentration of 4.5 mM, the geometric mean of extra- and intracellular concentrations. The curve therefore lies within 0.35 log units of the 2 and 10 mM concentrations for extra- and intracellular phosphate. Near pH 8.3 the free Al^{3+} permitted by 4.5 mM inorganic phosphate exceeds that allowed by the solubility of an amorphous $Al(OH)_3$. The two dotted curves in Figs. 2 and 3 are not limited by the 1 μM total Al(III) concentration assumed for all the other curves.

TABLE 2 Al^{3+} binding to phosphate ligands

Ligand	pK_a[a]	Log K_1	Log K_2
ATP	6.4	7.9	4.6
ADP	6.3	7.8	4.3
AMP	6.1	6.2	4.2
2,3-DPG	6.1, 7.2	6.0 ($AlLH^+$)	4.2 ($AlL_2H_2^-$)[b]
$HOPO_3^{2-}$	6.77	6.3^c	4.4^c

[a]Phosphate group values adjusted to usual activity pH scale.
[b]2,3-DPG (2,3-diphosphoglycerate) 2:1 complex loses protons with $pK_a = 5.7$ and 6.4, so the main species in neutral solutions is AlL_2^{3-}.
[c]Estimated.

Al^{3+} will frequently form soluble complexes with phosphate groups in biological systems. We compare the binding strengths of various phosphate-containing ligands. Stability constant logarithms recently determined for Al^{3+} binding to adenosine 5′-nucleotides (Kiss et al 1991a) and 2,3-diphosphoglycerate (DPG) (Sovago et al 1990) appear in Table 2. The strongest stability constants occur for ADP and ATP, where chelation occurs. At about 3 mM there is three times as much DPG as ATP in the red blood cell but, since ATP binds Al^{3+} 80 times more strongly, most Al^{3+} binds to ATP rather than to DPG. For comparison, the stability constant for Mg^{2+} binding to ATP and other nucleoside triphosphates is log $K_1 = 4.3$ (Sigel et al 1987), 4000 times weaker than for Al^{3+}. Thus, 0.2 µM Al^{3+} competes with 1 mM Mg^{2+} for ATP.

Because of the insolubility of $AlPO_4$, the soluble monophosphate complexes are difficult to study. We have used the pK_a and log K_1 results in Table 2 to predict log $K_1 = 6.3$ for Al^{3+} binding to inorganic phosphate, $HOPO_3^{2-}$, in the last row of Table 2. This complex dominates in acidic solutions, mainly as AlO_3POH^+. It loses a proton to give either AlO_3PO ($AlPO_4$), or more likely an alternative microform, $HOAlO_3POH$. The latter in turn may lose a proton to yield $HOAlO_3PO^-$ or the more likely microform $(HO)_2AlO_3POH^-$. Water molecules occupy the other coordination positions. The extent to which cooperativity (Martin 1991b) might favour Al^{3+}-bound OH^- in this system is unknown. The net negatively charged microforms probably predominate in neutral solutions. By assigning $pK_a = 5.5$ and 6.0 to the successive deprotonations, we obtain the pAl values listed in the eighth row of Table 1. At pH 6.6 the pAl value compares with that allowed by insoluble $AlPO_4$. The estimates suggest that though inorganic phosphate is an effective Al^{3+} binder, its middle position in Table 1 indicates that it is less effective than other ligands found in a cell and the plasma.

Many suppose that Al^{3+} binds to DNA in the cell nucleus. However, Al^{3+} binding to DNA is so weak that a quantitative study was limited to a high pH = 5.5, because of metal ion hydrolysis and precipitation (Dyrssen et al 1987).

Therefore, DNA cannot compete with ATP, $HOPO_3^{2-}$ and other ligands in Table 1 for Al^{3+}. With very weakly basic phosphates of $pK_a \approx 1$, DNA serves merely as a polyelectrolyte interacting with Al^{3+} weakly and non-specifically. Residing at the very top of Table 1 with the lowest pAl values or highest allowed free Al^{3+} concentrations, DNA loses Al^{3+} to all other entries in the table. We deduce that Al^{3+} binding to double-helical DNA is so weak under intracellular conditions that it fails by several orders of magnitude to compete with either metal ion hydrolysis or insolubility of even an amorphous $Al(OH)_3$. No matter what other ligands are present, this competition with aqueous solvent components remains. *Therefore, we conclude that the observation of aluminium with nuclear chromatin is due to its coordination not to DNA but to other ligands.*

What ligands might bind Al^{3+} in the cell, especially in the nuclear chromatin region? ATP and ADP are comparably strong Al^{3+} binders (Table 2). A crucial Al^{3+} binding site in chromatin promises to be phosphorylated proteins, perhaps phosphorylated histones. Phosphorylation and dephosphorylation reactions normally accompany cellular processes. The phosphate groups of any phosphorylated protein provide the requisite basicity, and, in conjunction with juxtaposed carboxylate or other phosphate groups, become strong Al^{3+}-binding sites. Abnormally phosphorylated proteins have been found in brains of patients with Alzheimer's disease (Grundke-Iqbal et al 1986, Sternberger et al 1985). Al^{3+} aggregates highly phosphorylated brain cytoskeletal proteins (Diaz-Nido & Avila 1990). Ternary Al^{3+} complexes have received little study, and Al^{3+} has been used as a tanning or cross-linking reagent. Very possibly, Al^{3+} cross-links proteins, and proteins and nucleic acids in the nucleus.

Ligands for Al^{3+}

The most likely Al^{3+} binding sites are oxygen atoms, especially if they are negatively charged. Carboxylate, catecholate and phosphate groups are the strongest Al^{3+} binders. Even when part of a potential chelate ring, sulphydryl groups do not bind Al^{3+} (Toth et al 1984). Amines do not bind Al^{3+} strongly except as part of multidentate ligand systems such as nitrilotriacetate (NTA) and EDTA. The nitrogenous bases of DNA and RNA do not bind Al^{3+} (Martin 1986b, 1988a). Fluoride binds Al^{3+} avidly, and, at the 1 p.p.m. level at which fluoride is added to acidic drinking water, most Al(III) appears as AlF_2^+ and neutral AlF_3 (Martin 1988b). In mixed complexes of ADP and F^-, the ternary complex appears with the frequency expected statistically on the basis of binary complex stabilities (Nelson & Martin 1991).

Reported observations of facilitated transfer of Al(III) into erythrocytes when whole blood is made 10 mM in glutamate, and into rat brains by the subcutaneous injection of glutamate, have been interpreted as the passage of a neutral Al(III)–glutamate complex across the erythrocyte membrane and blood–brain barrier (Deloncle et al 1990). It is also argued that glutamate and aspartate complexes of Al^{3+} play roles in Alzheimer's disease and other Al(III)-related conditions (Deloncle & Guillard 1990). From the stability constants, these ideas are not tenable. The quoted stability constants are unreliable; because of the weakness of binding, mainly Al^{3+} hydrolysis is being measured. Careful allowance for the highly competitive hydrolysis yields only weak stability constants of log $K_s = 7.69$ for glutamate and 7.77 for aspartate, with no 2:1 nor 3:1 complexes (Dayde 1990). With ligand pK_a values of 9.18 and 9.27, the conditional stability constants for pH 6.6 become log $K_{6.6} = 5.1$ for both amino acids. This value is less than that for all other ligands in Table 1. We conclude that glutamate and aspartate are not competitive with numerous other ligands for Al^{3+} in physiological systems. Any role for glutamate must be indirect and not as a non-competitive Al^{3+} complex.

There is ample proof that citrate facilitates the incorporation of Al(III) into mammalian tissues. Al(III) levels were elevated in both the brain and bones of rats fed a diet containing aluminium citrate, or even just citrate (Domingo et al 1991, Slanina et al 1984, 1985). Evidently citrate alone chelates trace Al(III) in the diet. Dosing lambs with aluminium citrate promotes the absorption of Al(III) and alters the balance of other minerals (Allen et al 1990, 1991). Increased levels of serum Al(III) were found in patients with chronic renal failure who were taking an Al(III)-containing phosphate binder with citrate (Hewitt et al 1988). A rapidly fatal encephalopathy in human patients with chronic renal failure has been attributed to concomitant ingestion of $Al(OH)_3$ and citrate (Bakir et al 1986, Kirschbaum & Schoolwerth 1989). Moreover, healthy adults taking $Al(OH)_3$-based antacids along with citric acid, citrate salts or citrus fruits showed substantial increases in Al(III) levels in blood (Slanina et al 1986, Weberg & Berstad 1986) and urine (Bakir et al 1989, Coburn et al 1991, Walker et al 1990).

Citrate exists mainly in the form of the tricarboxylate anion (L^{3-}) at pH > 6, and at 0.1 mM in the blood plasma it is the leading small molecule Al^{3+} binder (Martin 1986b,c). In neutral solutions the main species is $HOAlH_{-1}^{2-}$, followed by $AlLH_{-1}^{-}$ with $pK_a \simeq 6.5$ for the loss of a proton from metal ion-bound water (Martin 1988a). Even though much of the citrate in plasma occurs as a Ca^{2+} complex, Al^{3+} easily displaces Ca^{2+} from citrate. Even when alkaline earth cations in the plasma are taken into consideration, there is an almost 10^8 mole ratio of citrate-bound to unbound Al^{3+} (Martin 1991a). Because of the citrate binding of Al^{3+}, calcium citrate increases intestinal Al(III) absorption (Nolan et al 1990) and should not be given to uraemic patients; calcium acetate appears to be a better choice (Emmett et al 1991).

As indicated in Table 1 and Fig. 3 and previously (Martin 1986b,c, 1988a), citrate solubilizes Al^{3+} from both insoluble $Al(OH)_3$ and $AlPO_4$. On the basis of information from equilibrium constants, it was strongly urged in 1986 that people should not take Al(III) and citrate together (Martin 1986b,c). A more recent warning suggests that uraemic patients should avoid citrate compounds (Molitoris et al 1989). However, since a healthy diet always contains citrate, all deliberate ingestion of Al(III) compounds might be better avoided. The amount of citrate present should always be considered as a variable in Al(III) ingestion studies.

Transferrin is the main protein carrier of Al^{3+} in the plasma. Displacement of the 10^9 times stronger binding Fe^{3+} is unnecessary because plasma transferrin is about 50 µM in unoccupied sites. Recent experiments confirm the conclusion based on stability constants (Martin 1986b, Martin et al 1987) that citrate is the low molecular weight and transferrin the high molecular weight carrier of Al^{3+} in rat serum (Van Ginkel et al 1990). We emphasize again, however, that albumin does not bind Al^{3+} with anywhere near the strength of transferrin or citrate.

Upon binding of Al^{3+} to transferrin there are two scenarios. First, by removing Al(III) from the bloodstream and tightly complexing it, transferrin binding may be detoxifying and favourable to the body, though the final disposition of the Al(III) remains a source of concern. Second, transferrin binding may be dangerous if it results in incorporation of Al(III) at transferrin receptors. In this case it might be possible to detoxify the Al(III) by adding Fe(III) and saturating transferrin with this much more strongly binding metal ion to displace Al(III). The displaced Al(III) would be picked up by citrate and eliminated in the urine. In this scenario the antidote for Al(III) poisoning is Fe(III).

In body fluids low in citrate, transferrin and nucleotides, the catecholamines may well become important Al^{3+} binders (Kiss et al 1989). While dopa and noradrenaline (epinephrine) fail to bind Mg^{2+} at pH 7.4, they bind Al^{3+} at picomolar levels (Table 1). In neutral solutions the main species is a 3:1 complex with the catechol moiety chelating the Al^{3+} and the ammonium group remaining protonated (Kiss et al 1989, 1991b). The noradrenaline–Al^{3+} complex inhibits enzymic O-methylation but not N-methylation by catechol-O-methyltransferase (Mason & Weinkove 1983). This result conforms to that expected if Al^{3+} binds only to the catechol moiety of noradrenaline. When other metal ions are deficient, Al(III) decreases catecholamine levels in the rat brain (Wenk & Stemmer 1981). By binding to the catechol moiety of catecholamines, trace amounts of Al^{3+} may disrupt neurochemical processes.

Ligand exchange

In addition to the stability of metal ion complexes, an important and often overlooked feature is the rate of ligand exchange out of and into the metal ion coordination sphere. Ligand exchange rates take on special importance for Al^{3+} because they are slow, and systems may not be at equilibrium. The rate for the exchange of inner sphere water with solvent water is known for many metal ions, and the order of increasing rate constants in acidic solutions is given by

$$Al^{3+} \lll Fe^{3+} < Ga^{3+},\ Be^{2+} \lll Mg^{2+} < Fe^{2+} < Zn^{2+} < Ca^{2+}$$

Each inequality sign indicates an approximate 10-fold increase in rate constant from $1.3\ s^{-1}$ for Al^{3+} and increasing through eight powers of 10 to $10^8\ s^{-1}$ for Ca^{2+} at 25 °C (Martin 1991a). Though these specific rate constants refer to water exchange in aquo metal ions, they also reflect relative rates of exchange of other unidentate ligands. Reducing Fe^{3+} to Fe^{2+} gains a 10^4-fold increase in rate. Chelated ligands exchange more slowly, but the order remains. The slow ligand exchange rate for Al^{3+} makes it useless as a metal ion engaged in reactions at enzyme active sites. The 10^5 times faster rate for Mg^{2+} furnishes enough reason for the Al^{3+} inhibition of enzymes with Mg^{2+} cofactors. Processes involving rapid Ca^{2+} exchange would be thwarted by substitution of the 10^8-fold slower Al^{3+}.

References

Allen VG, Fontenot JP, Rahnema SH 1990 Influence of aluminium citrate and citric acid on mineral metabolism in wether sheep. J Anim Sci 68:2496–2505

Allen VG, Fontenot JP, Rahnema SH 1991 Influence of aluminium-citrate and citric acid on the tissue mineral composition in wether sheep. J Anim Sci 69:792–800

Bakir AA, Hryhorczuk DO, Berman E, Dunea G 1986 Acute fatal hyperaluminemic encephalopathy in undialyzed and recently dialyzed uremic patients. Trans Am Soc Artif Intern Organs 32:171–176

Bakir AA, Hryhorczuk DO, Ahmed S, Hessl SM, Levy PS, Spengler R, Dunea G 1989 Hyperaluminemia in renal failure: the influence of age and citrate intake. Clin Nephrol 31:40–44

Coburn JW, Mischel MG, Goodman WG, Salusky IB 1991 Calcium citrate markedly enhances aluminum absorption from aluminum hydroxide. Am J Kidney Dis 17:708–711

Dayde S 1990 Thesis, Université de Toulouse

Deloncle R, Guillard O 1990 Mechanism of Alzheimer's disease: arguments for a neurotransmitter–aluminium complex implication. Neurochem Res 15:1239–1245

Deloncle R, Guillard O, Clanet F, Courtois P, Piriou A 1990 Aluminum transfer as glutamate complex through blood–brain barrier. Biol Trace Elem Res 25:39–45

Diaz-Nido J, Avila J 1990 Aluminum induces the in vitro aggregation of bovine brain cytoskeletal proteins. Neurosci Lett 110:221–226

Domingo JL, Gomez M, Llobet JM, Corbella J 1991 Influence of some dietary constituents on aluminum absorption and retention in rats. Kidney Int 39:598–601

Dyrssen D, Haraldsson C, Nyberg E, Wedborg M 1987 Complexation of aluminum with DNA. J Inorg Biochem 29:67–75

Emmett M, Sirmon MD, Kirkpatrick WG, Nolan CR, Schmitt GW, Cleveland MV 1991 Calcium acetate control of serum phosphorus in hemodialysis patients. Am J Kidney Dis 17:544–550

Grundke-Iqbal I, Iqbal K, Tung Y, Quinlan M, Wisniewski HM, Binder LI 1986 Abnormal phosphorylation of the microtubule associated protein tau in Alzheimer cytoskeletal pathology. Proc Natl Acad Sci USA 83:4913–4917

Harris WR, Sheldon J 1990 Equilibrium constants for the binding of aluminum to human serum transferrin. Inorg Chem 29:119–124

Hewitt CD, Poole CL, Westervelt FB Jr, Savory J, Wills MR 1988 Risks of simultaneous therapy with oral aluminium and citrate compounds. Lancet 2:849

Kirschbaum BB, Schoolwerth AC 1989 Acute aluminum toxicity associated with oral citrate and aluminum-containing antacids. J Med Sci 297:9–11

Kiss T, Sovago I, Martin RB 1989 Al^{3+} binding to catecholamines and tiron. J Am Chem Soc 111:3611–3614

Kiss T, Sovago I, Martin RB 1991a Al^{3+} binding by adenosine-5'-phosphates: AMP, ADP, and ATP. Inorg Chem 30:2130–2132

Kiss T, Sovago I, Martin RB 1991b Comments on macrospecies and macroconstants versus microspecies and microconstants. Polyhedron 10:1401–1403

Macdonald TL, Martin RB 1988 Aluminum ion in biological systems. Trends Biochem Sci 13:15–19

Macdonald TL, Humphreys WG, Martin RB 1987 Promotion of tubulin assembly by aluminum ion in vitro. Science (Wash DC) 236:183–186

Martin RB 1986a Bioinorganic chemistry of metal ion toxicity. Metal Ions Biol Syst 20:21–65

Martin RB 1986b The chemistry of aluminum as related to biology and medicine. Clin Chem 32:1797–1806

Martin RB 1986c Citrate binding of Al^{3+} and Fe^{3+}. J Inorg Biochem 28:181–187

Martin RB 1988a Bioinorganic chemistry of aluminum. Metal Ions Biol Syst 24: 1–57

Martin RB 1988b Ternary hydroxide complexes in neural solutions of Al^{3+} and F^-. Biochem Biophys Res Commun 155:1194–1200

Martin RB 1990a Aluminosilicate stabilities under blood plasma conditions. Polyhedron 9:193–197

Martin RB 1990b Bioinorganic chemistry of magnesium. Metal Ions Biol Syst 26: 1–13

Martin RB 1991a Aluminum in biological systems. In Nicolini M, Zatta P, Corain B (eds) Aluminum in chemistry biology and medicine. Cortina international, Verona & Raven Press, New York, p 3–20

Martin RB 1991b Fe^{3+} and Al^{3+} hydrolysis equilibria. J Inorg Biochem 44:141–147

Martin RB, Savory J, Brown S, Bertholf RL, Wills M 1987 Transferrin binding of Al^{3+} and Fe^{3+}. Clin Chem 33:405–407

Mason L, Weinkove C 1983 Reversal of aluminium inhibition of enzymatic O-methylation by desferrioxamine. Ann Clin Biochem 20:105–111

Molitoris BA, Froment DH, Mackenzie TA, Huffer WH, Alfrey AC 1989 Citrate: a major factor in the toxicity of orally administered aluminum compounds. Kidney Int 36:949–953

Nelson DJ, Martin RB 1991 Speciation in systems containing aluminum, nucleoside diphosphates, and fluoride. J Inorg Biochem 43:37–43

Nolen CR, Califano JR, Butzin CA 1990 Influence of calcium acetate or calcium citrate on intestinal aluminium absorption. Kidney Int 38:937–941

Ohman L 1988 Stable and metastable complexes in the system $H^+ - Al^{3+}$ –citric acid. Inorg Chem 27:2565–2570

Ohman L, Sjoberg S 1983 A potentiometric study of aluminium(III) salicylates and aluminium(III) hydroxy salicylates. Acta Chem Scand A37:875–880

Sigel H, Tribolet R, Malini-Balakrishnan R, Martin RB 1987 Comparison of the stabilities of monomeric metal ion complexes formed with adenosine 5′-triphosphate (ATP) and pyrimidine-nucleoside 5′-triphosphates. Inorg Chem 26:2149–2157

Slanina P, Falkeborn Y, Frech W, Cedergren A 1984 Aluminium concentrations in the brain and bone of rats fed citric acid, aluminium citrate or aluminum hydroxide. Food Chem Toxicol 22:391–397

Slanina P, Frech W, Bernhardson A, Cedergren A, Mattson P 1985 Influence of dietary factors on aluminium absorption and retention in the brain and bone of rats. Acta Pharmacol Toxicol 56:331–336

Slanina P, Frech W, Ekstrom L, Loof L, Slorach S, Cedergren A 1986 Dietary citric acid enhances absorption of aluminum in antacids. Clin Chem 32:539–541

Sovago I, Kiss T, Martin RB 1990 2,3-Diphosphoglycerate binding of Zn^{2+} and Al^{3+}. Polyhedron 9:189–192

Sternberger NH, Sternberger LA, Ulrich J 1985 Aberrant neurofilament phosphorylation in Alzheimer disease. Proc Natl Acad Sci USA 82:4274–4276

Toth I, Zekany L, Brucher E 1984 Equilibrium study of the systems of Al(III), Ga(III) and In(III) with mercaptacetate, 3-mercaptopropionate and 2-mercaptobenzoate. Polyhedron 7:871–877

Van Ginkel MF, Van der Voet GB, Van Eijk HG, De Wolff FA 1990 Aluminium binding to serum constituents: a role for transferrin and for citrate. J Clin Chem Clin Biochem 28:459–463

Walker JA, Sherman RA, Cody RP 1990 The effect of oral bases on enteral aluminum absorption. Arch Intern Med 150:2037–2039

Weberg R, Berstad A 1986 Gastrointestinal absorption of aluminium from single doses of aluminium containing antacids in man. Eur J Clin Invest 16:428–432

Wenk GL, Stemmer KL 1981 The influence of ingested aluminum upon norepinephrine and dopamine levels in the rat brain. Neurotoxicology 2:347–353

Womack FC, Colowick SP 1979 Proton-dependent inhibition of yeast and brain hexokinases by aluminum in ATP preparations. Proc Natl Acad Sci USA 76:5080–5084

DISCUSSION

Kerr: If I have understood your pAl data, one would expect desferrioxamine to take aluminium off transferrin. Is that correct?

Martin: Yes, you would expect removal thermodynamically, but maybe not clinically. It's possible that it will be hard to get aluminium out of transferrin in a reasonable time.

Kerr: We generated conflicting results on the competition between desferrioxamine and transferrin for bound aluminium in Newcastle. In one series of unpublished studies by Dr Habibur Rahman, no aluminium was leached from transferrin during prolonged incubation against therapeutic concentrations of desferrioxamine. However, in a later study by Skillen & Moshtaghie (1986), substantial amounts of Al were removed during dialysis for 24 hours or more against Earle's medium at pH 7.4 containing 200 mg/l desferrioxamine. Certainly the binding to transferrin is tight and can only be broken by desferrioxamine under specific conditions of pH and bicarbonate concentration.

Martin: I would think that you need a small ligand to enter transferrin, proceed to the metal ion site, chelate the metal ion, remove it from the protein, and pass it to desferrioxamine (DFO). DFO itself is too large to serve as a mediating ligand. Fe^{3+} transfer from transferrin to DFO is slow, but is promoted by small ligands such as citrate and more effectively by nitrilotriacetate (Pollack et al 1976). Citrate also facilitates removal of Al^{3+} from transferrin (Marques 1991).

Candy: Crystallographic analysis of human apolactoferrin shows that this molecule is analogous to a Venus fly-trap (Anderson et al 1990). If this analogy is extended to transferrin, the iron cations bound by each of the protein lobes will be locked inside the molecule and will not be available for extraction by desferrioxamine.

Zatta: Going back to Professor Williams' Introduction, besides the influence of aluminium outside the cell and inside the cell, there is another universe which is in between, namely the cell membrane. This universe is not yet fully understood with respect to aluminium biochemistry.

A second comment is on the simplicity and complexity of aluminium chemistry. The chemistry of aluminium is apparently 'very simple', as Dr Martin said. This is because the background of a chemist is to try to simplify the model for studying these kinds of problems. On the contrary, the biological effects of aluminium are indeed very complex. As a biologist, I feel there is something more than just the metal speciation effects. Let me give you an example.

Transferrin is considered to be the major aluminium carrier in the serum. However, in the metal internalization, several situations must be carefully considered, such as—for instance—the interaction between transferrin and its receptor, or the implication of the modified glycosylation of either transferrin or its receptor in the intracellular movement of the metal, and the responsibility of the metal for modifying the permeability of the blood–brain barrier. This 'universe' is very complicated and highly articulated. My colleagues and I are trying to encourage a more interdisciplinary approach to these problems, otherwise we may have an over-simplification or, perhaps, an over-complication of the problem.

Greger: There is also the question of the amounts of iron, manganese and other elements that compete for sites on transferrin. Transferrin saturation varies with an individual's iron status; this could affect how much aluminium could be bound.

Candy: This will not make much difference, since under normal physiological conditions plasma transferrin is only 30% saturated with iron and there is therefore a vast excess of unoccupied metal-ion binding sites at a concentration of 50 μmol/l (Martin 1986).

Greger: In an anaemic person, or even a person with an infection, saturation can be 15% or less. In iron overload, transferrin saturation can be greater than 80% (Subcommittee on Iron 1979). So you could get quite a bit of variation in saturation.

Harrison: Were you suggesting that there might be an inverse relationship between aluminium toxicity and iron overload? Is there any evidence for that?

Greger: No, I was not. Factors that affect aluminium binding in the plasma may or may not affect aluminium levels in the brain, and brain function.

Blair: When we consider the binding to transferrin of aluminium, gallium and iron, it's important to realize that when one moves away from physiological conditions (pH 7.0; bicarbonate, about 25 µM), the binding falls. Low bicarbonate levels occur in some physiological conditions, such as anoxia.

Aluminium goes into the unoccupied site on transferrin and will not displace iron, as would be expected from stability constants. The same would be true using gallium, where stability constants are a factor of 100 less than for iron. If transferrin is treated with gallium, the unoccupied sites are filled and iron is not displaced.

You spoke of the binding of Al to DNA as being very weak, Professor Martin. Nevertheless binding to DNA will occur, given high levels of aluminium in the cell?

Martin: I think the point is that where DNA is, there is also ATP, and other strong aluminium binders are present, some of which are listed in Table 1 of my paper (p 9).

Blair: Yes; then as the concentration of aluminium inside the cell rises, you would see it filling up layer after layer of potential complexing agents, and ultimately it binds to DNA?

Martin: That would take a lot of aluminium.

Blair: Would it not correspond to the situation in renal dialysis patients, where plasma levels can rise to 200 µmol per litre? I am thinking of the extreme end of the spectrum, not the normal situation.

Martin: It may be so. To answer this question I would need reliable concentrations for the other species that are present in the plasma, in a cell, or in compartments of a cell.

Blair: I recognize the problem you face, because people take very simple binding calculations and apply them to complex interactions within a cell; nevertheless, the possibility *could* exist of aluminium binding to DNA.

McLachlan: Using human or mammalian brain nuclei, and fractionating the nuclear chromatin into linker histone-free (transcribable) DNA and heterochromatized (condensed) DNA, we found 18–20 times more Al/g DNA on the condensed, heterochromatized, component of DNA than on the linker histone-free DNA (Crapper et al 1980). The partitioning of Al relates to the protein–DNA complex rather than to DNA alone. The more complete description of aluminium binding characteristics in nuclear compartments requires measures involving chromatin.

Martin: I agree with that. My point is that double-helical DNA itself is unlikely to be a strong aluminium binder.

Birchall: Can your calculations take account of the fact that although binding might be weak, the product is insoluble, so the equilibrium is taken away? Consider say, polyacrylic acid; the amount of calcium needed to bring it out of solution is quite small.

Martin: Yes, all you need is a solubility product. I have used that to estimate pAl for $Al(OH)_3$, kaolinite, and $AlPO_4$ in Table 1 of my paper.

Birchall: With polymers, in which metal bindings may induce cross-linking and precipitation, the calculations may not be straightforward?

Martin: It might be awkward, but if someone has worked on it there may be a number.

Strong: Are there solubility constants for Al–histone binding?

Martin: Not that I know of.

McLachlan: It is the Al–histone–DNA binding that is the critical issue. Al in chromatin is not accessible to commonly used chelating agents such as EDTA or desferrioxamine. This suggests that aluminium is in a steric conformation within chromatin which is not accessible to chelating agents. One could defend the argument that Al has some biological role in nuclei, perhaps the repression of genes. For instance, Al would displace Mg^{2+} from the histone–DNA complex, resulting in conformational changes in chromatin which result in hetero-chromatization and gene repression. In human brain nuclei, after the first six months of life, the ratio of Al to DNA in chromatin remains constant until the ninth decade (Lukiw et al 1991). Alzheimer's disease is the only condition in which this fixed ratio changes. The Al/DNA ratio for the brains of various species of animals remains constant also. Bulk aluminium in brain ranges between 1.1 and 1.9 µg/g dry weight in all mammals examined (Crapper et al 1976).

We speculate that in post-differentiated cells, such as neurons, aluminium is an efficient cross-linker of proteins to DNA and a repressor of genes which will never be utilized in the remaining life of the neuron. Thus Al, with its very slow time of dissociation, would be an efficient electrostatic cross-linker which represses genes on a long-term basis. To conclude that Al has no biological function may be premature.

Martin: In a cell nucleus, I think Al would interact with ATP preferentially, and then with the phosphorylated proteins, especially the histones; so I agree with you.

Blair: So we would be looking at a long-term effect of exposure to a pulse dose of Al?

McLachlan: Yes. Despite long-term exposure to aluminium, healthy brain tissue concentrations remain constant. This supports the argument that the Al brain concentration is regulated. The molecular basis for this regulation is unknown.

Williams: Professor Martin was saying that with double-stranded (double-helical) DNA, he observes very weak Al binding. But with RNA, single-stranded, in crucifix form, the Mg binding constant inside tRNA is at least 10^4. The site there is special. Similar types of 'cruciform' structures are now also known in DNA; so there are special sites in DNA where three phosphates come together to form a cavity. It is perfectly possible, I believe, for Al to get in there—although we know nothing about that yet. So we have to be careful about specific sites of DNA in these crucifix-like forms where we know that a metal is very important (von Kitzing et al 1990).

Petersen: Dr Martin, you mentioned that Al^{3+} has a much smaller radius than Ca^{2+}, about the same as that of Mg^{2+}, and therefore would be expected to have binding properties rather more like those of Mg^{2+} than those of Ca^{2+}. You also mentioned the chelator, EDTA, which binds both Mg^{2+} and Al^{3+}. Another chelator, EGTA (ethylene glycol tetra-acetic acid), has a high affinity for Ca^{2+}. According to your hypothesis, one would expect that EGTA should have no higher affinity for Al^{3+} than EDTA, whereas it has a much higher affinity for Ca^{2+}. Is this the case?

Martin: EGTA exhibits a marginally higher stability constant than EDTA for Ca^{2+}. However, owing to the greater basicity of EGTA, in neutral solutions it binds Ca^{2+} slightly less strongly (lower conditional stability constant) than EDTA. But, as you surmise, both Al^{3+} and Mg^{2+} bind much more strongly to EDTA than to EGTA. These stability constants are tabulated in one of my reviews (Martin 1988). In my view, EGTA should be used only for Ca^{2+} binding; the published titration curve for its binding to Al^{3+} is poorly defined.

Williams: However, Ca binding to a flexible ligand, ATP, is very nearly the same as Mg binding to ATP, and with such flexible ligands, the hole size becomes less critical. It is only with a rigid hole that you have the possibility of making a comparison between Mg and Al based on size alone.

Now consider the chemistry of calcium outside the cell, where its concentration is high, 10^{-3} M (i.e. in the extracellular fluid); the competition is somewhat loaded in favour of Ca relative to Mg and protons at pH 7. Calcium binds strongly but Al binds poorly, owing to precipitation. There are certain regions of the body (e.g. in parts of the digestive tract and even in some vesicles in cells) where the acidity drops to pH 5. Free calcium changes little and while the proton concentration increases with the first power of $[H^+]$, the Al concentration goes up by the $[H^+]$ power 3 with a drop of pH. The competition between aluminium and calcium is now greatly changed, and Al can get to Ca sites perfectly well. In the cell, where [Ca] is so low (10^{-7} M), and the acidity is kept at 7, there is no problem of calcium/aluminium competition; outside the cell, where pH is below 7, and where Ca is high, I think the competition can be Al against Ca.

Martin: The strong pH dependence of the Al^{3+} concentration is illustrated in all three figures of my paper. Where there are specific Ca binding sites, Al fits poorly and rattles around. But, as you point out, Ca sites are often not very specific, and may adapt to the smaller Al. Many Ca sites are more adaptable than those of Mg (Martin 1990).

Zatta: So the interaction between Al and calmodulin inside the cell is a non-specific interaction.

Martin: Yes. Though there are papers reporting strong Al^{3+} binding to calmodulin, I have questioned them (Martin 1986, 1988). Only weak binding of Al^{3+} to calmodulin has been found by other investigators (Richardt et al 1985, You & Nelson 1991), the latter using some of the same techniques as the early studies. With one quarter of its amino acid residues bearing carboxylate side chains, calmodulin is an acidic protein that should bind multiply charged metal ions as a polyelectrolyte. When it does so, magnitudes in the circular dichroism spectrum decrease in the direction of denaturation. It is not through interactions with calmodulin directly, but through the calmodulin-regulated proteins that contain phosphate groups, that I expect Al^{3+} will exert any influence.

Blair: May I mention haemochromatosis? Grootveld et al (1989) have shown that in haemochromatosis, when transferrin saturation reaches up to about 75% or more, you get low molecular weight iron species increasing in concentration in the plasma. These were identified as iron–citrate complexes by high performance liquid chromatography. They are transient in the plasma.

Williams: Of course, in plants, citrate is the main carrier for iron.

The reactions of Al with fluoride are well known. If one is using 'fluoride' toothpaste with, say, Al around, could you imagine, Dr Martin, that the Al would be carried by the fluoride into the body, because the exchange rates are so slow?

Martin: They are not as slow as all that. Exchange rates for Al^{3+} with unidentate ligands are of the order of a second.

Blair: Using gallium as a model for Al, when fluoride is given to the rat, you get less gallium transport in both the starved and fed state; so when the fluoride concentration is sufficiently high to form the aluminium–fluoride complex, which is negatively charged and relatively insoluble, the intestinal transport rates will decrease (Farrar et al 1988).

Williams: We know of course that there's a problem about the biochemistry of AlF_4^- in the way it inhibits enzymes. I just wanted to make the point that there are dangers in all chemicals, and when you combine two elements like aluminium and fluoride, you have to think hard before making general statements that there is no associated risk. It may turn out to be trivial, but we need to raise these biological questions which we can't answer as chemists. It has to be a physiological answer, not a chemical answer. (See p 104 for further discussion on aluminium–fluoride complexes.)

Day: In relation to this point, I would like to refer to the kinetics of Al reactions. It seems to me that we need to make a distinction between the kinetics of the inorganic reactions of Al and the kinetics of Al when its bound, say, in transferrin. It is true that in its inorganic reactions, Al is slower and more inert than Ca, but this is relative. The rates of Al reactions in simple inorganic reactions are still very high. Thus the substitution of aluminium hydroxy species with fluoride and so on has time constants of much less than a second, even though they are slower than for Ca or Mg. So we shouldn't regard Al as inert in straightforward inorganic reactions. When Al is bound to something like transferrin, then I think the complex is inert (that is, it's kinetically stable) because of the properties, particularly the shape, of the organic ligand, and not because of the properties of the metal ion *per se*.

Martin: I agree. The exchange rate depends critically upon the rate-limiting step. For unidentate ligands the rate-limiting step of exchange of ligands into and out of the coordination sphere is mainly a property of the metal ion, while for large, multidentate ligands like transferrin the rate limiting becomes crucially a property of the ligand. Polynuclear complex formation is also slow, sometimes taking months for equilibrium to be attained with Al^{3+} complexes.

My purpose in raising the exchange rate issue relates to cases where Al^{3+} substitutes for Mg^{2+} (or Ca^{2+}) at an enzyme active site. The substitution of the relatively slowly reacting Al^{3+} may so retard a key step in the reaction sequence as even to render the enzyme non-functional.

Day: Yes, I agree. In such cases, where the metal ions were competing in a catalytic cycle, the relative rates of substitution would determine the outcome.

References

Anderson BF, Baker HM, Norris GE, Rumball SV, Baker EN 1990 Apolactoferrin structure demonstrates ligand-induced conformational change in transferrin. Nature (Lond) 344:784–787

Crapper DR, Krishnan SS, Quittkat S 1976 Aluminum, neurofibrillary degeneration and Alzheimer's disease. Brain 99:67–80

Crapper DR, Quittkat S, Krishnan SS, Dalton AJ, De Boni U 1980 Intranuclear aluminum content in Alzheimer's disease, dialysis encephalopathy, and experimental aluminum encephalopathy. Acta Neuropathol 50:19–24

Grootveld M, Bell JD, Halliwell B, Aruoma OI, Bomford A, Sadler PJ 1989 Non-transferrin-bound iron in plasma or serum from patients with idiopathic haemochromatosis. J Biol Chem 264:4417–4422

Farrar G, Morton AP, Blair JA 1988 The intestinal speciation of gallium: possible models to describe the bioavailability of aluminium. In: Brätter P, Schramel P (eds) Trace element analytical chemistry in medicine and biology. Walter de Gruyter, Berlin, vol 5: 342–347

Lukiw WJ, Krishnan B, Wong L, Kruck TPA, Bergeron C, McLachlan DRC 1991 Nuclear compartmentalization of aluminum in Alzheimer's disease (AD). Neurobiol Aging 13:115–121

Marques HM 1991 Kinetics of the release of aluminum from human serum dialuminum transferrin to citrate. J Inorg Biochem 41:187–193

Martin RB 1986 The chemistry of aluminum as related to biology and medicine. Clin Chem 32:1797–1806

Martin RB 1988 Bioinorganic chemistry of aluminum. Metal Ions Biol Syst 24:1–57

Martin RB 1990 Bioinorganic chemistry of magnesium. Metal Ions Biol Syst 26:1–13

Pollack S, Aisen P, Lasky FD, Vanderhoff G 1976 Chelate mediated transfer of iron from transferrin to desferrioxamine. Br J Haematol 34:231–235

Richardt G, Federolf G, Habermann E 1985 The interaction of aluminum and other metal ions with calcium-calmodulin-dependent phosphodiesterase. Arch Toxicol 57:257

Skillen AW, Moshtaghie AA 1986 The binding of aluminium by human transferrin and the effect of desferrioxamine—an equilibrium dialysis study. In: Taylor A (ed) Aluminium and other trace elements in renal failure. Ballière Tindall, London, p 81–85

Subcommittee on Iron 1979 Iron. University Park Press, Baltimore, MD, p 100

You G, Nelson DJ 1991 Al^{3+} versus Ca^{2+} ion binding to methionine and tyrosine spin-labeled bovine brain calmodulin. J Inorg Biochem 41:283–291

von Kitzing E, Lilley DMJ, Diekmann S 1990 The stereochemistry of a four-way DNA junction: a theoretical study. Nucleic Acids Res 18:2671–2683

Dietary and other sources of aluminium intake

J. L. Greger

Department of Nutritional Sciences, University of Wisconsin, 1415 Linden Drive, Madison, WI 53706, USA

Abstract. Aluminium in the food supply comes from natural sources including water, food additives, and contamination by aluminium utensils and containers. Most unprocessed foods, except for certain herbs and tea leaves, contain low (<5 µg Al/g) levels of aluminium. Thus most adults consume 1–10 mg aluminium daily from natural sources. Cooking in aluminium containers often results in statistically significant, but not practically important, increases in the aluminium content of foods. Intake of aluminium from food additives varies greatly (0 to 95 mg Al daily) among residents in North America, with the median intake for adults being about 24 mg daily. Generally, the intake of aluminium from foods is less than 1% of that consumed by individuals using aluminium-containing pharmaceuticals. Currently the real scientific question is not the amount of aluminium in foods but the availability of the aluminium in foods and the sensitivity of some population groups to aluminium. Several dietary factors, including citrate, may affect the absorption of aluminium. Aluminium contamination of soy-based formulae when fed to premature infants with impaired kidney function and aluminium contamination of components of parenteral solutions (i.e. albumin, calcium and phosphorus salts) are of concern.

1992 Aluminium in biology and medicine. Wiley, Chichester (Ciba Foundation Symposium 169) p 26–49

Oral exposure to aluminium is from foods, water and pharmaceutical products. The aluminium in the food supply comes from natural sources, water used in food preparation, food additives, and contamination by aluminium utensils and containers.

Unprocessed foods as sources of aluminium

Most foods, except for herbs and tea leaves, contain less than 5 µg aluminium per gram (Table 1). Furthermore, only small quantities of herbs are consumed by most individuals and most of the aluminium in tea leaves is insoluble. Thus most individuals consume only 1 to 10 mg of aluminium from natural sources daily.

TABLE 1 Estimated aluminium concentrations of selected foods

Foods	Aluminium concentration (µg/g)	Foods	Aluminium concentration (µg/g)
Animal products		**Vegetables & legumes**	
Beef, cooked[a]	0.2[bc]	Asparagus	4.4[bg]
Cheese, cheddar	0.2[d]	Beans, green, cooked[a]	3.4[c]
Cheese, cottage, creamed	0.1[d]	Beans, navy, boiled	2.1[d]
Cheese, processed	297.0[be]	Cabbage, raw	0.1[c]
Chicken w/skin, cooked[a]	0.7[c]	Cauliflower, cooked	0.2[c]
Eggs, cooked[a]	0.1[c]	Corn	0.1[d]
Fish (cod), cooked[a]	0.4[c]	Cucumber, fresh, pared	0.1[d]
Ham, cooked[a]	1.2[c]	Lettuce	0.6[g]
Milk, whole	0.06[d]	Peanut butter	5.8[d]
Salami	1.1[d]	Peas, green, cooked	1.9[c]
Yoghurt, plain low-fat	1.1[d]	Potatoes, unpeeled, boiled[a]	0.1[c]
		Potatoes, unpeeled, baked	2.4[c]
Fruits		Spinach, cooked[a]	25.2[bg]
Apple	0.1[d]	Tomatoes, cooked[a]	0.1[c]
Banana, fresh	0.05[d]		
Grapes	0.5[bf]	**Herbs & spices**	
Orange juice, frozen reconstituted	0.06[d]	Basil	3082.0[bf]
Peaches	0.4[bf]	Celery seed	465.0[bf]
Raisins, dried	3.1[d]	Cinnamon	82.0[bf]
Strawberries, fresh	2.2[d]	Oregano	600.0[bf]
		Pepper, black	143.0[bf]
Grains		Thyme	750.0[bf]
Biscuits, baking powder, refrigerated	16.3[d]		
Bran, wheat	12.8[bg]	**Other**	
Bread, white	3.0[f]	Baking powder	2300.0[bf]
Bread, whole wheat	5.4[f]	Beer, canned	0.7[d]
Corn chips	1.2[d]	Candy, milk chocolate	6.8[d]
Cornbread, homemade	400.0[d]	Cocoa	45.0[g]
Muffin, blueberry	128.0[d]	Cola, carbonated	0.1[d]
Oatmeal, cooked	0.7[d]	Cream substit., powdered	139.0[d]
Rice, cooked[a]	1.7[c]	Pickles w/Al additives	39.2[be]
Spaghetti, cooked[a]	0.4[c]	Salt w/Al additives	164.0[be]
		Tea, steeped	4.3[e]

[a]Food reported to *not* be stored or cooked in aluminium pans, trays or foil.
[b]Value is an average of several values reported in the reference.
[c]Greger et al 1985.
[d]Pennington & Jones 1989.
[e]Greger 1985.
[f]Sorenson et al 1974.
[g]Schlettwein-Gsell & Mommsen-Straub 1973.

Averages, such as those reported in Table 1, can be misleading for several reasons. First, during the last 20 years estimates of the amount of aluminium in serum samples have decreased 50-fold because of improvements in methodology (Versieck & Cornelis 1980). Similarly, many values reported in the literature for the aluminium content of foods also appear to reflect the inadequacies of analytical methodology. Second, the aluminium content of some foods, especially those plants known as 'aluminium accumulators', can vary greatly because of differences in plant varieties and soil conditions, including pH (Greger 1988). For example, Eden (1976) estimated that tea leaves contained on average 1000 µg Al/g, but some contained as much as 17 000 µg Al/g. Third, the adhesion of soil to vegetable matter can greatly increase its aluminium content. The aluminium content of soil is variable, but on average soil is estimated to contain 7.1% aluminium (Jones & Bennett 1986). ✗

Water as a source of aluminium

The amount of aluminium in surface and groundwater is variable; concentrations of 0.012 to 2.25 mg/l have been reported in North American rivers (Jones & Bennett 1986). When the pH of water is < 5, the amount of soluble aluminium in water tends to increase.

Aluminium-containing flocculants are used to clarify municipal water supplies. Miller et al (1984) reported that the concentrations of aluminium in the finished water at 186 water utilities in the United States ranged from < 0.014 to 2.67 mg Al/l. They estimated that there was a 40–50% chance that aluminium coagulants increased the aluminium concentration of finished water above that naturally present in water. However, the median level of aluminium in the finished water samples studied was very low (< 0.017 mg/l). Thus, individuals consuming two litres of water daily would take in less than 0.04 mg aluminium from water.

as already noted etc.

✗ Food additives as a source of aluminium

Food additives are a major source of dietary aluminium in the United States, although aluminium-containing additives are present in only a limited number of foods (Greger 1985, Pennington 1987). Variability in the usage of these foods make it very difficult to estimate aluminium intake from these sources.

In 1982 approximately four million pounds weight of aluminium was used in food additives in the United States (Committee on Food Additives Survey Data 1984). This means that the average US citizen theoretically consumed 21.5 mg aluminium daily in food additives (Greger 1985). The use of industrial production figures in this manner tends to overestimate intakes. However, calculations based on standardized menus such as those used in the Total Diet Study of the Food and Drug Administration (FDA) may underestimate the use of certain processed foods, such as processed cheese which can contain very high levels of aluminium (Table 1).

The data compiled by the Committee on the GRAS List Survey—Phase III (1979) indicated that individual usage of aluminium-containing food additives varied greatly. About 5% of adults in the United States consumed more than 95 mg aluminium in food additives daily; 50% consumed 24 mg or less aluminium daily in food additives. The most commonly used aluminium-containing food additives are: acidic sodium aluminium phosphate (leavening agent in baked goods); the basic form of sodium aluminium phosphate (emulsifying agent in processed cheese); aluminium sulphates (acidifying agents); bentonite (materials-handling aid); aluminium lakes of various food dyes and colours; and aluminium silicates (anti-caking agents) (Greger 1985, Pennington 1987).

Pennington & Jones (1989) noted that the three foods that contributed the most aluminium in the Total Diet Study menus were processed American cheese, homemade cornbread, and yellow cake with white icing. Food additives were the major source of aluminium in these three foods. Only 9–15 g of each of these foods was included in the standardized menu. An individual consuming standard servings of processed cheese (28 g), cornbread (45 g), and cake (60 g) would consume proportionally more aluminium from additives.

Packaging and utensils as sources of aluminium

Many foods accumulate statistically significant amounts of aluminium when cooked or stored in aluminium pans, trays or foil as compared to similar batches of food processed in stainless steel containers (Table 2). However, most foods accumulate less than 2 µg Al/g food during preparation and storage (Greger et al 1985).

Several factors appear to influence the accumulation of aluminium by foods cooked in aluminium pans, such as pH, length of cooking period, and the use of new pans or pressure cookers (Greger et al 1985, Inoue et al 1988). Thus, investigators have found that tomato sauces cooked for several hours in aluminium pans accumulated 3–6 mg aluminium per 100 g serving and Chinese noodles accumulated 2.6 mg aluminium per serving.

Organic acids, including citric acid, and copper in foods may increase the solubilization of aluminium from pans and foil (Baxter et al 1988, Flaten & Ødegård 1989, Ellen et al 1990). The addition of fluoride increased slightly the amount of aluminium leached from pans when citric acid solutions were heated in the pans. However, this effect of fluoride was not observed when fluoride was added to acidic foods (Savory et al 1987, Baxter et al 1988).

TABLE 2   **Concentrations of aluminium in foods cooked in aluminium or stainless steel pans** [a]

Food	Pan type	Cooked in aluminium pans [b]	Cooked in stainless steel pans
		(μg Al/g wet weight)	
Apple sauce [c]	Saucepan	7.1	0.12
Beef, roast [c]	Pressure cooker	0.85	0.21
Cabbage [c]	Pressure cooker	3.6	0.20
Chicken, with skin	Frying pan	1.00	0.66
Eggs [c]	Frying pan	1.6	0.13
Pudding, chocolate	Saucepan	0.42	4.0
Rice	Saucepan	1.7	1.7
Tomato sauce	Saucepan	57.1	0.16

[a]Greger et al 1985.
[b]Pans conditioned through standardized cooking procedures.
[c]Products cooked in aluminium pans contained significantly ($P < 0.05$) more aluminium than products cooked in stainless steel pans, as determined by Student's t-tests.

Dietary aluminium

Most Americans probably consume 2–25 mg aluminium daily, with 1–10 mg from natural sources including water, 0–20 mg from food additives, and 0–2 mg from contamination from pans and utensils. Pennington & Jones (1989) reported that the daily menu for 25- to 30-year-old males in the FDA's Total Diet Study contained 13.66 mg aluminium. Shiraishi et al (1988) found 2.2 to 8.1 mg aluminium in duplicate portions of foods consumed by Japanese adult males daily, but estimated that 'model' daily diets consumed by 40-year-old Japanese males contained 2.3 mg aluminium. Ellen et al (1990), using duplicate portion techniques, reported that Dutch adults consumed 0.60 to 33.3 mg Al per day. Average intakes of Dutch adults in 1984–1985 were estimated to be 2.8 mg. The higher aluminium levels in duplicate portions than in 'model' diets may reflect differences in food choices and aluminium contamination from cooking utensils (Ellen et al 1990). The higher intakes of aluminium by North Americans probably reflect the composition and popularity of chemically leavened baked goods and of processed cheeses.

The oral load of aluminium tends to be higher for children than adults. If the average body weights of three-year-old and 40-year-old males in Japan are assumed to be 13 and 75 kg respectively, oral aluminium loads were greater for children than adults (0.28 versus 0.03 mg aluminium/kg body weight). Calculations based on the FDA's Total Diet Study suggest that two-year-olds (13 kg body weight) in the USA consumed almost three times as much aluminium per kg body weight as adult males (75 kg body weight) and females (60 kg body weight), respectively (0.48 vs. 0.18 and 0.15 mg aluminium/kg body weight).

Similarly, oral aluminium loads of 3.9-year-old Hungarians appeared to be two-fold greater than those of 14-year-old males (0.32 vs. 0.15 mg aluminium/kg body weight) (Schamschula et al 1988).

Infant formulae

Recently a number of investigators have become concerned about the aluminium content of infant formulae (Koo et al 1988, McGraw et al 1986, Simmer et al 1990). The aluminium content of human or cow's milk is negligible (<0.05 µg/ml) (Koo et al 1988). Dabeka & McKenzie (1990) reported that ready-to-use milk-based and soy-based formulae contained 0.01–0.36 and 0.40–6.4 µg Al/g, respectively. Thus 1–3-month-old infants consuming certain soy-based formulae could take in as much as 2.1 mg Al daily, whereas infants fed human or cow's milk would consume only 3 µg Al daily. The infants believed to be most at risk would be preterm infants with impaired renal function because they would be less able to excrete absorbed aluminium (Greger 1987, Koo et al 1988).

Pharmaceutical products as sources of aluminium

The typical quantities of aluminium consumed in foods and beverages amount to less than 1% of the quantities that can be consumed in pharmaceutical products, such as antacids, buffered analgesics, antidiarrhoeal agents, and certain anti-ulcer drugs (Lione 1983, 1985). Lione (1985) estimated that 126–728 mg and 840–5000 mg were possible daily doses of aluminium in buffered analgesics and antacids, respectively.

Aluminium-containing phosphate-binding gels have been used for years to treat hyperphosphataemia in patients with chronic renal failure. Unfortunately, very high doses (100 mg/kg per day) were needed to control hyperphosphataemia in some paediatric cases, and symptoms of aluminium-associated bone diseases were sometimes observed. Thus the use of these aluminium-containing pharmaceutical agents is now discouraged among children with chronic renal failure (Committee on Nutrition, American Academy of Pediatrics 1986).

Of special concern are parenterally delivered pharmaceuticals, because the protective barrier of the gut has been bypassed (Greger 1987). Several substances administered intravenously, including albumin and calcium and phosphate salts, can contain significant quantities of aluminium (Table 3) (Sedman et al 1985). Parenterally administered products intended for repeated use should be tested for aluminium contamination, and patients, especially those with impaired renal function, should be monitored for aluminium intoxication (Hoiberg 1989).

TABLE 3   **Aluminium concentrations of substances added to intravenously administered solutions**[a]

Additive	Aluminium concentration ($\mu g/g$)
Calcium gluconate (10%)	5056
Heparin (1000 units/ml)	684
Potassium phosphate (3000 mmol/l)	16 598
Sodium phosphate (3000 mmol/l)	5977
Normal serum albumin	1822
Total parenteral nutrition solution (2.125% essential amino acids)	72

[a]Sedman et al 1985.

Bioavailability of aluminium

The potential negative effects of exposure to aluminium reflect not only the dose, but also the percentage of the dose that is absorbed and retained. The effectiveness of the gut as a protective barrier is illustrated by the observation that only 0.01–0.05% of oral doses of aluminium appears to be retained in rats (Ecelbarger & Greger 1991).

The bioavailability of aluminium is dependent on the total dietary milieu, not just on the chemical form in which aluminium is ingested (Dayde et al 1990, Greger 1988). For example, the elevation of dietary calcium levels has been found to decrease the retention of aluminium in bone in rats (Ecelbarger & Greger 1991). Citrate is another dietary factor that can influence aluminium retention in tissues.

Investigators have reported that citrate increased the tissue retention of aluminium in animals that were gavage fed aluminium daily (Slanina & Falkeborn 1984, Slanina et al 1985) or given aluminium in their drinking water (Fulton & Jeffery 1990). Recently, we (Ecelbarger & Greger 1991) demonstrated that the addition of citrate (10–31 µmol/g diet) to the diets of rats increased their zinc absorption and the retention of aluminium in their bones (Table 4, Study 1). The molar ratio of added citrate to added aluminium ranged up to 0.9. However, the increased retention of aluminium was not linearly related to dietary citrate levels, as might be expected if citrate–aluminium complexes were absorbed. These data suggest that citrate decreased the pH in the gut and increased the solubility of trace elements, such as zinc and aluminium, in the gut. This increased solubility resulted in greater absorption. Moreover, in a second study, rats with one kidney removed and reduced kidney function were more sensitive to the effects of dietary citrate on aluminium retention than intact rats (Table 4, Study 2). These modest but consistent effects of dietary citrate on

TABLE 4 Aluminium concentrations in tibia of rats fed various levels of aluminium and citrate, with and without impaired kidney function[a]

Study 1				Study 2		
Dietary addition		Tibia			Dietary addition		Tibia
Al	Citrate	Al[b]			Al	Citrate	Al[d]
(μmol/g diet)		(nmol/g)	Renal status[c]		(μmol/g diet)		(nmol/g)
0.5	0	31 ± 2	2		0.3	0	31 ± 2
0.5	10	47 ± 8	2		39	0	39 ± 2
0.5	21	38 ± 4	2		0.3	21	37 ± 3
0.5	31	45 ± 4	2		39	21	49 ± 3
34	0	46 ± 7	1[c]		0.3	0	32 ± 3
34	10	59 ± 14	1		39	0	48 ± 4
34	21	52 ± 5	1		0.3	21	46 ± 3
34	31	74 ± 13	1		39	21	51 ± 3

[a]Ecelbarger & Greger 1991.
[b]The effects of dietary aluminium ($P < 0.005$) and dietary citrate ($P < 0.05$) were significant in a two-way analysis of variance.
[c]One kidney was removed surgically at the initiation of Study 2; this resulted in 20–30% reduction in kidney function as judged by decreased creatinine clearance and increased blood urea nitrogen levels.
[d]The effects of kidney function ($P < 0.05$), dietary aluminium ($P < 0.0001$) and dietary citrate ($P < 0.0001$) were significant in a three-way analysis of variance.

aluminium retention are probably important in practical situations. Americans are estimated to consume 4 g of preformed citrate in foods daily.

The assessment of aluminium exposure is complex. Scientists need to define the dietary, physiological and pathological factors that affect the absorption and retention of dietary aluminium, because any assessment of dietary aluminium exposure is incomplete if it does not include an assessment of the bioavailability of aluminium. Ultimately, these variables are more important than the absolute amounts of aluminium in foods and in the environment.

Acknowledgements

Supported by College of Agricultural and Life Sciences Project 2623 and National Institutes of Health grant DK41116.

References

Baxter M, Burrell JA, Massey RC 1988 The effects of fluoride on the leaching of aluminum saucepans during cooking. Food Addit Contam 5:651–656
Committee on Food Additives Survey Data 1984 Poundage update of food chemicals 1982. PB 84-16214. National Academy Press, Washington, DC

Committee on the GRAS List Survey—Phase III. 1979 The 1977 survey of industry on the use of food additives. National Academy of Sciences, Washington, DC

Committee on Nutrition, American Academy of Pediatrics 1986 Aluminum toxicity in infants and children. Pediatrics 78:1150–1154

Dabeka RW, McKenzie AD 1990 Aluminum levels in Canadian infant formulae and estimation of aluminum intakes from formulae by infants 0–3 months old. Food Addit Contam 7:275–282

Dayde S, Filella M, Berthon G 1990 Aluminum speciation studies in biological fluids, part 3. Quantitative investigation of aluminum–phosphate complexes and assignment of their potential significance in vivo. J Inorg Biochem 38:241–259

Ecelbarger CA, Greger JL 1991 Dietary citrate and kidney function affect aluminum, zinc and iron utilization in rats. J Nutr 121:1755–1762

Eden T 1976 Tea. Longman, London, p 8–15 (chapter on climate and soils)

Ellen G, Egmond E, Van Loon JW, Sahertian ET, Tolsma K 1990 Dietary intakes of some essential and non-essential trace elements, nitrate, nitrite and N-nitrosamines, by Dutch adults: estimated via a 24-hour duplicate portion study. Food Addit Contam 7:207–221

Flaten TP, Ødegård M 1989 Dietary aluminum and Alzheimer's disease—a reply. Food Chem Toxicol 27:496–498

Fulton B, Jeffery EH 1990 Absorption and retention of aluminum from drinking water. 1. Effect of citrate and ascorbic acids on aluminum tissue levels in rabbits. Fundam Appl Toxicol 14:788–796

Greger JL 1985 Aluminum content of the American diet. Food Technol 9(5):73, 74, 76, 78–80

Greger JL 1987 Aluminum and tin. World Rev Nutr Diet 54:255–285

Greger JL 1988 Aluminum in the diet and mineral metabolism. In: Sigel H, Sigel A (eds) Metal ions in biological systems, vol 24: Aluminum and its role in biology. Marcel Dekker, New York, p 199–215

Greger JL, Goetz W, Sullivan D 1985 Aluminum levels in foods cooked and stored in aluminum pans, trays and foil. J Food Prot 48:772–777

Hoiberg CP 1989 Aluminum in parenteral products: overview of chemistry concerns and regulatory actions. J Parenter Sci Technol 43:127–131

Inoue T, Ishiwata H, Yoshihara K 1988 Aluminum levels in food-simulating solvents and cooked in aluminum pans. J Agric Food Chem 36:599–601

Jones KC, Bennett BG 1986 Exposure of man to environmental aluminum—an exposure commitment assessment. Sci Total Environ 52:65–82

Koo WWK, Kaplan LA, Krug-Wispe SK 1988 Aluminum contamination of infant formulas. J Parenter Enteral Nutr 12:170–173

Lione A 1983 The prophylactic reduction of aluminum intake. Food Chem Toxicol 21:103–109

Lione A 1985 Aluminum intake from non-prescription drugs and sucralfate. Gen Pharmacol 16:223–228

McGraw ME, Bishop N, Jamison R et al 1986 Aluminum content of milk formulae and intravenous fluids used in infants. Lancet 1:157

Miller RG, Kopfler FC, Kelty KC, Stober JA, Ulmer NS 1984 The occurrence of aluminum in drinking water. J Am Water Assoc 76:84–91

Pennington JAT 1987 Aluminum content of foods and diets. Food Addit Contam 5:161–232

Pennington JAT, Jones JW 1989. Dietary intake of aluminum. In: Gitelman HJ (ed) Aluminum and health: a critical review. Marcel Dekker, New York, p 67–100

Savory J, Nicholson JR, Wills MR 1987 Is aluminium leaching enhanced by fluoride? Nature (Lond) 327:107–108

Schamschula RG, Sugár E, Un PSH, Duppenthaler JL, Toth K, Barmes DE 1988 Aluminum, calcium and magnesium content of Hungarian foods and dietary intakes by children aged 3.9 and 14 years. Acta Physiol Hung 72:237–251

Schlettwein-Gsell D, Mommsen-Straub S 1973 Spurenelemente in Lebensmitteln. XII. Aluminium. Int Z Vitamin Ernaehrungsforsch 43:251–263

Sedman AB, Klein GL, Merritt RJ et al 1985 Evidence of aluminum loading in infants receiving intravenous therapy. N Engl J Med 312:1337–1343

Shiraishi K, Yamagami Y, Kameoka K, Kawamura H 1988 Mineral contents in model diet samples for different age groups. J Nutr Sci Vitaminol 34:55–65

Simmer K, Fudge A, Teubner J, James SL 1990 Aluminum concentrations in infant formulae. J Pediatr Child Health 26:9–11

Slanina P, Falkeborn Y 1984 Aluminum concentrations in the brain and bone of rats fed citric acid, aluminum citrate or aluminum hydroxide. Food Chem Toxicol 33:391–397

Slanina P, Frech W, Berhardson A, Cedergren A, Mattsson P 1985 Influence of dietary factors on aluminum absorption and retention in the brain and bone of rats. Acta Pharmacol Toxicol 56:331–336

Sorenson JRJ, Campbell IR, Tepper LB, Lingg RD 1974 Aluminum in the environment and human health. Environ Health Perspect 8:3–95

Versieck J, Cornelis R 1980 Measuring aluminum levels. N Engl J Med 302:468

DISCUSSION

Blair: You spoke about aluminium pans used for cooking. However, these are of several kinds—plain aluminium, or anodized aluminium, or aluminium coated with teflon.

Greger: We didn't test the teflon-coated type, because Al is not going to leach through that surface in significant quantities. According to the Cookware Manufacturers' Association (Lake Geneva, Wisconsin), fewer than 10% of the aluminium pans sold in the United States are anodized. Accordingly we used a series of Al pans that were commercially available and non-anodized.

Blair: I raise this point because there is a difference between plain and anodized Al in the solubility of the metal. It is often assumed that the anodizing process reduces the Al solubility, but it in fact leads to much more vigorous pitting and etching, potentially leading to increased leaching of aluminium during cooking. It is important to distinguish between the two types.

Greger: In our studies we tested a variety of pans: ones that were 20 years old, brand new pans, and commercially available pans of aluminium alloys 3003 and 3004 that were conditioned through standardized cooking techniques. We could measure statistically significant amounts of aluminium being leached from the pans. But, practically, very little Al moved from the pans into servings of foods, except when brand new pans were used (Greger et al 1985).

Blair: There is a whole series of problems in Al availability, from ingestion, which involves the food mass and the processes in the gut, which are quite complex, to the problems of the chemical solution of Al. In Birmingham in the UK there are two classes of Al pan in use, plain and anodized, and there are considerable differences in leaching.

Martin: Dr Greger mentioned the high accumulation of Al by tomato sauce cooked for several hours in aluminium pans. Do Italians stew tomato sauce for spaghetti in steel or in aluminium pans?

Zatta: In Italy we don't use Al pans anymore, as we used to do years ago. Now we prefer steel pans, mainly for aesthetic as well as for practical reasons.

Schlatter: We have recently done a total diet study in Switzerland and can confirm other European data (e.g. the Dutch studies) that the daily Al intake is on average 2–5 mg for adults (Knutti & Zimmerli 1985). We also did migration studies from utensils, and we investigated plain, non-anodized Al pans (J. Müller, PhD thesis, in preparation). The surprising result was a considerable migration of Al in boiling tap water at pH 8.3, whereas the high migration rate in tomato sauce (pH 4.2) was expected (Fig. 1). There was a higher initial migration from new pans; after a cooking time of 50 min we recorded a migration of 2.5 mg/l, using a new Al pan.

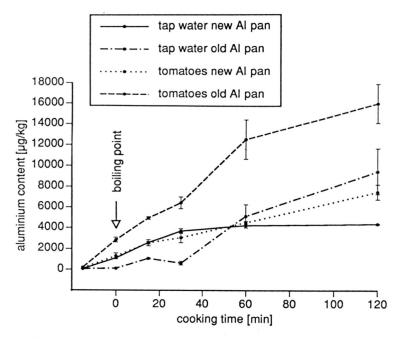

FIG. 1. (*Schlatter*) Aluminium concentrations (mean ± SD) in tap water (pH 8.3) and in blended tomatoes (pH 4.2) during cooking in a new and in an old aluminium saucepan.

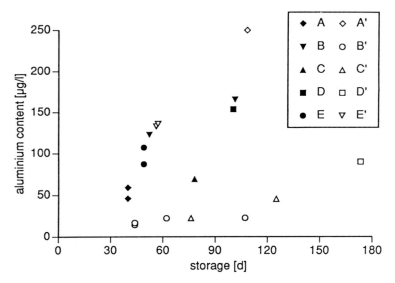

FIG. 2 (*Schlatter*) Aluminium concentrations in various batches of Coca Cola® (A,B,C,D,E) and Coca Cola® light (A',B',C',D',E') after storage in aluminium drink containers.

Since the general use of Al pans has declined in recent years and the prolonged boiling and cooking of acid food items is not regularly performed, the daily Al contribution from Al pans is almost negligible. We estimate less than 0.1 mg/day. Because of the coating of Al cans used for beer or soft drinks, such as Coca Cola, these beverages fail to make any contribution to Al intake, even after several weeks of storage (Fig. 2). Coffee prepared in the typical Italian aluminium mocha coffee-maker contains less Al than the water used for the preparation. Obviously, the Al is absorbed on the coffee powder. On the other hand, storage of acidic beverages in non-coated Al bottles must be discouraged, because of the high migration rate (Fig. 3).

Greger: Consumers often worry about Al from soda and beer cans, but this isn't a problem. The cans would dissolve, or at least be pitted, before they left the factory if they were not totally coated.

Flaten: As you indicated, the average intake of dietary Al seems to be higher in the USA than in Europe; from recently published studies I would say typically about three times higher. I agree with you that this difference is mainly due to food additives, but I believe that it is more related to legislation than to personal food choice. As far as I know, fewer Al additives are allowed in fewer foods in Europe than in the USA; this is definitely the case in Norway.

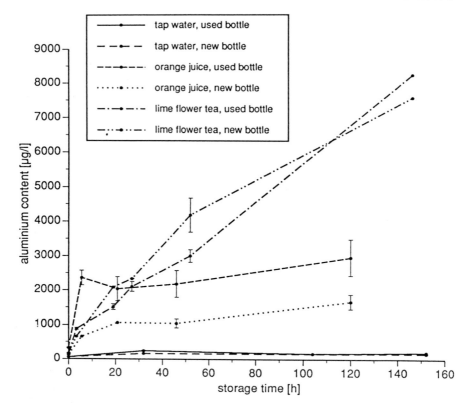

FIG. 3   (*Schlatter*) Aluminium concentrations measured in tap water (pH 8.2), orange juice (pH 3.8) and lime flower tea (pH 3.1), during storage in a new and in a used aluminium bottle.

Greger: It is a matter of different food choices in the sense that biscuits, cookies, quickbreads and cornbread that are made with baking powder are not as popular in Europe as they are in certain sections of the USA.

Flaten: I believe that Al-continuing baking powder is less commonly used in Europe than in the USA. Regarding your point that with improvement in analytical methodology, natural levels of Al are now considered to be very low in most foodstuffs, it is necessary to ingest some 10–100 kg of meat to reach an intake of 1 mg of Al; Al levels generally seem to be considerably higher in vegetables than in meat. Some studies also seem to indicate a discrepancy between Al intake calculated from the concentrations measured in individual foodstuffs and those actually measured in replicate food portions; the latter often given higher values for Al intake. Was not that the case in a study from Switzerland too, Dr Schlatter?

Schlatter: This was *not* the case. There is a good agreement between results from analyses of individual food items and the total daily Al intake of 2–5 mg.

Flaten: I would add that the average European, ingesting some 2–5 mg Al/day, would double the intake of Al by drinking a litre of tea daily.

You ended your talk, Dr Greger, by saying that the actual amount of Al intake is probably less important than the way Al is handled in the body. In Norway, Knut Nordal and coworkers are carrying out a study of 200 65-year-old women representative of the general population of that age in Oslo, none of whom had markedly reduced kidney function. They were given a single dose of 972 mg elemental Al in the form of an aluminium hydroxide-containing antacid, together with a citrate solution. Serum Al rose from a baseline level of 2–10 μg Al/l to between 40 and more than 300 μg Al/l, 2–4 hours after this challenge. Thus, in healthy elderly women there seems to be large individual differences in the absorption of aluminium.

Wisniewski: Dr Greger, can you elaborate on the use of flocculants in relation to the water content of Al?

Greger: I cited the work of Miller et al (1984). They reported that the Al content of natural and finished water in the United States at 186 utilities varied 100-fold. They stated that there was a 40–50% chance that Al flocculants added Al to the water, as I said. But a National Academy document (Safe Drinking Water Committee 1982) stated that 'In modern purification practice, aluminum-based coagulants usually result in the presence of lower concentrations of aluminum in drinking water than in raw water'. On the other hand, the report noted that aluminium intoxication occurred in Chicago among long-term haemodialysis patients after the city adopted a water purification using aluminium sulphate. Using US data, I can't give a better answer. I suspect that with some of the problems in Britain in the last few years you may have better data on this than we do!

Fawell: There are occasions when the use of Al sulphate as a flocculant will add Al to the water, but there are other occasions where the use of Al sulphate as a flocculant *reduces* the level of Al present naturally, and if you don't use Al sulphate you will finish up with a higher concentration of Al than if you had used it. It is a good way of removing natural Al, under certain circumstances, but there are occasions (it depends how good the water treatment works is) where Al will be left behind, and the resultant Al concentration may be higher than that originally present.

Edwardson: However, the naturally occurring Al which you are removing by flocculation may be in colloidal form and not bioavailable to the same extent as the more soluble Al species that are introduced during flocculation.

Fawell: That is a different question; we were talking about Al intake, as opposed to bioavailability. The whole bioavailability problem in terms of Al used as a flocculant by the water companies is open to question. There is Al bound to organic matter; if you look at the Al in the raw water, it does appear, by chemical techniques such as fast oxime extraction or dialysis, that the Al is in a form which is less chemically labile and of higher molecular weight than that present after treatment. It is the form of Al in the water before and after treatment that is important, but at present we have not actually investigated what this means for bioavailability in a biological system.

Kerr: If you look back in the literature, the fluctuations in the Al content of tap water in Britain at the time of the epidemics of dialysis-induced encephalopathy were almost entirely determined by whether Al flocculation was used. The differences were enormous; occasionally as much as a 1000 µg of residual Al was being added to the water per litre (Parkinson et al 1981). I don't know whether water authorities are now much better at it, but I would like to be convinced about that.

Fawell: It's not what is added, it's what comes out at the end of the treatment process that is important, because the added Al sulphate is mostly removed. It is the Al remaining in the water when the flocculation process has not been operating properly that reaches the tap. It is certainly true that there was considerable variation in different areas of the UK. That has improved tremendously, but a lot will still depend on the water treatment plant and its operation. It is worth noting that the drinking water standard for Al is 200 µg/l, based on aesthetic considerations.

Day: Our experience is that the water Al levels in the Manchester area are strongly related to whether Al treatment is used, and very little related to the Al content of the raw water that goes into the reservoirs in those areas. We have been looking at Al from 600 home-dialysis patients for 12–13 years, and that is one very clear conclusion (O'Hara et al 1987).

The second conclusion is that there hasn't been much change in the Al level of these patients' water supplies, over those years. We have reduced the Al load to dialysis patients by treating the water, but the water as supplied to the house has hardly changed in its Al content over that period. This is probably a reflection of the treatment works in the Manchester area, and the possibility that they haven't changed their technology very much over that time. But we observe that for water supplies that are *not* treated by the flocculation process, the water Al concentration ranges from five to 50 µg/litre, possibly up to 100 µg/l on rare occasions. Whereas for flocculation-treated water, the level is almost invariably over 50 µg/l and may exceptionally even reach 2000 µg/l. We would regard normal concentrations of Al in Al-flocculated water as being 100–300 µg/l. This reflects the actual practicalities of what's received at the household tap in normal circumstances. And of course it may reflect not only the treatment, but also what has got into the pipes over a long period, and may subsequently be leached out from the pipework.

Blair: Dr Greger, I gather that you have used the absorption of radiolabelled gallium as a marker for Al. This is of great interest, because one of the deficiences of Al studies is the lack of a proper marker. There are a whole series of observations you can't make. But it does seem that every time I listen to comparisons with gallium, it is in fact a good marker!

Greger: We fed ^{67}Ga along with a dose of Al to rats and looked at its excretion and its appearance in tissues. We also injected both together. Generally, aluminium and ^{67}Ga were distributed in tissues pretty similarly in both cases, but ^{67}Ga was not a perfect marker. However, other factors, such as citrate, appeared to affect both ^{67}Ga and aluminium absorption (Ecelbarger & Greger 1991).

Farrar: We also have data on the tissue distribution of ^{67}Ga and aluminium after oral administration. We showed similar areas of tissue accumulation with both metals. Citrate administered with ^{67}Ga increased absorption in fed rats only (Farrar et al 1987, 1988).

Blair: I had just wondered if you found any major discrepancies, Dr Greger, and you do not. It does seem that if you can use ^{67}Ga in intact human subjects, you can get rid of the necessity for balance studies, which you cannot really do because you don't have access to the tissues. You can measure the amount of gallium in the brain, for example.

Williams: There is a chemical problem. If you look at Al, Ga and Fe, then Ga comes somewhere in between in complex ion chemistry, and often nearer to iron than to Al. You might expect gallium as a marker for Al to confuse you, because it will lean to the side of interfering with Fe and away from Al.

Blair: I am not saying that Ga is an exact analogue, but it is a model which you use in the absence of a proper tracer experiment with Al. And even using [26]Al, you cannot for example take out lumps of liver tissue.

Edwardson: When the patient dies, you can. [26]Al has a half-life of 0.75 million years, so this is a feasible experiment!

Blair: I agree that you get excellent value for money when you use [26]Al, but, nevertheless, we want to look at people in life.

Klinowski: Nobody is using the expression 'ionic products' here. But it's a very different proposition to take one teaspoon of the food additive called Coffeemate, which includes 150 mg of readily available Al^{3+} cation, which moves rapidly through cell channels, to worrying about how much comes off the pan, the surface of which is covered with aluminium oxide. The oxide has a very low ionic product, and is not therefore readily solubilized. Talking about Al impurity, or intake, or loading, or content is chemically unhelpful.

Williams: The problem is that these are almost all the data we have. The Al moves in the system and you must follow it as Al species analytically as best you can. If you ask about its speciation, which is what Bruce Martin talked about, it is in chemically defined fluids, and Al is as ionized Al, or Al in specific forms. Once you are outside the context of a test-tube or a chemically defined system, it's very hard for anybody to describe anything but total aluminium. That is what we are up against—the bioavailability problem against the nature of the Al compound we start with.

Klinowski: We have been studying gallium compounds, and they have a very different chemistry from Al. This may be a trap. I know of someone who made his reputation using cadmium, which has a very nice NMR isotope, instead of Ca which doesn't. He says that Cd is just like Ca. My response is: if so, why is cadmium poisonous but calcium isn't? And gallium, incidentally, is also much more toxic than aluminium!

Greger: One point which is often forgotten is that the forms of Al in foods are not necessarily the forms present in the duodenum. The forms of aluminium in food will be affected by the digestive state and by other foods present. That was why I stressed in my paper the effects of other dietary components on aluminium absorption.

Birchall: In a recent study of the excretion of Al by people who work in the Al industry, breathing in a lot of dust and presumably ingesting some, there was evidence for quite a long delay between the consumption and excretion of Al, so it remains in the body for a while, somewhere (Rollin et al 1991).

Greger: That's an interesting point. A number of investigators have attempted to monitor aluminium bioavailability by measuring urinary Al excretion (Ganrot 1986). However, our work demonstrates that some aluminium is retained in tissue. But how much Al is retained? How long is the aluminium retained? Some Al may be retained in tissues for a while and then excreted. There is not enough information on the turnover of Al in tissues.

Birchall: In addition, there are data showing that patients with generalized septicaemia start to excrete Al. It is mobilized from some store, presumably on tissue destruction (Davenport et al 1988).

Greger: We have preliminary data along that line too. We were studying ageing rats. We assumed that if we fed the rats 1 mg Al/g diet for as long as nine months, they would accumulate ever-increasing amounts of Al, especially in bone; but it wasn't so. The Al content of bone went up to a certain level and then levelled off; then soft tissue levels of Al increased. We measured hydroxyproline excretion as an index of bone turnover. We hypothesized that if bone turnover was increased, more Al would be released and hence would move into other tissues. There are several questions. When Al leaves bone, is it bound to transferrin or to some other protein? Does this affect where it goes? Moreover, would enough Al be released from bone during turnover to be practically significant?

Birchall: In other words, we haven't got a mass balance, have we?

Greger: No!

Perl: It is very nice that we are beginning to get some of these data, not only in terms of the availability of Al, but in relation to its retention. Most of these studies are done on young animals, or young adult animals; but I would be interested in the Al retention of really old animals, and what effect ageing has on this process. It may be very different.

Greger: The data are limited. McDermott et al (1979) and Markesbery et al (1981) demonstrated that Al accumulates in brains of humans with age. But Massie et al (1988) did not find that mice accumulated much aluminium in tissues as they aged, even when fed aluminium (270 µg Al/g diet). Similarly, in continuing studies that we are now in the process of doing, we found that rats fed aluminium accumulated aluminium in tissues with age but the effect was not dramatic.

Garruto: You discussed the intake of aluminium in infants. If you calculate the amount of Al ingested per kg of body weight, what figures do you come up with, and how do they compare to those for children and adults?

Greger: A 3 kg infant (newborn) who was consuming 3 µg Al/day in human milk would get a dose of 1 µg/kg. This would be lower than the estimates that I gave for adults (0.15 mg Al/kg) or for two-year-olds (0.48 mg Al/kg). The higher intake of aluminium by two-year-old children may reflect the foods that American children like, such as processed cheeses or the brightly coloured drinks, such as cherry sodas, in which the colour is from Al lakes.

Candy: In relation to Al intake in infants, it has to be remembered that in the rat there appears to be a rapid increase in iron uptake by the brain after birth which peaks at 15 days and then declines (Taylor & Morgan 1990).

Greger: That is why I think it would be interesting to study low birth weight infants with perhaps impaired kidney function, who would be unable to excrete absorbed Al effectively. With these very small infants you often have a so-called 'leaky' gut, which allows greater absorption of at least some minerals. Admittedly, Litov et al (1989) didn't find changes in plasma aluminium levels of term infants fed soy-based infant formulas. But they looked only at plasma Al levels; it's possible that Al could clear from the plasma very rapidly but could accumulate in bone or some other tissue.

Candy: We have been investigating the possible accumulation of Al in the human brain during development, using imaging dynamic secondary ion mass spectrometry (SIMS). We feel that SIMS imaging provides a considerably more sensitive index of aluminium exposure than bulk chemical methods, since we have evidence that significant focal accumulation of aluminium may occur in the brain in the absence of any change in the bulk concentration (Candy et al 1992). The increased discrimination using SIMS is probably due to the focal nature of the aluminium deposits within the tissue, which will be 'diluted out' in bulk samples for chemical analysis.

We have studied 27 human subjects, ranging in age from premature infants to teenagers. Of the 27, only five showed a complete absence of focal accumulations of Al in unfixed, unstained cryostat sections from the frontal cortex. Our series included eight premature infants, three of whom had received total parenteral nutrition (TPN). The three infants who had received TPN, for up to six days, didn't show any increased accumulation of Al in the frontal cortex, compared to the other premature infants. The very high proportion of our cases in which we had evidence of focal Al accumulation in the brain may be due to increased uptake and transport of Al, immediately postnatally. Alternatively, there is the possibility that maternal transfer of Al may occur, so aluminium in the brain might not be coming only from dietary sources after birth.

Blair: What was the source of these children's brains? I am also unclear what general conclusion you draw from them.

Candy: The source was post mortem material from Newcastle. The tentative conclusions are that a very high proportion of the cases studied showed evidence of focal Al accumulations in the frontal cortex, and that there was no clear increase in either the density or the intensity of focal aluminium accumulations in the frontal cortex in infants that had been given total parenteral nutrition, compared to pre-term infants that hadn't received TPN.

Blair: Was the number of foci greater in 5- or 10-year-olds?

Candy: It is difficult at this stage of our investigation to draw hard conclusions. All we can say is that in 22 out of 27 cases examined, we found evidence of focal Al accumulation at a very low level—in the low p.p.m. range— and that there didn't appear to be any clear age-related increase in either the density or intensity of the focal Al accumulations, in this group of patients.

Perl: Dr Greger talked about the vulnerability of the infant exposed to Al. One other aspect relates to exposure to Al in the infant via the DPT vaccine (diphtheria, pertussis and tetanus), which contains aluminium hydroxide as an adjuvant, and is injected intramuscularly, on multiple occasions.

Martyn: Some preparations of DPT vaccine are adsorbed on to an Al adjuvant, and some are not. If the adsorbed version is used, the dose is several milligrams of Al by the time three injections have been given.

Perl: That's a rather substantial dose?

Martyn: Yes. We are giving rather small children a parenteral dose of Al when we immunize them.

Greger: I agree; the experts have not considered enough the importance of pharmaceutical products, particularly injected ones, as sources of aluminium.

Day: In relation to medications, I can give you an interesting example. A patient in our dialysis unit had increasing monthly plasma Al levels. His doctors took him off Al-containing antacids and managed to get his Al levels down again. Then, for three months in succession, we noticed that his plasma Al was again increasing. We looked at every possible route of exposure. We eventually found that his main Al exposure was from the iron capsules he was taking as iron supplements. These were coloured capsules; the Al was in the red coating, to bind the dye, and he was taking 3 mg/day of Al from this unsuspected source. And, of course, the uptake was increased because the iron pills contained an enhancer, intended to increase the uptake of iron, and it assisted Al uptake. So, sources of Al are not always obvious!

Garruto: With respect to total parenteral nutrition (TPN) in non-uraemic infants, it appears that Al loading occurs in bone and liver in these infants. There are also reports that the nervous system in some cases is compromised as well. I think this is an important area that should be pursued.

Greger: Both Sedman et al (1985) and Koo et al (1989) have published data on Al exposure of infants through intravenous therapy. In some of the early work, aluminium-contaminated protein sources were used. The drug companies and Food and Drug Administration were concerned with Sedman's data because they already knew there were problems with Al in renal dialysis patients (Hoiberg 1989). Thus some of the protein sources were contamined with Al and have been taken off the market, and the aluminium contents of parenteral products are often monitored now.

Garruto: I am personally aware that Dr Gordon Klein of the University of Texas Medical Branch at Galveston is currently following a series of TPN patients (both infants and adults, uraemic and non-uraemic) who are Al loaded, so obviously there is still a problem with Al loading during TPN.

Greger: That is why physicians should monitor the aluminium content of parenteral products that are given chronically, particularly to infants and to patients with renal impairment. Ca and P sources are very prone to Al contamination and should be checked.

Strong: In collaboration with Dr Gordon Klein, we have performed the neuropathological analysis of the neonatal cases (Klein et al 1992). The interest was originally generated by tissues that were sent to us from five neonatal infants who had received TPN. In three of these cases we found neuroaxonal spheroids in locations in the CNS where we would not have expected them to occur in infants; they can normally occur in areas of myelination in neonates, but this was outside that territory. We have now looked at a total of 11 children (10 neonates; one six-year-old) who have been exposed to TPN for intervals of six days to about 11 months. The presence or absence of sepsis has not been looked at specifically, but should be.

Gestational ages 27–40 weeks were examined, with an additional case at age six years. Several conclusions arose from this analysis. Firstly, the handling of the tissue for analysis is of paramount importance. When you handle the fresh tissue sample extremely carefully, ensuring that Al-contaminated solutions or utensils are not used, Al deposition is not observed in the brain. Aluminium deposition is however uniformly present at the osteoid–bone interface.

Additionally, neuropathological changes in the CNS that would traditionally be attributed to Al toxicity are not observed. So, at least in the premature infant, the pathological substrate for aluminium accumulation is bone and not the nervous system, and the preferential site of adsorption is to the osteoid–bone interface. But this doesn't address what happens when you load bone with Al and allow the infant to survive: does the Al mobilize from the osteoid–bone interface and, if so, where does it go to thereafter? This might be particularly important for young infants on long-term parenteral nutrition; are we giving them, at a vulnerable point in life, a large dose of Al? It may well be more bioavailable, and if Al is binding at the histone-DNA level, and it's *not* released from there, what are the long-term consequences for gene transcription? That has not yet been addressed.

Edwardson: What technique are you using to localize Al in the brain?

Strong: This is at the moment done simply by immunohistochemistry or by tinctorial staining (using Solochrome azurine).

Edwardson: This would not give sufficient sensitivity to reveal intracellular accumulation of Al at concentrations known to be neurotoxic.

Strong: I agree, but, given the threshold of sensitivity at which we are looking, Al is preferentially adsorbed to the osteoid–bone interface. Unless we actually probe the nucleus, we shall not be able to answer the question of whether Al is not there at all.

Garruto: Are these uraemic or non-uraemic infants?

Strong: Non-uraemic infants.

Kerr: I was very interested in Dr Greger's data on citrate and Al absorption, which seem to suggest that citrate is much less important that we are teaching at the moment. Since the editorial review by Molitoris et al (1989), we have been scared of the ingestion of Al and citrate in renal failure. For instance, we teach our students that they mustn't give sodium citrate to correct acidosis with aluminium hydroxide to prevent phosphate absorption. The data of Molitoris and others (e.g. Bakir 1989) seem to contradict yours, in the context of uraemic patients.

You looked at rats with one kidney and two kidneys. Have you gone further and looked at uraemic rats, to see whether they have this much greater Al absorption with citrate?

Greger: We didn't study uraemic rats. We created a rodent model with some impairment in kidney function but normal growth, because we felt this would be parallel to premature infants with some reduced kidney function. Moreover, uraemic animals generally have reduced growth and food intake. These confounding factors can be overcome by gavage feeding, but gavage feeding is certainly not a normal physiological state. We have done about five studies now and we always find citrate significantly affects Al retention in the tibia; but the effect is not as impressive as in Slanina's data (Slanina & Falkeborn 1984). We think this is because our animals were consuming Al in a nutritionally balanced diet, with 5 mg of Ca/g and 35 μg Fe/g. We showed in another study that if we lowered Ca intake, Al retention increased (Ecelbarger & Greger 1991).

Edwardson: We have data to suggest that there may be significant changes in the absorption of Al in older subjects (Taylor et al 1992). We gave an oral dose of Al citrate, based on absorption studies carried out in young volunteers aged 20–30; we were aiming to double the blood levels of Al after one hour.

When we gave this challenge to a group of 10 elderly controls (age range, 69–76 years) their mean blood Al level rose from 8 µg/l to 38 µg/l at 60 minutes, a response twice as high as that seen in the younger subjects. When we gave the same challenge to a group of 10 age-matched patients with Alzheimer's disease, the blood levels went up on average to 104 µg/l. So there was also a significant difference between the disease group and the control patients in this small study.

The challenge was repeated with 10 older controls (77–88 years) and 10 older Alzheimer's disease patients (79–89 years). There was no significant difference between the peak blood levels in these groups, but this was because the very elderly control group showed evidence of a further increase in absorption with age, peak blood levels being 101 µg/l. This suggests that there is an increase throughout life in the amount of Al absorbed, at least following an Al citrate challenge. In this symposium we shall be discussing possible insults to the ageing brain, which is compromised by a variety of other environmental and age-related factors. Increased accumulation of Al in old age may be very much more important than differences in exposure at birth.

Greger: I agree that age is likely to affect Al retention, and that changes in the gut and in kidney function could make more Al bioavailable.

Birchall: Were your rats fed standard rat chow?

Greger: No, we fed a semi-purified diet, because chow has a lot of Al contamination, as much 400 µg Al/g diet.

Birchall: We also find a lot of silica (as SiO_2) in it, which could influence availability.

Blair: We were talking earlier about transferrin and the binding of iron or Al in the active centre of the transferrin molecule (p 20). This involves the carbonate anion. What is known about how this complex breaks up? Which ion in fact first leaves the active centre—the carbonate anion? Does that then release the iron? Or does the ferric iron leave first, leaving carbonate behind?

Williams: The structure of transferrin is known (Anderson et al 1987). In the iron binding by this structure, there a third event, that the protein folds around the metal site when the metal is finally incorporated; so it's not a single problem of carbonate against iron alone with this protein, since it does not have a static structure. It is an open loop at first and then it closes around the iron and the carbonate. It may be that once the protein opens, the iron and the carbonate come out very fast, one after the other.

Harrison: In the binding of iron to serum transferrin, spectroscopic and kinetic evidence suggests that the synergistic bicarbonate (or the carbonate) binds before iron, and indeed the protein resembles the periplasmic sulphate-binding protein (Anderson et al 1987, Quiocho 1990). It may be that these proteins evolved to bind anions, and then subsequently have developed the capacity to bind iron. But in relation to the release of the metal and delivery to the tissue, you are talking about a different situation in which the protein is bound to the receptor. Although it has been postulated that protonation of the anion is a likely first step in release, there is no real evidence on whether metal or anion comes out first, *in vivo*.

References

Anderson BF, Baker HM, Dodson EJ et al 1987 Structure of human lactoferrin at 3.2-Å resolution. Proc Natl Acad Sci USA 84:1769–1772

Bakir AA 1989 Acute aluminemic encephalopathy in chronic renal failure: the citrate factor. Int J Artif Organs 12:741–743

Candy JM, Oakley AE, Mountfort SA et al 1992 The imaging and quantification of aluminium in the human brain using dynamic secondary ion mass spectrometry (SIMS). Biol Cell 74:109–118

Davenport A, Williams PS, Roberts NB, Bone JN 1988 Sepsis: a cause of aluminium release from tissue stores associated with acute neurological dysfunction and mortality. Clin Nephrol 30:48–51

Ecelbarger CA, Greger JL 1991 Dietary citrate and kidney function affect aluminum, zinc and iron utilization in rats. J Nutr 121:1755–1762

Farrar G, Morton AP, Blair JA 1987 Gallium and scandium as models for aluminium in studies of intestinal absorption and tissue distribution. In: Heavy metals in the environment, vol 2. GEP Consultants, Edinburgh, p 14–16

Farrar G, Morton AP, Blair JA 1988 The intestinal speciation of gallium: possible models to describe the bioavailability of aluminium. In: Brätter P, Schramel P (eds) Trace element analytical chemistry in medicine and biology. Walter de Gruyter, Berlin, vol 5:342–347

Ganrot PO 1986 Metabolism and possible health effects of aluminum. Environ Health Perspect 65:363–441

Greger JL, Goetz W, Sullivan D 1985 Aluminum levels in foods cooked and stored in aluminum pans, trays and foil. J Food Prot 48:772–777

Hoiberg CP 1989 Aluminum in parenteral products: overview of chemistry concerns and regulatory actions. J Parenter Sci Technol 43:127–131

Klein GL, Hodsman AB, Campbell GA et al 1992 Early histochemical detection of aluminum in bone but not in central nervous system of infants on total parenteral nutrition. J Bone Miner Res, in press

Knutti R, Zimmerli B 1985 Untersuchung von Tagesrationen aus schweizerischen Verpflegungsbetrieben. III. Blei, Cadmium, Quecksilber, Nickel und Aluminium. Mitt Geb Lebensmittelunters Hyg 76:206–232

Koo WWK, Kaplan LA, Krug-Wispe SK, Succop P, Bendon R 1989 Response of preterm infants to aluminum in parenteral nutrition. J Parenter Enteral Nutr 13:516–519

Litov RE, Sickles VS, Chan GM, Springer MA, Cordano A 1989 Plasma aluminum measurements in term infants fed human milk or a soy-based infant formula. Pediatrics 84:1105–1107

Markesbery WR, Ehmann WD, Hossain TIM, Alauddin M, Goodin DT 1981 Instrumental neutron activation analysis of brain aluminum in Alzheimer's disease and aging. Ann Neurol 10:511–516

Massie HR, Arello VR, Tuttle RS 1988 Aluminum in organs and diet of ageing C57BL/6J mice. Mech Ageing Dev 45:145–156

McDermott JR, Smith AI, Iqbal K, Wisniewski HM 1979 Brain aluminum in aging and Alzheimer's disease. Neurology 29:809–814

Miller RG, Kopfler FC, Kelty KC, Stober JA, Ulmer NS 1984 The occurrence of aluminum in drinking water. J Am Water Assoc 76:84–91

Molitoris BA, Froment DH, Mackenzie TA, Huffer WH, Alfrey AC 1989 Citrate: a major factor in the toxicity of orally administered aluminum compounds. Kidney Int 36:949–953

O'Hara M, Day BJ, Day JP, Ackrill P 1987 Water aluminium concentrations in renal dialysis. In: Grandjean P (ed) Trace elements in human health and disease. Extended abstracts of the second Nordic symposium, August 1987. WHO Environ Health Ser No. 20, Copenhagen

Parkinson IS, Ward MK, Kerr DNS 1981 Dialysis encephalopathy, bone disease and anaemia: the aluminum intoxication syndrome during regular haemodialysis. J Clin Pathol 34:1285–1294

Quiocho FA 1990 Atomic structure of periplasmic binding proteins and the high-affinity active transport systems in bacteria. Philos Trans R Soc Lond Ser B Biol Sci 326:341–351

Rollin HB, Theodorou P, Kilroe-Smith TA 1991 The effect of exposure to aluminium on concentrations of essential metals in serum of foundry workers. Br J Ind Med 48:243–246

Safe Drinking Water Committee 1982 Drinking water and health, vol 4. National Academy Press, Washington, DC, p 155–167

Sedman AB, Klein GL, Merritt RJ et al 1985 Evidence of aluminum loading in infants receiving intravenous therapy. N Engl J Med 312:1337–1343

Slanina P, Falkeborn Y 1984 Aluminum concentrations in the brain and bone of rats fed citric acid, aluminum citrate or aluminum hydroxide. Food Chem Toxicol 33:391–397

Taylor EM, Morgan EH 1990 Developmental changes in transferrin and iron uptake by the brain in the rat. Dev Brain Res 55:35–42

Taylor GA, Ferrier IN, McLoughlin IJ et al 1992 Gastrointestinal absorption of aluminium in Alzheimer's disease: response to aluminium citrate. Age Ageing 21:80–89

The interrelationship between silicon and aluminium in the biological effects of aluminium

J. D. Birchall

ICI plc, PO Box 11, The Heath, Runcorn, Cheshire WA7 4QE, UK

Abstract. It is well established that aluminium is toxic at the cellular level and that pathological symptoms follow its entry into organisms (plants, fish, humans) when the normal exclusion mechanisms fail or are bypassed, as for example in renal dialysis. The present debate concerns the *availability* of environmental aluminium and the possible impact of its slow and insidious absorption and accumulation in vulnerable individuals. Silicon is considered an essential element but the mechanisms underlying its essentiality remain unknown and binding of the element (through oxygen) with biomolecules has not been demonstrated. There is, however, a unique affinity between aluminium and silicon, not only in solid state chemistry ($[AlO_4]^{5-}$ and $[SiO_4]^{4-}$ are isostructural), but also in aqueous solution chemistry as illustrated by the synthesis of zeolite from aluminate and silicate anions at high pH and under hydrothermal conditions. This affinity exists also in very dilute solution ($<10^{-5}$ M) at near-neutral pH when hydroxyaluminosilicate species form. These species mediate the bioavailability and cellular toxicity of aluminium. The observed effects of silicon deficiency can be attributed to consequential aluminium availability. There are important implications for the epidemiology and biochemistry of aluminium-induced disorders and any consideration of one element must include the other.

1992 Aluminium in biology and medicine. Wiley, Chichester (Ciba Foundation Symposium 169) p 50–68

There are three threads to the story outlined in this paper. One is the cellular toxicity of aluminium; a second is the 'essentiality' of silicon; the third is the question of what determines the bioavailability of aluminium and the cellular response to its presence. With aluminium being so ubiquitous, why, given its cellular toxicity, is there life on the planet? Aluminium is for the most part locked up in the aluminosilicates of rock and soil minerals from which it can be leached by acidity resulting from human activity ('acid rain'). It is released from its oxide (bauxite) in metallic aluminium production, and many aluminium salts and compounds are used in a wide variety of industries.

In 1982, 4×10^6 lb of aluminium were used in food additives in the USA (Greger 1988); the amount of aluminium salts used in water treatment in the UK is of the order of 10^5 tonnes per year. In a near-neutral aquatic environment the availability of aluminium is limited by the precipitation of gibbsite (the crystalline form of aluminium hydroxide, $Al(OH)_3$), at about pH 6; solubility rises at pH lower than this (the 'acid rain' problem) and at more alkaline pH, owing to the formation of the aluminate anion, $Al(OH)_4^-$ (see this volume: Martin 1992). An important limitation to the bioavailability of aluminium is the interaction of aqueous aluminium species with 'dissolved silica'—silicic acid, $Si(OH)_4$— generating hydroxyaluminosilicates (Birchall & Chappell 1988). These species mediate the bioavailability of aluminium and, early indications suggest, the cellular response to aluminium. Soil scientists (Farmer & Frazer 1982) and ocean chemists (Wiley 1975) have known for some time that the presence of dissolved silica limits the solubility of aluminium. Non-dialysable hydroxyaluminosilicate species form at pH 4 and upwards when silicon concentrations exceed 6 mg SiO_2 per litre (Farmer 1986). What has not been recognized is the significance of this control mechanism to biology and its relevance to understanding the epidemiology and biochemistry of aluminium-induced pathology.

Aluminium in biology

When aluminium becomes available to organisms through the acidification of surface waters, it is toxic to plants, affecting root development (Taylor 1988), and to fish, affecting gill function (Driscoll et al 1980). Its toxicity to humans was first clearly recognized in renal medicine, when the element was identified as the causal agent in the neurological and bone disorders observed in patients dialysed with aluminium-containing water (this volume: Kerr et al 1992). The use of reverse osmosis and de-ionization has removed this problem, although the continued use of oral aluminium hydroxide as a phosphate binder still produces aluminium overload in some patients (Fleming et al 1982).

In dialysis-induced aluminium overload, when plasma aluminium levels can rise from less than 1 μmol/litre to above 5 μmol/litre, aluminium accumulates in many tissues, including kidney, liver, skeletal muscle, heart, brain and bone (Roth et al 1984). It gives rise to encephalopathy, osteomalacia, and an anaemia responsive neither to iron therapy nor to erythropoietin; it may also be responsible for the myocardial dysfunction observed in many dialysis patients (London et al 1989). Symptoms can arise within months. The cellular toxicity of aluminium is undoubted, but the mechanisms of its toxicity are not fully understood. The question of current interest is whether exposure to environmental aluminium in food, water, medication and in other ways is a causal agent in various diseases, in particular Alzheimer's disease. The daily human intake of aluminium is estimated to be about 20 mg (Jones & Bennett 1985) and absorption is influenced *inter alia* by dietary constituents such as citrate and maltol. Iron status may influence aluminium absorption (Fernandez Menendez et al 1991). Is a slow and insidious accumulation of aluminium in tissues responsible for disease?

The possible relationship between aluminium and Alzheimer's disease was first suggested by the demonstration of neurofibrillary degeneration in rabbits after direct exposure of the central nervous system to aluminium salts (Klatzo et al 1965). Subsequently, increased aluminium levels were found in the brains of Alzheimer's disease patients; later came the identification of aluminium in tangle-bearing hippocampal neurons (Perl & Brody 1980; this volume: Perl & Good 1992) and at the core of senile plaques (Candy et al 1986; this volume: Edwardson et al 1992).

Silicon in biology

The fact that silicon is present in small concentrations in living organisms has for years prompted questions about a possible functional role. Silicon is considered to be essential for diatoms, not only for the construction of the siliceous frustule, but also for the maintenance of major metabolic processes (Werner 1977). Grasses accumulate silicon, depositing opaline silica as phytoliths in aerial parts, and some plant species (notably rice) are thought to require the element. The experiments that caused silicon to be listed as an essential element in at least some mammalian species were reported in 1972 (Schwarz & Milne 1972, Carlisle 1972). These workers independently showed reduced weight gain and pathological changes to bone and connective tissue in rats and chickens maintained on a silicon-depleted diet—symptoms that could be reversed by silicon (silicate) supplementation. The development of bone was particularly affected, both the formation of the collagenous organic matrix and its mineralization. A role for silicon in osteogenesis had already been suggested by the finding that silicon was uniquely localized at the mineralization front in the bones of young rats and mice (Carlisle 1970). This suggestion was reinforced by the detection of silicon in connective tissue cells and in isolated osteocytes (Carlisle 1982).

There is a striking similarity between the pathological features of bone reported in silicon deficiency and those observed in aluminium-induced osteodystrophy, in which aluminium is localized at the mineralization front (Denton et al 1984) and is found in the osteogenic cells congregated at that front (Schmidt et al 1989). These observations suggested a physiological interaction between these two elements, comparable to their geochemical interaction.

Mechanism underlying the essentiality of silicon

The essentiality of silicon and the possible reasons for it were discussed at an earlier Ciba Foundation Symposium (1986). It was concluded that no Si—C bonds existed in biology, that the existence of stable $Si-O-C$ bonds had not been demonstrated and, indeed, that there was no convincing evidence for any organic binding of silicon in biological systems, in which the element exists as the neutral silicic acid, $Si(OH)_4$. How then does silicon deficiency produce pathological changes?

At that symposium a relationship with aluminium was first proposed (Birchall & Espie 1986) and the effect of silicic acid on an 'aluminium-poisoned' enzyme was demonstrated. It had been suggested that silicon was essential for maximal prolyl hydroxylase activity (Carlisle & Alpenfels 1980), with reduced activity accounting for impaired collagen production in silicon deficiency. It is difficult to envisage any chemical interaction between silicic acid and the known cofactors of this enzyme or the protein itself. The cofactor metal is iron; it was shown that aluminium could act as a weak inhibitor of hydroxyproline production and that this inhibition was eliminated when sufficient silicic acid was present. Although it is almost certainly incorrect to suppose that aluminium will generally interfere with iron-dependent systems, the described effect is reminiscent of the inhibition of hexokinase activity by aluminium and reactivation of the system by citrate, which displaces aluminium from ATP and allows Mg^{2+} to bind (Viola et al 1980). Silicic acid selectively removes aluminium from the aluminium-contaminated prolyl hydroxylase system; preliminary work suggests that a similar effect can be observed in the aluminium-contaminated hexokinase system (C. Exley, personal communication 1991).

The first demonstration that silicic acid prevents a toxic effect of aluminium in an organism was in experiments on Atlantic salmon fry (*Salmo salar*), which are susceptible to low concentrations (6–7 µmol/l) of aluminium in acidic water. Gill damage was induced, with loss of ionoregulatory and osmoregulatory function (Exley et al 1991). The fish were exposed to water at about pH 5 containing various concentrations of aluminium and silicon (Birchall et al 1989). Figure 1 shows that water containing 6.26 µmol/l Al with only 0.60 µmol/l Si was acutely toxic, whereas water containing 7.15 µmol/l Al and 93 µmol/l Si was innocuous. Furthermore, the average whole-body aluminium content of the fish was >2 µmol/g dry mass in the toxic water, but only 0.40 µmol/g dry mass in the silicon-rich water, 10% less even than that accumulated in the control water (0.85 µmol/l Al, 0.66 µmol/l Si). The water containing 6.26 µmol/l Al and 0.60 µmol/l Si caused extensive gill damage with mucus production and cellular sloughing: gills remained intact in the silicon-rich water and no aluminium was detected at gill surfaces. In the presence of a high level of silicic acid, the adsorption of aluminium onto gill membranes and systemic absorption had both been inhibited.

This effect of silicic acid in promoting the exclusion of aluminium (and manganese) appears also to apply to plants (Foy et al 1978).

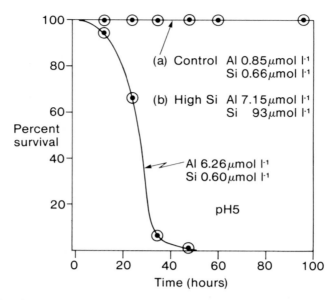

FIG. 1.   Survival curves for Atlantic salmon fry (*Salmo salar*) in water at about pH 5 containing various concentrations of aluminium and silicic acid. Note that the 'control' (a) and 'high Si' (b) conditions are represented by the same (zero death) curve.

Interaction of aqueous aluminium species with silicic acid

The speciation of aluminium in aqueous solution is strongly pH dependent, basic species being produced by the hydrolysis of $[Al(H_2O)_6]^{3+}$ above pH 4 (this volume: Martin 1992). The speciation of silicic acid is relatively simple; at pH <9 and concentrations less than 2 mM, silicic acid exists as the neutral monomer, $Si(OH)_4$. This is the physiologically relevant species. At concentrations and pH values found in aquatic environments (<300 μmol/l) and in biological systems (5–200 μmol/l, say) stable reactions with organic molecules have not been demonstrated. Silicic acid, however, reacts with basic metal ions at a pH just below that at which the hydroxide is precipitated. Thus, at physiological pH there can be no interactions with Ca^{2+} or Mg^{2+}, but interactions with Fe^{3+} and Al^{3+} are possible. Al—O—Si bonds are formed readily at pH 5 and above, whereas Fe—O—Si complexes exist only below pH 2 (Schenk & Weber 1968). Above pH 2, finely divided iron-oxy-hydroxide coated with adsorbed silica is formed. Thus the chemistry suggests that the activity of silicic acid in biology will be restricted to reactions with aluminium.

When dilute (around 10^{-4} M) acidic (about pH 5) solutions of silicic acid and aluminium are heated, the mineral imogolite is precipitated. This tubiform structure has the ideal composition $(HO)_3Al_2O_3SiOH$ and can be considered as a single gibbsite sheet with inner hydroxyls replaced by orthosilicate

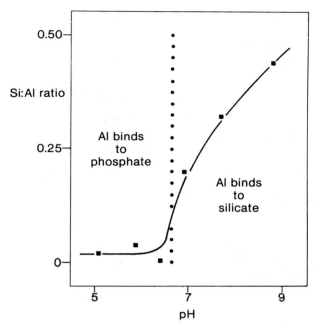

FIG. 2. The Si:Al ratio of hydroxyaluminosilicate species bound to the functional groups on aminodiphosphonate resin, as a function of pH.

(Farmer & Frazer 1982). Unheated, such solutions remain clear and stable for many weeks, but have been shown to contain hydroxyaluminosilicate species by the infrared examination of solids recovered after freeze-drying solutions (Farmer et al 1979) and by ion-exchange experiments (Birchall & Chappell 1988a, Chappell & Birchall 1988). These species have an Si:Al ratio of 0.25–0.5 and can be regarded as fragments of or precursors to the imogolite structure. Such 'protoimogolite' species form above pH 7, even in the presence of citric acid, an aluminium chelator. For example, solutions containing $AlCl_3$ (0.10 µmol/l) with equimolar citrate and silicic acid (0.5 µmol/l) adjusted to pH 7.4 yield no filterable solids after 12 weeks, yet solutions left for 20 hours can be shown by ion-exchange to contain hydroxyaluminosilicate species with a Si:Al ratio of about 0.5.

These interactions of aluminium with silicic acid are complicated by the presence of phosphate. Solutions containing 0.1 mM aluminium, 0.5 mM silicic acid and 0.5 mM phosphate gave solids in which the Si:Al ratio varied with pH—at pH 7.4 it was 0.44, whereas at pH 6.4 the silicon content was negligible and the Si:Al ratio was 0.02. The switch in the binding of aluminium between phosphate and silicic acid is illustrated in Fig. 2. Solutions with the above composition were passed over an aminodiphosphonate resin and the retained species were analysed. The situation is summarized as follows.

Aluminium   Hydroxyaluminosilicate
  phosphate   pH | $Al^{(3-x)+}$ | pH   species (Si:Al ratio
  species plus <6.6 | PO_4^{3-} | >6.6 0.25–0.5) plus free
  free silicic acid   ← | $Si(OH)_4$ | →   phosphate

The formation of protoimoglite species with a Si:Al ratio of 0.5 requires a silicic acid concentration of 100 µmol/l or more. Hydroxyaluminosilicate species with lower Si:Al ratios (0.1–0.5) are generated at lower concentrations.

The formation of hydroxyaluminosilicates clearly reduced the toxicity of aluminium to fish and many of the reported biological effects of silicon can be interpreted in terms of the chemistry outlined above. For example, the beneficial effects of silicic acid on plant growth are likely to be due to its demonstrated abilities to limit the uptake of toxic metals. Above pH 6.6, the ability of silicic acid to prevent aluminium from binding phosphate aids plant (and diatom) growth by maintaining phosphate availability.

Biological consequences of aluminium–silicic acid interactions

We need now to consider two possibilities: firstly, the effect of silicic acid on the availability of ingested aluminium and, secondly, the effect of plasma and tissue silicic acid on the cellular response to aluminium.

Availability of ingested aluminium

Does the effect of silicic acid observed on fish gills also apply to the mammalian intestine? In normal human subjects, aluminium ingestion is about 20 mg/day. Silicon ingestion has been estimated at 20–50 mg/day, most (60%) of which is derived from cereals. Water and other drinks can provide 19%, presumably in a highly absorbable form (Pennington 1991). The contribution from water will vary geographically; silicon concentrations in water range from 10 µmol/l to 300 µmol/l, being generally high in hard water areas and low in soft water areas (Dobbie 1982). Absorbed silicon is rapidly excreted in urine, as is shown by studies of patients taking magnesium trisilicate (Dobbie & Smith 1986). Few studies have been reported on the effect of a high silicic acid intake on aluminium absorption in the gastrointestinal tract. Aluminium dietary supplementation increased brain aluminium levels in 28-month-old rats on a low silicon diet, but no such increase was observed in the brains of rats on a silicon-supplemented diet (Carlisle & Curran 1987). A study of aluminium absorption in patients on oral aluminium hydroxide therapy in areas with high or low silicion levels in water is warranted.

Plasma silicic acid concentrations are normally in the range 5–20 µmol/l, but can rise to above 100 µmol/l in renal disease. An important question is whether a sustained high level of silicon in plasma protects against aluminium-induced disorders such as osteomalacia and encephalopathy. Experiments to investigate this are planned.

Cellular response to aluminium and effect of silicic acid on it

A major cause of aluminium's toxicity is its ability to displace Mg^{2+} from key sites at which this metal is catalytic. Aluminium will also bind to phosphate groups, especially when two or more phosphate groups can cooperate. Therefore aluminium is commonly bound to chromatin and to phosphorylated polymers (this volume: Perl & Good 1992). This drew attention to the possible effects of aluminium on the inositol phosphate second messenger system and hence on intracellular Ca^{2+} homeostasis (Birchall & Chappell 1988b).

In isolated pancreatic cells, acetylcholine provokes an oscillatory increase in cytoplasmic Ca^{2+} which can be monitored by measuring the Ca^{2+}-dependent chloride current. Intracellular infusion of an $AlCl_3$-containing solution using a fine tube inserted into a patch-clamp pipette eliminated or attenuated the response to acetylcholine (Wakui et al 1990; this volume: Petersen et al 1992). The Ca^{2+} release channel is activated by micromolar Ca^{2+}, but inhibited at millimolar levels. It is possible that aluminium has a high affinity for this site and blocks Ca^{2+} release. It is considered that aluminium can successfully compete for Mg^2-binding sites even in the presence of a 10^8-fold molar excess of Mg^{2+}. This demonstrated effect on Ca^{2+} homeostasis supports the perceptive comment made by Kruck & McLachlan (1988) that: 'aluminium is implicated as an agent involved in disruption of Ca^{2+} dependent electrophysiologic functions and of intra-cellular Ca^{2+} regulation'.

Preliminary experiments (this volume: Petersen et al 1992) have shown that the injection of silicic acid (100 µM solution) alone into pancreatic cells *stimulates* the response to acetylcholine. This might be due to 'neutralization' of traces of aluminium introduced during cell isolation (such contamination has been shown in our laboratory to be inevitable). The co-injection of aluminium and silicic acid greatly reduced the inhibitory effect of aluminium. If this latter effect is confirmed, the potential protective effect of elevated levels of plasma/tissue silicic acid will be strongly indicated.

Recent *in vitro* work (Wiegland 1990) has indicated that silicic acid stimulates the activity of adenylate cyclase in several tissues, including bone, kidney and liver. It has been suggested that this effect may be due to silicic acid binding excess Ca^{2+} which would otherwise inhibit the enzyme. At physiological pH this is an unlikely explanation. A more plausible mechanism is competition for binding sites between trace aluminium (introduced during tissue manipulation) and Mg^{2+}, followed by removal of aluminium by silicic acid.

Implications of aluminium–silicon interaction

Epidemiological studies relating aluminium in water to the incidence of Alzheimer's disease (Martyn et al 1989; Martyn 1992: this volume) raise several problems, notably a lack of correlation between dose and effect, and the fact that the intake of aluminium from water is but a fraction of that from food. Water that contains significant amounts of aluminium (soft water and that from upland areas) contains little silicic acid, whereas water high in silicic acid (hard) contains little aluminium. The epidemiological studies therefore suggest an inverse relationship between *silicon* in water and Alzheimer's disease, with a high silicic acid intake restricting the absorption of aluminium from food (Birchall & Chappell 1989).

The ingestion of aluminium in food will be relatively constant throughout a population. Superimposed on this will be the dietary silicon intake (mainly from cereals and vegetables), but this silicon may be less available for reaction with ingested aluminium than the monomeric silicic acid in water. The silicon concentration in domestic water can vary from $10\,\mu mol/l$ to $>200\,\mu mol/l$ and any exclusion of aluminium would be likely only when water contains more than $100\,\mu mol/l$ silicic acid. Fresh epidemiological studies may reveal such a relationship.

Recent results indicate that a high plasma silicon content may protect against the detrimental effects of absorbed aluminium. Studies of aluminium and silicon excretion in patients after kidney transplantation suggest that the two elements are excreted in tandem (N. Roberts, personal communication 1991). Are they combined together as hydroxyaluminosilicates? It appears that much of the absorbed aluminium is temporarily (and some permanently) stored in tissue, so there is a delay before excretion (Schlatter & Steinegger 1991). Preliminary work (J. P. Belia & N. Roberts, personal communication 1991) suggests that a high silicic acid intake may mobilize aluminium and reduce its tissue storage. There is therefore a puzzle.

Solid silica (for example, phytoliths in cereal foods) will adsorb aluminium and so prevent its gastrointestinal absorption. A high intake of silicic acid may, through the formation of hydroxyaluminosilicates of low molecular mass, increase gastrointestinal availability (to high affinity sites), but compensate by increasing aluminium excretion. Although most (95%) aluminium in plasma is bound to high molecular mass ($>5\,kDa$) components (Day et al 1991), the rest is contained in a low molecular mass form. Citrate (0.1 mM in plasma) is a candidate carrier, but plasma silicic acid may contribute, because hydroxyaluminosilicates form in the presence of citrate above pH 7 and remain of low molecular mass. These would be expected to be rapidly excreted, prompting the mobilization of aluminium.

Finally, the switch in the binding of aluminium from phosphate to silicic acid above about pH 6.6 (Fig. 2, p 55) suggests that intracellular aluminium will be bound to phosphate groups, with binding to silicic acid being possible only in the extracellular milieu. Intracellular aluminium will cause cell death (by interference with Ca^{2+} homeostasis—this will have consequences for phosphorylation, for example); after which the export of aluminium (bound to phosphate groups on macromolecules, which would thereby be cross-linked) into the extracellular environment would allow its binding to silicic acid and 'co-deposition' in senile plaques. This is seen as a 'late stage' event, following primary cell damage.

In summary, it is proposed that the role of silicon (silicic acid) is to aid the exclusion of aluminium from organisms, to 'sequester' the metal, so reducing its effect on enzymic and Mg^{2+}-dependent processes, and to reduce tissue retention, promoting the excretion of aluminium. Much work is required to confirm these ideas, which could have far-reaching implications in renal medicine and in the understanding of (and possibly the control of) aluminium-related disorders, including Alzheimer's disease, if it is shown that aluminium has a role in this condition.

Acknowledgements

I would like to thank colleagues who over the years have contributed to this research, especially Dr A. Espie, Dr J. S. Chappell and Dr C. Exley.

References

Birchall JD, Chappell JS 1988 The solution chemistry of aluminium and silicon and its biological significance. In: Thornton I (ed) Geochemistry and health. Science Reviews Ltd, Northwood, Middlesex, UK, p 231–342

Birchall JD, Espie AE 1986 Biological implications of the interaction of silicon (via silanol groups) with metal ions. In: Silicon biochemistry. Wiley, Chichester (Ciba Found Symp 121) p 140–159

Birchall JD, Chappell JS 1988a The chemistry of aluminium and silicon in relation to Alzheimer's disease. Clin Chem 34:265–267

Birchall JD, Chappell JS 1988b Aluminium, chemical physiology and Alzheimer's disease. Lancet 1:1008–1010

Birchall JD, Chappell JS 1989 Aluminium, water chemistry and Alzheimer's disease. Lancet 1:953

Birchall JD, Exley C, Chappell JS, Phillips MJ 1989 Acute toxicity of aluminium to fish eliminated in silicon-rich acid waters. Nature (Lond) 338:146–148

Candy JM, Oakley AE, Klinowski J et al 1986 Aluminosilicates and senile plaque formation in Alzheimer's disease. Lancet 1:354–357

Carlisle EM 1970 Silicon: a possible factor in bone calcification. Science (Wash DC) 167:179–180

Carlisle EM 1972 Silicon: essential element for the chick. Science (Wash DC) 178:154–156

Carlisle EM, Alpenfels WF 1980 A silicon requirement for normal growth of cartilage in culture. Fed Proc 37:1123 (abstr)

Carlisle EM 1982 The nutritional essentiality of silicon. Nutr Rev 40:193–198

Carlisle EM, Curran MJ 1987 Effect of dietary silicon and aluminium on silicon and aluminium levels in rat brain. Alzheimer Dis Assoc Disord 1(2):83–89

Chappell JS, Birchall JD 1988 Aspects of the interaction of silicic acid with aluminium in dilute solution and its biological consequences. Inorgan Chim Acta 153:1–4

Ciba Foundation 1986 Silicon biochemistry. Wiley, Chichester (Ciba Found Symp 121)

Day JP, Barker J, Evans LJA et al 1991 Aluminium absorption studied by [26]Al tracer. Lancet 337:1345

Denton J, Freemont AJ, Ball J 1984 Detection and distribution of aluminium in bone. J Clin Pathol 37:136–142

Dobbie JW 1982 Silicate nephrotoxicity in the experimental animal. Scot Med J 27:10–16

Dobbie JW, Smith MJB 1986 Urinary and serum silicon in normal and uraemic individuals. In: Silicon biochemistry. Wiley, Chichester (Ciba Found Symp 121) p 194–213

Driscoll CT, Baker JP, Bisigni JJ, Schofield CL 1980 Effect of aluminium speciation on fish in dilute acidified waters. Nature (Lond) 284:161–164

Edwardson JA, Candy JM, McArthur FK et al 1992 Aluminium accumulation, β-amyloid deposition and neurofibrillary changes in the central nervous system. In: Aluminium in biology and medicine. Wiley, Chichester (Ciba Found Symp 169) p 165–185

Exley C, Chappell JS, Birchall JD 1991 A mechanism for acute aluminium toxicity in fish. J Theor Biol 151:417–428

Farmer VC 1986 Sources and speciation on aluminium and silicon in natural waters. In: Silicon biochemistry. Wiley, Chichester (Ciba Found Symp 121) p 4–23

Farmer VC, Frazer AR 1982 Chemical and colloidal stability of sols in the Al_2O_3-Fe_2O_3-SiO_2-H_2O system: their role in podzolization. J Soil Sci 33:737–742

Farmer VC, Frazer AR, Tait JM 1979 Characterisation of the chemical structures of natural and synthetic aluminosilicate gels and sols by infra-red spectroscopy. Geochim Cosmochim Acta 4:1417–1420

Fernandez Menendez MJ, Fell GS, Brock JH, Cannata JB 1991 Aluminium uptake by intestinal cells: effect of iron status and precomplexation. Nephrol Dial Transplant 6:672–674

Fleming LW, Stewart WK, Fell GS, Halls DJ 1982 The effect of oral aluminium therapy on plasma aluminium levels in patients with chronic renal failure in an area with low water aluminium. Clin Nephrol 17:222–227

Foy CD, Chaney RL, White MC 1978 The physiology of metal toxicity to plants. Annu Rev Plant Physiol 29:511–566

Greger JL 1988 Aluminum in the diet and mineral metabolism. In: Sigel H, Sigel A (eds) Metal ions in biological systems, vol 24: Aluminum and its role in biology. Marcel Dekker, New York, p 199–215

Jones KC, Bennett BG 1985 Exposure commitment assessments of environmental pollutants. Monitoring and Assessment Research Centre Report No. 3, vol 4, King's College, University of London

Kerr DNS, Ward MK, Ellis HA, Simpson W, Parkinson IS 1992 Aluminium intoxication in renal disease. Wiley, Chichester (Ciba Found Symp 169) p 123–141

Klatzo I, Wisniewski H, Streicher E 1965 Experimental production of neurofibrillary degeneration. J Neuropathol Exp & Neurol 24:187–199

Kruck TPA, McLachlan DR 1988 Mechanisms of aluminum neurotoxicity—relevance to human disease. In: Sigel H, Sigel A (eds) Metal ions in biological systems, vol 24: Aluminum and its role in biology. Marcel Dekker, New York, p 285–314

London GM, de Vernejoul MC, Fabiani F, Marchais S, Guerin A et al 1989 Association between aluminum accumulation and cardiac hypertrophy in haemodialysed patients. Am J Kidney Dis 12:75–83

Martin RB 1992 Aluminium speciation in biology. In: Aluminium in biology and medicine. Wiley, Chichester (Ciba Found Symp 169) p 5–25

Martyn CN 1992 The epidemiology of Alzheimer's disease in relation to aluminium. In: Aluminium in biology and medicine. Wiley, Chichester (Ciba Found Symp 169) p 69–86

Martyn CN, Barker DJP, Osmond C, Harris EC, Edwardson JA, Lacey RF 1989 Geographical relationship between Alzheimer's disease and aluminium in drinking water. Lancet 1:59–62

Pennington JAT 1991 Silicon in foods and drinks. Food Addit Contam 8(1):97–118

Perl DP, Brody AR 1980 Alzheimer's disease: x-ray spectrometric evidence of aluminum accumulation in neurofibrillary tangle-bearing neurons. Science (Wash DC) 208:297–299

Perl DP, Good PF 1992 Aluminium and the neurofibrillary tangle: results of tissue microprobe studies. In: Aluminium in biology and medicine. Wiley, Chichester (Ciba Found Symp 169) p 217–236

Petersen OH, Wakui M, Petersen CCH 1992 Intracellular effects of aluminium on receptor-activated Ca^{2+} signals in pancreatic acinar cells. In: Aluminium in biology and medicine. Wiley, Chichester (Ciba Found Symp 169) p 237–253

Roth A, Nogues C, Galle P, Drueke T 1984 Multiorgan aluminium deposits in a chronic haemodialysis patient. Virchows Arch A Pathol Anat Histopathol 405:131–140

Schenk JE, Weber WJ 1968 Chemical interactions of dissolved silica with Fe(II) and Fe(III). J Am Water Works Assoc 60:199–212

Schlatter C, Steinegger AF 1991 Messung der Aluminiumexposition an Arbeitsplatzen in der Aluminiumprimarindustrie. Erzmetall 44(6):326–331

Schmidt PF, Zumkley H, Barckhaus R, Winterberg B 1989 Distribution patterns of aluminum accumulations in bone tissue from patients with dialysis osteomalacia determined by LAMMA. In: Russel PE (ed) Microbeam analysis. San Francisco Press, San Francisco, p 50–54

Schwarz K, Milne DB 1972 Growth promoting effects of silicon in rats. Nature (Lond) 239:333–334

Taylor GJ 1988 The physiology of aluminium phytotoxicity. In: Sigel H, Sigel A (eds) Metal ions in biological systems, vol 24: Aluminum and its role in biology. Marcel Dekker, New York, p 123–163

Viola RE, Morrison JF, Cleland WW 1980 Interaction of metal(III)–adenosine 5'-triphosphate complexes with yeast hexokinase. Biochemistry 19:3131–3137

Wakui M, Itaya K, Birchall JD, Petersen OH 1990 Intracellular aluminium inhibits acetylcholine- and caffeine-evoked Ca^{2+} mobilization. FEBS (Fed Eur Biochem Soc) Lett 267:301–304

Werner D 1977 Silicate metabolism. In: Werner D (ed) The biology of diatoms. Blackwell, Oxford, p 110–149

Wiegland KE 1990 Modulation of adenylate cyclase response (Patent): International Publication Number W090/07574

Wiley J 1975 Silica–alumina interactions in sea water. Mar Chem 3:241–251

DISCUSSION

Garruto: You referred to the possible 'co-deposition' of Al and Si in senile plaques. In 1984 we reported the first chemical mapping (imaging) of the intraneuronal co-deposition of Ca and Al, and two years later Si, in neurofibrillary tangle-bearing hippocampal neurons in patients with amyotrophic lateral sclerosis (ALS) and parkinsonism-dementia (PD) of Guam (Garruto et al 1984, 1986). Our order-of-magnitude estimates for the highest concentrations of these elements in neurons were 500 p.p.m. for Al, 3000 p.p.m. for Si and 7000 p.p.m. for Ca.

Williams: Can I just clarify that these deposits were *inside* the cells? How did you make the analysis?

Garruto: Yes, these were intracellular deposits. We used a novel application of computer-controlled X-ray microanalysis and wavelength dispersive spectrometry to produce chemical maps of these intracellular deposits (see p 230).

Williams: So the cells were dead?

Garruto: The studies were conducted using both fixed and frozen autopsy tissue and the elemental deposits were found in neurofibrillary tangle-bearing neurons. The deposits appear to be 'ghost' neurons as well as in intact neurons— a point which I shall amplify later (p 230–231).

Birhcall: One problem has been that people who have analysed for Si in tissue have seldom analysed for Al, and vice versa. It is important to know if the two elements occur together.

Garruto: I agree.

Candy: Dr Garruto refers to the co-localization of Si and Al in the Guam cases. In contrast, in renal dialysis patients, we find, using SIMS, numerous focal accumulations of Al in the frontal cortex in the absence of any evidence of Si accumulation, even in cases with the highest Si concentrations determined using graphite furnace atomic absorption spectrometry. We find co-localization of Al and Si only in the core of senile plaques. Whether silicic acid actually gets inside neurons is another question.

Birchall: There is a pH-sensitive competition between phosphate and silicic acid binding to aluminium, with Al–phosphate bonds being favoured at acidic pH and Al–silicic acid bonds being favoured at pH values above 6.6 (Birchall & Chappell 1988a) (Fig. 2, p 55). At around pH 6.6, mixed binding is possible ('silicophosphate'). This pH sensitivity will dictate the intracellular and extracellular binding of aluminium.

Martin: The cross-over in solubilities described by Derek Birchall, with $AlPO_4$ more insoluble in acidic acid and aluminosilicates more insoluble in basic solutions, is evident from the results in Table 1 of my paper (p 9) (this volume: Martin 1992). At pH 6.6, $AlPO_4$ exhibits the greater pAl (lower $[Al^{3+}]$), while at pH 7.4, kaolinite shows a greater pAl. The cross-over has also been illustrated graphically (Martin 1990).

Blair: Dr Candy told us earlier (p 43–44) about studies on focal Al deposition in the neonate and newborn. Was there any co-deposition of Si there?

Candy: We didn't look at Si in those cases; we have only studied Si, using SIMS, in renal dialysis patients, so far.

Perl: We too have probed the brains of the Guam ALS/parkinsonism-dementia complex cases (Perl et al 1986), and in *intact* tangle-bearing neurons we *don't* find these high levels of Si in the tangled neurons that Dr Garruto and his colleagues found. Whether that's true of cells that have died and have the remaining so-called 'tombstone' tangles is another story. But, as you say, the concept of Al and Si being in an intact cell is difficult to accept.

Birchall: One must be careful to distinguish between finding Al and Si 'co-localized' and the existence of aluminosilicate species. In the former, the two elements may be at the same site but they need not be chemically combined. When phosphate is present, aluminosilicates require a pH above 6.6 to form.

Perl: In the intact neurons that we have probed, we don't find Si accumulating with Al.

Garruto: Theoretically we cannot differentiate whether neurons are intact or not, using X-ray microanalysis.

Williams: It is important to discover whether the living cell is allowing these ions to come in, and if you just do post mortem analysis, of course, you won't know whether the Al is coming in after the death of the cell. This leads on to what could be mechanistically important in the nature of aluminium deposits.

Klinowski: Professor Birchall, you are saying that silicate ions are produced by solubilization of silica, which is measurable, even at room temperature. Aluminosilicate species form when you introduce Al into the system. These are insoluble, but are also difficult to crystallize, and therefore they form a 'subcolloid', which is completely harmless to man or beast, I gather?

Birchall: Certainly, these species are harmless to fish and to plants, because the aluminium in them is unavailable.

Klinowski: This process explains the increased solubility of Al, because you are removing it from the solution by producing the subcolloid. By the law of mass action, you can introduce more and more Al. But in this case, why don't the water companies, having added aluminium sulphate to water, titrate the excess of Al with silicic acid?

Birchall: That is a very good point!

Fawell: This would require making considerable changes in water treatment procedures and water treatment works, and Professor Birchall's suggestions are relatively recent. Also, in the UK, when you look at the northwest–southeast distribution pattern of silica in water, you can also correlate it with a range of other things. I am not denying what Derek Birchall says. We need to look more closely and see how some of these other things (such as hardness) correlate with Al, silica, and the incidence of Alzheimer's disease as well.

Birchall: I agree that this is a complex problem. One of the curious paradoxes is that the areas in the UK which have low Si levels in water happen to be the areas where most often you need to use aluminium salts, in order to clarify the water. Whereas in areas with high Si, you don't need to use Al in water treatment; so everything works against you, in this sense. People living in areas with Al in water will, I suggest, be exposed to Al in water *and* in food with no 'neutralizing' silicic acid, or too little of it.

Kerr: When Jim Dobbie described the Si levels in tap water and in uraemic patients he showed, as you would expect from your data, higher serum Si levels in London than in Glasgow or Edinburgh (Dobbie & Smith 1986). The levels correspond to the Si concentrations in the dialysis fluid, because in those days we were not treating the tap water by de-ionization or reverse osmosis and Si was readily taken up by the patients in the course of one dialysis. Nowadays most renal units remove all solutes from water used for dialysis, yet you are still finding similar differences in serum Si. Are they therefore traceable to Si in the drinking water?

Birchall: Yes, in most cases. There might be a difference of 20-fold in silicon levels in drinking water from different regions, sufficient to produce quite significant differences in plasma silicon.

Kerr: And you absorb enough Si from drinking water to raise the plasma level substantially?

Birchall: Yes; you absorb almost all the silicon in water.

Martin: Dr Birchall, would you say more about the postulated Al-citrate-silicic acid ternary complex?

Birchall: It used to be said by the soil scientists that hydroxyaluminosilicates, such as imogolite, didn't form in the presence of citrate. In fact, the truth seems to be that the imogolite solid phase doesn't form in the presence of citrate, but its precursors seem to form, so that it 'wants' to crystallize; there are aluminosilicate units of the crystal in solution which don't aggregate into a crystal when citrate is present. Imogolite can be considered as a single gibbsite sheet with $Si(OH)_4$ adsorbed on the surface. We seem to have citrate adsorbed as well, sitting on that plane. This complex can't grow to form imogolite solid phase, but subcolloidal or solution species exist, containing aluminium, silicic acid and citric acid (Lou & Huang 1989).

Blair: The key to your work seems to be the ratio of Si to Al, not the absolute concentrations of each, in an aqueous phase. How do these ratios conform to the ratios of Al and Si in human plasma?

Birchall: Al in plasma is normally less than $1.0 \mu M$. Silica in normal plasma varies between 5 and $20 \mu M$. It can rise to $100 \mu M$. Hydroxyaluminosilicate species with low Si:Al ratio (0.1) can be detected at silicic acid concentrations as low as $10^{-6} M$ at pH 7.2. The Si:Al ratio of the species rises with silicic acid concentration to that of imogolite precursors (0.5, the ideal composition of imogolite being $(HO)_3Al_2O_3SiOH)$ at $10^{-4} M$ silicic acid. At plasma pH, silicic acid can be a significant ligand for aluminium, its potential for binding rising with concentration, perhaps in cooperation with citrate.

We looked at the effect of silicic acid and citrate on the competitive binding of Fe^{3+} and aluminium by desferrioxamine at pH 7.2 (Birchall & Chappell 1988b). The binding of Fe^{3+} was *promoted* in the presence of silicic acid/citrate, which cooperate in keeping Fe^{3+} hydroxide of low molecular mass and dispersed. The binding of aluminium was *inhibited* in the presence of silicic acid, at least in the early stages of the reaction, due to hydroxyaluminosilicate formation. This suggests that silicic acid could influence the kinetics of ligand exchange of aluminium *in vivo*. Silicic acid (from magnesium trisilicate, or beer, which can contain $> 500\,\mu M$ Si, etc.) is rapidly absorbed and excreted in urine. A high plasma level resulting from continuous ingestion (as in a hard water area) might be expected to influence not only the availability of ingested aluminium, but also the destination of absorbed aluminium—that is to say, its entry into cells, mobilization, excretion, and so on.

Day: In the early days of purifying water for renal dialysis at home, de-ionizers were used; more recently it has become fashionable to use reverse osmosis. There is a big difference in the effect on silicon. The de-ionizer doesn't remove Si from water; in fact, our experience is that it often increases it, because in most de-ionizers the resins are in fibre-glass containers and the Si concentration rises enormously as the water goes through. Whereas reverse osmosis removes all the solute components, ions or molecules, and Si levels will be very low. So your observations could be very important for the practical treatment of renal patients.

Birchall: I am aware of the difference between the early de-ionization, which put silica into water and took cations out, and what is done now, using reverse osmosis. We are looking retrospectively at that, and hoping to find a difference in patients exposed to the two conditions. In other words, in the old days, in high-silicon water areas, one might expect no obvious aluminium toxicity.

We have variable silicon levels in tap water in Liverpool, with an enormous variation of $10–200\,\mu g$ per litre in different areas. But if it's right that the effects of Al are neutralized by high plasma levels of Si, that would be extremely important.

Fawell: How important do you think high Si levels are in relation to the availability of Al from other environmental sources such as food? And how does that compare to the importance of Si actually being absorbed and providing silica which will combine with Al in the blood plasma, and help to remove the Al? In other words, Si has two roles, stopping Al going in and helping it to come out. Which is more significant?

Birchall: A good question! In the fish, the gill is rather a curious organ; on one side of the gill is the lake, and the other side is the fish! It is clearly the case that Si stops Al adsorbing to the gill membranes and to mucus. It equally stops it getting absorbed, so we don't find Al in the chloride cells of the gill. So Si stops Al getting into the fish itself. Also, Si stops Al getting into root cell membranes in plants. One might think that this happens in the gut too. However, that has yet to be demonstrated. We don't know enough about the mechanism of aluminium absorption to predict with certainty, and it is difficult to extrapolate from fish gill and plant root cell membranes to intestinal epithelia. I expect absorption to be reduced. There is an argument for increased aluminium absorption in the gut because hydroxyaluminosilicates are subcolloidal (not precipitates). However, in this case, I would expect rapid clearance by the kidneys and decreased 'hang up'.

Edwardson: What do you mean by this 'hang-up' in the excretion of Al?

Birchall: One recent paper (Rollin et al 1991) suggests that if you absorb Al and measure the amount excreted there is a delay in excretion, and there is retention.

Edwardson: If you are inhaling dust which contans Al into the lungs, and then expectorating and swallowing it and subsequently absorbing Al from the gastrointestinal tract, there will be a delay between exposure and excretion. Similarly, with a large Al exposure, there will be uptake into bone, which acts as a sink, with subsequent slow release and excretion. However, the evidence suggests that most circulating Al is excreted rapidly.

Birchall: We have not got a mass balance; as a simple chemist, I don't understand anything until I have a mass balance! So we cannot say what happens, but there is strong evidence of Al retention.

Blair: Intestinal absorption represents a dynamic system, consisting not only of flow out of the lumen of the gut into the body, but the flow from the stomach, through the lumen of the small intestine and into the lower colon. According to Professor Birchall, if you have Si and phosphate competing for Al, the pH level is important: phosphate is the dominant species at acidities greater than pH 6.6. Ingested material enters the stomach, where it reaches pH 2, so the Al will all bind to phosphate. It then leaves the stomach, moves into the duodenum, where it goes into pH 6 at the top of the duodenum, fluctuating a bit with the various fluid secretions. Then it passes down the gut and becomes more alkaline, so that Al binds to silicate. What is the rate of change from Al bound to phosphate to Al bound to silicate? Is it a slow rate? If it is (and the passage down the gut will be relatively fast), silicate species will not have any significance. With a fast rate of exchange, my argument doesn't hold. We need to know the kinetics of the change in Al binding as pH changes.

Birchall: I don't know the *exact* kinetics: I think that exchange is fast.

Blair: It would be interesting to know, because it could vitiate your argument. This is the kind of variable we need to think about when we consider intestinal absorption of aluminium. It is a very fine and complicated piece of plumbing!

Williams: We have some evidence, from *plants*, which supports the type of argument that Derek Birchall is making. There are two types of flowering plant, in terms of leaf structure. The grasses (monocotyledonous plants) take up an enormous amount of silica; they don't make what you describe as imogolite; they make amorphous silica, with a random distribution of Si, O and OH groups, i.e. $SiO_{2n}(OH)_{4-2n}$ where n goes from 0 to 1, in little globules, like opals (phytoliths). These 'silicates' in the grasses never contain any Al, to my knowledge.

Most broad-leaved plants do not take up much Si. An example is the tea plant, which does not take up Si, but its leaves are often heavily laden with Al. They put this Al in combination with phenolate groups for example, in the tannins. But in a certain chemical sense this phenolate OH surface is exactly the same as the surface of $SiO(OH)_2$, except that it is carbon hydroxyl instead of Si hydroxyl. These hydroxyls have about the same acidity constant (pK_a). So, higher plants such as the broad-leaved plants (dicotyledonous plants) have devised a trick to get rid of Al, based on phenols, whereas the grasses must have got rid of Al earlier within uptake mechanisms, and can lay down silica later absolutely free of Al. The chemistry nicely matches these different types of plant.

It would be interesting to look further at all these plant systems; obviously they must all be taking some Al across into the roots, but plants that make silica (the grasses) never get it into the leaf. Perhaps Al uptake is prevented because they are taking up so much silica in the root, by deliberately pumping Si in, that they protect themselves from Al, following Birchall's recipe. And that isn't the end, because the grass-eating animals then take in the phytoliths, which could well protect sheep and cows from Al in their first digestive system.

Klinowski: In support of what you say, in certain areas of India and China there are bamboo plants, which contain enormous amounts of Si inside the stem: it rattles when it dries. There is not a trace of Al in there. Where does it go? The white product, a hydrated silicic acid known locally as 'tabasheer', is purer than a silicon chip from Sinclair Research!

Williams: Some broad-leaved plants can handle silica, for example pears, but it would be interesting to investigate this further, because the plants may be able to help us in resolving the Al–Si story before we need to go to animal experiments.

References

Birchall JD, Chappell JS 1988a The chemistry of aluminium and silicon in relation to Alzheimer's disease. Clin Chem 34:265–267

Birchall JD, Chappell JS 1988b The solution chemistry of aluminium and silicon and its biological significance. In: Thornton I (ed) Geochemistry and health. Science Reviews Ltd, Northwood, Middlesex, UK, p 231–242

Dobbie JW, Smith MJB 1986 Urinary and serum silicon in normal and uraemic individuals. In: Silicon biochemistry. Wiley, Chichester (Ciba Found Symp 121) p 194–209

Garruto RM, Fukatsu R, Yanagihara R, Gajdusek DC, Hook G, Fiori CE 1984 Imaging of calcium and aluminum in neurofibrillary tangle-bearing neurons in parkinsonism-dementia of Guam. Proc Natl Acad Sci USA 81:1875–1879

Garruto RM, Swyt C, Yanagihara R, Fiori CE, Gajdusek DC 1986 Intraneuronal co-localization of silicon with calcium and aluminum in amyotrophic lateral sclerosis and parkinsonism with dementia of Guam. N Engl J Med 315:711–712

Lou G, Huang PM 1989 Nature and charge properties of X-ray non-crystalline aluminosilicates as affected by citric acid. Soil Sci Soc Am J 53:1287–1293

Martin RB 1990 Aluminosilicate stabilities under blood plasma conditions. Polyhedron 9:193–197

Martin RB 1992 Aluminium speciation in biology. In: Aluminium in biology and medicine. Wiley, Chichester (Ciba Found Symp 169) p 5–25

Perl, DP, Muñoz-Garcia D, Good PF, Pendlebury WW 1986 Calculation of intracellular aluminium concentration in neurofibrillary tangle (NFT) bearing and NFT-free hippocampal-neurons of ALS/Parkinsonism-dementia (PD) of Guam using laser microprobe mass analysis (LAMMA). J Neuropathol & Exp Neurol 45:379 (abstr)

Rollin HB, Theodorou P, Kilroe-Smith TA 1991 The effect of exposure to aluminium on concentrations of essential metals in serum of foundry workers. Br J Ind Med 48:143–246

The epidemiology of Alzheimer's disease in relation to aluminium

Christopher N. Martyn

MRC Environmental Epidemiology Unit, Southampton General Hospital, Tremona Road, Southampton, Hants SO9 4XY, UK

Abstract. The combination of an ageing population, an exponential increase in the incidence of dementing illness with age, and the high demands that demented patients place on health care resources makes Alzheimer's disease a major public health issue. So far, epidemiologists have made better progress in quantifying the frequency of the disease than in identifying strong risk factors, but evidence is accumulating to implicate environmental exposure to aluminium in the aetiology. The finding of a geographical correlation between death rates from dementia and water aluminium concentrations in Norway has since been replicated in several other surveys. Although ecological studies of this type should be interpreted cautiously, the association between Alzheimer's disease and aluminium in drinking water may prove to be an example of a potentially important biological effect of aluminium.

1992 Aluminium in biology and medicine. Wiley, Chichester (Ciba Foundation Symposium 169) p 69–86

The presence of aluminium in the cores of senile plaques and in neurofibrillary tangle-bearing neurons in the brains of patients with Alzheimer's disease (Candy et al 1986, Perl & Brody 1980) suggests that this element, which, in higher animals at least, has no known biological function, may be involved in the aetiology of the disease. As yet, only a handful of population-based studies that address the question of whether exposure to environmental sources of aluminium affects the risk of developing Alzheimer's disease have been carried out. This chapter attempts to review these endeavours, to discuss the methodological problems that limit their interpretation, and to indicate ways in which epidemiological methods might be used in future investigations.

Geographical relation between Alzheimer's disease and aluminium in drinking water

Two surveys undertaken in Norway in the early 1980s compared geographical variations in mortality with a number of measured variables in drinking water

TABLE 1   Age-adjusted death rates per 100 000 (1974–1983), 95% confidence intervals and relative risks for three groups of Norwegian municipalities, grouped according to aluminium concentrations in drinking water

Disease and sex	Aluminium in drinking water (mg/l)	No. of cases	Rate	95% confidence intervals	Relative risk
Dementia, males	<0.05	1559	25.1	23.8–26.4	1
	0.05–0.2	3550	28.8	27.8–29.7	1.15
	>0.2	533	33.2	30.3–36.0	1.32
Dementia, females	<0.05	2177	38.3	36.7–39.9	1
	0.05–0.2	5995	45.7	44.5–46.8	1.19
	>0.2	913	54.4	50.8–58.0	1.42

(Modified from Flaten 1990.)

(Vogt 1986, Flaten 1990). Many of the data used were common to both surveys and the conclusions reached by the investigators did not differ very much. The results described here originate from Flaten's work which is the only study to have been published in English.

One of the few statistically significant correlations revealed by this study was between concentrations of aluminium in water and mortality from dementia (Table 1). There are, however, as the investigator frankly acknowledged, some problems in interpreting this result. The observed correlation was not very strong and, because during the analysis multiple comparisons were made between many different causes of death and many variables in drinking water, the apparent association may represent nothing more than chance. Further, the association observed was between aluminium concentrations and dementia in general rather than Alzheimer's disease as a specific cause of dementia.

There is also doubt about the validity of using mortality data, collected routinely by a central office of health statistics, as an indicator of rates of dementia. Quite apart from the fact that dementia is under-recorded on death certificates, the pattern may be distorted by adventitious factors. A review of deaths from dementia in England and Wales showed that areas of high mortality from this cause correspond very closely geographically with the locations of large psychiatric hospitals (Martyn & Pippard 1988). The explanation is simply that many people who are demented eventually move to such institutions for long-term care. The Office of Population Censuses and Surveys considers these hospitals as the patients' usual address six months after the date of first admission. On death, therefore, the person's place of residence is recorded as the psychiatric hospital. Similar distortions have also been found in Australia; Jorm et al (1989) showed that the recorded local mortality from dementia could be strongly affected by the certification practices of a single medical practitioner. It is important to bear in mind that conclusions about the

TABLE 2 Relative risks (95% confidence intervals) of Alzheimer's disease, dementia from other causes, and epilepsy in patients aged 40–69 in county districts (UK) grouped according to water aluminium concentration (risks adjusted for distance from CT scanning unit and CT scanning rate)

Aluminium concentration (mg/l)	Probable Alzheimer's disease (n = 445)	Possible Alzheimer's disease (n = 221)	Other causes of dementia (n = 519)	Epilepsy (n = 2920)
0–0.01	1	1	1	1
0.02–0.04	1.5 (1.0–2.2)	1.1 (0.7–1.8)	1.2 (0.9–1.7)	0.9 (0.8–1.1)
0.05–0.07	1.4 (1.0–1.9)	1.1 (0.7–1.7)	1.1 (0.8–1.4)	0.9 (0.8–1.0)
0.08–0.11	1.3 (0.9–2.0)	0.8 (0.5–1.4)	1.0 (0.7–1.4)	0.9 (0.8–1.1)
>0.11	1.5 (1.1–2.2)	1.2 (0.7–1.9)	1.2 (0.8–1.6)	0.9 (0.8–1.1)

(From Martyn et al 1989 with kind permission of *The Lancet*.)

distribution of Alzheimer's disease that are based on mortality data are likely to be unreliable.

In England and Wales, unlike in Norway, most of the aluminium in water is derived not from natural sources but from treatment processes. The highly coloured water from upland catchments is often treated with aluminium sulphate in order to remove colour and particulate matter. Such waters are soft and contain little natural buffer. Their pH varies considerably over quite short periods of time. Because the solubility of aluminium in water is highly dependent on pH, high concentrations of aluminium may pass into supply. A survey of England and Wales exploited the geographical variation in the type of water (and hence its aluminium content) supplied to different parts of these countries to investigate a possible association between rates of Alzheimer's disease and the concentration of aluminium in drinking water (Martyn et al 1989).

Incident cases of Alzheimer's disease, other dementing illnesses and late-onset epilepsy in people between the ages of 40 and 70 years were identified from the records of neuroradiology centres. Seven areas of the country were included: South Wales, East Anglia, Hampshire, Merseyside, Northumbria, Nottinghamshire and Devon and Cornwall. In all of these areas the incidence of Alzheimer's disease, other forms of dementia and epilepsy was estimated for each of the county districts within the locality served by the CT scanner. These estimates of incidence were age-standardized and adjusted for the distance that patients had to travel to reach the neurological centre and for differing availability of CT scans in different parts of the country. The mean aluminium concentration of the water supplied to the county districts for the 10 years prior to the survey was obtained from records made available by local water companies and water authorities.

TABLE 3 Relative risks (95% confidence intervals) of Alzheimer's disease, dementia from other causes, and epilepsy in UK patients aged 40–64 (risks adjusted for distance from CT scanning unit and CT scanning rate)

Aluminium concentration (mg/l)	Probable Alzheimer's disease (n = 307)	Possible Alzheimer's disease (n = 153)	Other causes of dementia (n = 372)	Epilepsy (n = 2461)
0–0.01	1	1	1	1
0.02–0.04	1.4 (1.0–2.2)	0.9 (0.5–1.5)	1.2 (0.8–1.7)	1.0 (0.8–1.1)
0.05–0.07	1.4 (1.0–2.2)	1.1 (0.7–1.8)	1.1 (0.8–1.6)	0.9 (0.8–1.1)
0.08–0.11	1.6 (1.0–2.5)	0.6 (0.3–1.2)	1.2 (0.8–1.8)	1.0 (0.9–1.2)
>0.11	1.7 (1.1–2.7)	0.9 (0.5–1.6)	1.2 (0.8–1.8)	1.0 (0.8–1.1)

(From Martyn et al 1989 with kind permission of *The Lancet*.)

The results of this study are summarized in Table 2. The 88 county districts within the study are divided into five groups according to their water aluminium concentrations. Relative risks for probable Alzheimer's disease, possible Alzheimer's disease, other causes of dementia and late-onset epilepsy are shown with their associated 95% confidence intervals. The risk of probable Alzheimer's disease is raised in districts in which water aluminium concentration exceeds 0.01 mg/l when compared with districts where the water aluminium concentration is lower. There is no evidence of increased risk of possible Alzheimer's disease, other causes of dementia or epilepsy in districts with raised aluminium concentrations. When the analysis is restricted to the subgroup of cases under the age of 65 years, a stronger relation between water aluminium concentration and risk of probable Alzheimer's disease is seen (Table 3).

A French group subsequently attempted to replicate these findings in Aquitaine (Michel et al 1991). They used data collected in a population-based study of ageing and showed that prevalence rates of Alzheimer's disease correlated positively with aluminium concentrations in water (Table 4). The number of cases from which these ratios were calculated is small but the results are statistically significant. Since the publication of these results, the authors have presented a re-analysis of the data using a set of values for water aluminium concentrations that had been obtained more recently. The second analysis shows no relation between water aluminium concentrations and prevalence of Alzheimer's disease.

An association between Alzheimer's disease and aluminium in drinking water has also been found in a Canadian case–control study (Neri & Hewitt 1991). The methods of this study have only been given in outline; cases of Alzheimer's disease and presenile dementia were identified from discharge records of general hospitals in the province of Ontario and matched by age and sex to controls

TABLE 4 Prevalence (%) of dementia among those 65 years and over in areas of Aquitaine according to aluminium concentration in drinking water

	Aluminium concentration (mg/l)			
	≤0.01	0.02–0.04	0.05–0.07	≥0.08
Probable Alzheimer's disease	0	1.2	1.6	5.8
Possible Alzheimer's disease	0	0.3	0.6	1.3
Other dementia	0	0.4	1.4	0.6
Not demented	100	98.1	96.4	92.3
Total number	147	1984	506	155

(Modified from Michel et al 1991.)

with a non-psychiatric diagnosis. Aluminium exposure was estimated from water quality surveillance data corresponding to the locality of place of residence of the subject. The results are shown in Table 5. Unfortunately, the authors include little information about statistical significance in their report, but a χ^2 test using the numbers of cases of Alzheimer's disease and matched controls given in columns two and three of the table indicates that $P < 0.01$ ($\chi^2 = 12.3$, df = 3). This estimate, however, takes no account of the matched nature of the study design and must be regarded as provisional.

Aluminium from other sources

The daily dietary intake of aluminium is estimated as 5–10 mg for people living in the UK. Even among those people living in areas where the concentration of aluminium in drinking water is high, the proportion of the total dietary intake derived from water is unlikely to exceed 10%. And people who take aluminium-containing antacid preparations regularly have a daily intake of aluminium which is measured in grams rather than milligrams. Little is known about the relative

TABLE 5 Aluminium concentrations in drinking water for patients with Alzheimer's disease and matched controls, from a Canadian case–control study

Aluminium concentration (mg/l)	Alzheimer's disease	Matched controls	Case/control ratio	Estimated relative risk
<0.01	14	17	0.82	1.00
0.01–0.099	1261	1361	0.93	1.13
0.10–0.199	442	425	1.04	1.26
≥0.200	515	429	1.20	1.46

(From Neri & Hewitt 1991 with kind permission of *The Lancet*.)

bioavailability of aluminium from different dietary sources and it is possible that the aluminium in drinking water, which is largely uncomplexed, makes a disproportionate contribution to the amount absorbed from the gastrointestinal tract. Despite these uncertainties about bioavailability, studies of the risk of Alzheimer's disease associated with the use of aluminium-containing products are of great interest.

A brief report of a long-term mortality follow-up study of nearly 10 000 patients taking cimetidine, who were assumed to have taken large quantities of antacids before entry to the study, found that, out of the 20% of the cohort who had died, eight had some mention of Alzheimer's disease (ICD 9, code 331.0) or presenile dementia (ICD 9, code 290.1) (Colin-Jones et al 1989). The investigators did not calculate the expected numbers of deaths from this cause for an unexposed population of similar age and sex distribution but concluded that patients taking antacids were not at greatly increased risk of Alzheimer's disease. In view of the inadequacy of death certificates in identifying deaths from dementia, and the lack of an expected number of deaths with which to compare the number of deaths occurring in the cohort, this conclusion may not be warranted.

A Norwegian cohort of more than 4000 patients who had surgery for gastro-duodenal ulcer disease between 1911 and 1978 has been subjected to a complete analysis of mortality from dementia (Flaten et al 1991). It should be noted that aluminium-containing antacids were not used extensively in Norway before 1963. In this cohort, the standardized mortality ratio (SMR) for dementia was 1.10 for all patients, while that for patients operated on in the period 1967–1978 was 1.25. These estimates are based on small numbers of deaths ($n = 64$ for the whole cohort; $n = 13$ for those operated on in 1967–1978) and although both SMRs are increased, neither is statistically significantly different from unity.

Two case–control studies carried out in the mid 1980s included questions about regular use of antacids (Heyman et al 1984, Amaducci et al 1986), but the statistical power of both was so low that no useful conclusions can be drawn from their failure to find any association. More recently, two larger studies have produced conflicting results. In a pair-matched case–control study of 130 patients with Alzheimer's disease from North America (Graves et al 1990), surrogate informants for both cases and controls were interviewed by telephone using a structured, standardized questionnaire that included questions about the use of anti-perspirants, antacids and analgesics. Brand names of products used were enquired about and were used to distinguish between those that did and did not contain aluminium. The results are shown in Table 6. There was a statistically significant increase in relative risk of Alzheimer's disease associated with the use of aluminium-containing anti-perspirants and with the taking of any antacid preparation. A dose–response gradient was present for the use of antacids. However, when the analysis was restricted only to the use of aluminium-containing antacids, no statistically significant increase in relative

TABLE 6 Relative risks (95% confidence intervals) for Alzheimer's disease associated with the use of antacids in a case–control study from the USA (adjusted for age, family history and history of head trauma)

	No use	Low dose	Moderate dose	High dose
Any antacid	1.00	1.09 (0.32–3.76)	6.06 (1.20–30.64)	11.66 (1.28–106.4)
Aluminium-containing antacid	1.00	0.32 (0.07–1.36)	3.11 (0.42–23.03)	1.54 (0.25–9.60)
Any anti-perspirant	1.00	1.25 (0.55–2.86)	1.31 (0.57–2.98)	1.00 (0.38–2.67)
Aluminium-containing anti-perspirant	1.00	3.10 (0.74–12.97)	3.10 (0.94–10.47)	3.20 (0.76–13.20)

(Modified from Graves et al 1990.)

risk was found. An Australian study (Broe et al 1990) that was similar in size and design failed to find any increased risk associated with the use of antacids. These results are perplexing and can probably only be resolved by a further study with more reliable data on exposure.

Industrial exposure to aluminium

A common method of investigating possible health effects of industrial exposures is to form a cohort of workers in the industry retrospectively from records of employment, to trace them through a central office of health statistics, and to examine whether their experience of mortality from specific causes differs from that expected for those of similar age and sex in the general population. Such an approach is not likely to be useful in exploring a possible link between industrial exposure to aluminium and the development of Alzheimer's disease in most countries because of the deficiencies of death certification of dementia that were discussed earlier. Industrial cohort studies of aluminium reduction workers have shown no excess of mortality from dementia (Milham 1979), but little should be concluded from this negative finding.

Between 1944 and 1979, miners in northern Ontario were deliberately exposed to finely ground aluminium powder (McIntyre Powder) in an attempt to protect against silicotic lung disease. Recent chemical analysis of stored samples of McIntyre Powder shows it to contain particles of aluminium tri-hydroxides, elemental aluminium and traces of iron oxide. The long-term effects of this exposure have been investigated (Rifat et al 1991). A sample of 1353 subjects was drawn from a cohort of 6604 men, born between 1918 and 1928, who first started mining between 1940 and 1959 and who had been examined in one of three Ontario chest clinics. Data on exposure to McIntyre Powder were obtained from records of miners' annual chest clinic examinations and the McIntyre Research Foundation's records of mining companies licensed to dispense their powder. 607 men in this cohort were successfully traced and interviewed. At the interview they were given three tests of cognitive function: the mini-mental state examination, Raven's coloured progressive matrices test, and the symbol digit modalities test.

No significant differences between exposed and non-exposed miners in reported diagnoses of neurological disorders were found but exposed miners performed less well on the battery of cognitive tests than unexposed miners. These differences persisted after adjustment for age, number of years spent mining, age of starting mining, immigrant status, years of education, reported head injury, systemic hypertension at the time of the interview and an interviewer rating of subject co-operation. The results are shown in Table 7.

Without clinical or neuropathological follow-up of the subjects in this cohort, the relevance of these findings to the aetiology of Alzheimer's disease is not

TABLE 7   Indicators of cognitive function in Canadian miners according to exposure to McIntyre Powder

	Mean sum of three test scores		Proportion impaired on at least one test	
	Unadjusted	Adjusted	Unadjusted	Adjusted
Exposed ($n = 261$)	77.4	78.6	0.14	0.13
Unexposed ($n = 346$)	83.7	82.8	0.04	0.05
Difference (SE)	6.3 (1.1)	4.2 (1.1)	0.10 (0.02)	0.08 (0.02)

Values in the Adjusted column have been adjusted for age, number of years spent mining, age of starting mining, immigrant status, years of education, reported head injury, systemic hypertension at the time of the interview, and an interviewer rating of subject co-operation.
(Modified from Rifat et al 1991.)

clear. And there are no data that would directly answer the important question of whether exposure to McIntyre Powder results in an increased tissue burden of aluminium. The results do, however, provide prima facie evidence of the neurotoxicity of chronic aluminium exposure.

Conclusion

The evidence that links aluminium in drinking water to Alzheimer's disease is derived to a large extent from ecological studies that compare rates of the disease in populations that are supplied with water containing different amounts of aluminium. This type of study is notoriously vulnerable to confounding. One possible confounder, as Birchall (1992) has pointed out, is silicon. Waters that are high in aluminium tend to be low in silicon, and vice versa. He has suggested that what appears to be a toxic effect of aluminium may be better explained by a deficit in the protective effects of silicon. No epidemiological data are currently available that would allow his hypothesis to be tested, but a case–control study is in progress that is collecting data on exposure to silicon in water. These studies have also been criticized because of inaccuracy both in the diagnostic classification of cases and in the measurement of population exposure to aluminium. While it is true that such inaccuracies are likely to be present, this criticism is less valid. Random error, either in classification of disease or exposure, tends to obscure associations and to bias estimates of relative risk towards unity.

The results of the studies of antacid use are hard to evaluate. Leaving aside those investigations whose statistical power is low, there are two well-designed case–control studies that have produced very different risk estimates. No conclusion can be reached at present.

It is inevitable that the interpretation of studies that are less than ideal (or whose methods are not fully described) will be, to some extent, a matter of individual judgement. My own view is that, even if one takes into account a possible bias against the publication of studies that found no association between an environmental exposure to aluminium and risk of Alzheimer's disease, there is more than enough positive evidence to justify further epidemiological investigation, in spite of the obvious difficulties of obtaining reliable data on past environmental exposures of patients who are demented. In designing the next generation of studies, efforts should be concentrated on improving assessment of exposure to specific sources of aluminium, including measurement of possible confounding variables such as silicon and applying case–control methods to investigate the relation between exposure to aluminium and Alzheimer's disease in individuals.

References

Amaducci LA, Fratiglioni L, Rocca WA et al 1986 Risk factors for clinically diagnosed Alzheimer's disease: a case–control study of an Italian population. Neurology 36:922–931

Birchall JD 1992 The interrelationship between silicon and aluminium in the biological effects of aluminium. In: Aluminium in biology and medicine. Wiley, Chichester (Ciba Found Symp 169) p 50–68

Broe GA, Henderson AS, Creasey H et al 1990 A case–control study of Alzheimer's disease in Australia. Neurology 40:1698–1707

Candy JM, Oakley AE, Klinowski J et al 1986 Aluminosilicates and senile plaque formation in Alzheimer's disease. Lancet 1:354–357

Colin-Jones D, Langman MJS, Lawson DH, Vessey MP 1989 Alzheimer's disease in antacid users. Lancet 1:1453

Flaten TP 1990 Geographical associations between aluminium in drinking water and death rates with dementia (including Alzheimer's disease), Parkinson's disease and amyotrophic lateral sclerosis in Norway. Environ Geochem Health 12:152–167

Flaten TP, Glattre E, Viste A, Soreide O 1991 Mortality from dementia among gastroduodenal ulcer patients. J Epidemiol Community Health 45:203–206

Graves AB, White E, Koepsell TD et al 1990 The association between aluminium-containing products and Alzheimer's disease. J Clin Epidemiol 43:35–44

Heyman A, Wilkinson WE, Stafford JA et al 1984 Alzheimer's disease: a study of the epidemiological aspects. Ann Neurol 15:335–341

Jorm A, Henderson A, Jacomb P 1989 Regional differences in mortality from dementia in Australia: an analysis of death certificate data. Acta Psychiatr Scand 79:179–185

Martyn CN, Pippard EC 1988 Usefulness of mortality data in determining the geography and time trends of dementia. J Epidemiol Community Health 42:134–137

Martyn CN, Barker DJP, Osmond C, Harris EC, Edwardson JA, Lacey RF 1989 Geographical relation between Alzheimer's disease and aluminium in drinking water. Lancet 1:59–62

Michel P, Commenges D, Dartigues JF et al 1991 Study of the relationship between aluminium concentration in drinking water and risk of Alzheimer's disease. In: Iqbal K, McLachlan DRC, Winblad B, Wisniewski HM (eds) Alzheimer's disease: basic mechanisms, diagnosis and therapeutic strategies. Wiley, Chichester

Milham S 1979 Mortality in aluminium reduction plant workers. J Occup Med 21:475–480

Neri LC, Hewitt D 1991 Aluminium, Alzheimer's disease, and drinking water. Lancet 338:390

Perl DP, Brody AR 1980 Alzheimer's disease: x-ray spectrometric evidence of aluminum accumulation in neurofibrillary tangle-bearing neurons. Science (Wash DC) 208:297–299

Rifat SL, Eastwood MR, McLachlan DRC, Corey PN 1990 Effect of exposure of miners to aluminium powder. Lancet 336:1162–1165

Vogt T 1986 Water quality and health—study of a possible relationship between aluminium in drinking water and dementia. (Sosiale og okonomiske studier 61; English abstract) Central Bureau of Statistics of Norway, Oslo

DISCUSSION

Martin: How long-term was the usage of antacids in the two case–control studies you outlined? If Alzheimer's disease is a long-term neurodegenerative process, we should be talking about many years.

Martyn: In one study antacids had to be used daily for a year or more, and, in the other, daily for six months or more. But we don't really know how prolonged exposure to aluminium needs to be. I agree that there is a feeling that it has to be low amounts of Al over a long period of time, but we don't actually know that.

Klinowski: Would it be correct to say that the antacid you referred to was polycrystalline gamma alumina? This is one of the least soluble substances. I wouldn't be surprised if it was totally harmless, because I don't think you can dissolve a pound of it in the Atlantic Ocean! So whereas Al in drinking water that was perhaps complexed by humic acid would be bioavailable, antacids wouldn't be.

Martyn: Except that we know that patients with renal failure given aluminium-containing antacids have raised plasma concentrations of aluminium; so they must be bioavailable to some extent.

Williams: I don't think one should make such firm statements about whether Al would be bioavailable. Al is not easily available from compounds such as humic acid. I don't think we know enough. We just know that Al is there in some form.

Martin: The Al compound enters a very acid stomach, where it likely dissolves.

Williams: If so, it will become free Al at that stage, but the question is whether, in that very acid region, it has already passed the place important for uptake.

Martin: It depends upon whether you have drunk some citrus juice!

Wisniewski: When we talk about stomach acidity and Al solubility, we forget that while stomach juices are very acid, between these acid juices and the gastric mucosa there is a mucosal epithelium covered with mucus which is not acid. Therefore, if we want to better understand the condition of gastrointestinal absorption of Al we must take into consideration the pH of the mucus covering the mucosal epithelium.

Fawell: It is clear to me from the epidemiological studies that something is going on with water that needs to be followed up, but I'm not sure whether the Al part of the water story that's emerging in these studies is not a red herring, and that perhaps other aspects may be important.

What is important is that we do follow these findings up, because there may be valuable clues from these epidemiological studies as to what is happening in relation to Al exposure and Al uptake in populations. Something that we at the Water Research Centre are planning to do, with Christopher Martyn's help, is to look at the uptake of Al from different waters in individuals, examining serum and urinary Al levels, to see if there is a difference in the availability of Al in different parts of the UK. We also want to look at the different waters from the studies that Dr Martyn did, to see if there are other factors that ought to be brought into the equation. We are seeing something happening, and it's an opportunity to try to find out what it is. I think Derek Birchall's ideas are important, and we have to go on from there and try to take account of the different factors.

Flaten: Personally, I doubt that studies of populations with different drinking water sources will show any differences in serum and urine Al levels, although I agree that such studies are necessary to follow up the results of the epidemiological studies. Another theoretical possibility is that some drinking water sources contain Al complexes that not only are effectively absorbed from the gut, but also are stable enough to be transported and delivered to target organs. The levels of such compounds need not lead to a marked increase in serum Al to make a big difference in e.g. the brain Al burden when ingested over a lifetime.

In direct relation to Dr Martyn's paper, the French epidemiological study found a very clear association between Alzheimer's disease and their first set of measurements of Al in drinking water. However, when they reanalysed the water sources for Al, the association disappeared, as you pointed out. The later results are probably more valid; their first set of Al values consisted of results supplied from the different waterworks, some analyses dating back to 1965, and probably based on different analytical methods. When they presented their first results showing the strong association between Alzheimer's disease and drinking water Al, the local population and authorities wanted to reinvestigate the matter, so they resampled all the water sources and did all the determinations in the same laboratory with up-to-date methods. They found that, on average, the new analytical values for Al were about five times lower than the first results, so that almost all the samples were below 30 µg/l, which is very low. To me, there's no doubt which set of data is most trustworthy.

Martyn: It's not quite so clear-cut. We know that, in general, Al levels of water have fallen as treatment plants have improved.

Flaten: These French communities are mostly served by untreated ground water sources; no Al has been added, so I think the old values are erroneous, because of the inadequate analytical methods.

This leads to another point I would like to make: how could this rather dubious set of analytical values for Al correlate so closely with Alzheimer's disease in the first place? This strengthens my gut feeling about many kinds of epidemiological studies; no matter how they are carried out, there can always be some underlying confounding variables that operate to produce the observed association. This is especially critical when we are talking about relative risks that are only slightly elevated.

Martyn: The data of all the studies that I have described contain error, both in the measurement of exposure to Al in drinking water, and also in the classification of subjects as to whether they are cases of Alzheimer's disease or not. But error of this sort, if it's random, tends to militate *against* finding an association. A large amount of noise makes it hard to pick up a signal. I disagree with your interpretation; in my view, the presence of random error in data is likely to obscure associations.

Flaten: I agree; random errors *will* generally operate to disguise an existing association. However, I shall continue to be sceptical towards all epidemiological studies showing only slightly elevated relative risks, although it should be noted that for the studies of Alzheimer's disease/dementia and drinking water Al, the striking consistency between most studies greatly strengthens the evidence for an existing positive association.

Garruto: Dr Martyn, you mentioned rightly that epidemiologically we can no longer concentrate on looking solely at population risk factors; we have to move to another level of epidemiological investigation. You said that individual risk factors are important. I would add that household risk factors are equally important where there is a familial association with disease, as in ALS and parkinsonism-dementia. For example, geographical location (even within a town), lifetime residence histories, consumption of foodstuffs, activity patterns including sexual division of labour and so on, may provide important information in the assessment of overall risk.

Martyn: If you can sharpen that up, and tell me what things I should ask the subjects about, I would certainly do so. I don't think the general approach of asking as many questions as you can in the hope that something will turn up is a good strategy.

Greger: The Al content of water may vary with when you sample it. For example, we know that lead levels in water increase when water remains in the pipes overnight. Households vary. Some people turn on the water and immediately use it for cooking, or drinking; others flush the pipes by letting the water run first. I do not know the importance for aluminium levels in water.

Martyn: It would be hard to get that sort of information. People drink different amounts of water, and do different things with it. Some use it to make orange squash and some to make tea!

Day: We did an experiment in Manchester, enlisting the help of Greater Manchester Radio. We got nearly 500 people to sample their tap water more

TABLE 1 (*Day*) **Lead and aluminium concentrations in Manchester tapwater: (a) first flush sample obtained in the morning, and (b) obtained after running the tap for one minute after sample (a)**

	Lead (μg/l)	Aluminium (μg/l)
Sample (a)	13.7 (1–447)	27.6 (1–793)
Sample (b)	3.9 (1–87)	28.6 (1–690)

Values are the mean and range of 495 samples.
(Unpublished results of J. P. Day & J. Templar.)

or less simultaneously throughout Manchester, first thing in the morning, when it had been standing in the pipe over night. Then they let the tap run for a minute and took a second sample. We determined lead and Al (see Table 1). There was an enormous difference in the lead content between the first and second sample, which is well known. There was little difference in the Al content (unpublished results). So I think running the water doesn't affect the Al level. This result is consistent with the expected sources of the two metals—the lead by dissociation of lead pipes or soldered joints, the Al from the water source itself.

Zatta: In view of Professor Birchall's paper, it could be very important to re-examine the data published in your 1989 paper, Dr Martyn, taking account of the presence of Si in drinking water (Martyn et al 1989).

Martyn: That is one approach, but it would require an enormous amount of effort to collect from the water suppliers the necessary data concerning silicon concentrations. Instead, we are incorporating measurement of silicon in our current case–control study.

Day: It has always interested me that in your 1989 study you don't get a dose-response relation. Your relative risk starts at an Al concentration of 0.02 mg/ml, and remains almost constant above this level. I have never heard a convincing explanation for this.

Martyn: It is a valid criticism. In the places where aluminium concentrations were 0.01 mg/l or lower, we set the risk to be 1.0. Places where aluminium concentrations were 0.02 mg/l or greater were all associated with a risk of about 1.5. If our data for the baseline are wrong, the estimates of risk in the other groups will be wrong. On the other hand, these risks are not estimated very precisely. It may be that a dose–response relation is obscured by the imprecision. When we analysed the subgroup of cases under the age of 65 years, there was some indication that risk increased progressively with aluminium concentrations. The way forward is surely to replicate the study in other populations, or to attack the problem in a different way. If lack of a dose–response relationship makes you feel that the evidence is not compelling, I won't quarrel with you.

My view at the moment is only that it seems convincing enough to be worth pursuing.

Day: I wasn't suggesting that this invalidated the conclusions, but I think there may be additional reasons for the relationship which are not necessarily emerging. One might explain the lack of dose–response simply on the basis of analytical inadequacy; possibly people either don't detect any Al, or they do, and are not very good at measuring the exact concentration. In the era when your data were coming out, that was particularly the case. So your quantitative ranges for Al levels might not be a very accurate measure of exposure.

Martyn: That is true, but, as I have said before, inaccuracy will tend to obscure any relation.

Day: Another possibility is that you might find particularly high, or even particularly low, concentrations of Al in certain cases because of the presence of something else. You might get Al in your drinking water because something is there to bring it in. An example would be if the presence of Si was inversely related to the level of Al; then you might be looking at 'no Si' and 'a lot of Si' as the parameter. Alternatively, it could be that in certain areas the presence of humic acid in water results in an enhanced concentration of Al, through increased leaching of Al from soil, or whatever. So it's quite possible that you were looking not strictly at the Al dependence, but at the dependence on some other component, linked to aluminium, which we don't know about.

Birchall: The point I wish to emphasize is that the aluminium content of water may be irrelevant in terms of a major contribution to the total ingested and absorbed. There is far more aluminium in food and medications. Water that contains aluminium contains little silicic acid and vice versa. My suggestion (Birchall & Chappell 1989) is that drinking water or cooking with water that is high in silicic acid inhibits the absorption of the aluminium ingested in food. The reciprocal relationship between Si and Al in water could then give rise to an apparent relationship between Al in water and the incidence of Alzheimer's disease, assuming that Al has indeed a role in this disease.

Williams: If you were an enzymologist doing an activity test and obtained those data, however, you would have to say that the Al was *not* responsible for the activity!

Martyn: It may be that at the Al concentrations that we chose to measure, we are on a part of the curve where it has reached a plateau.

Williams: Absolutely; but you have then to go back and measure the other points before making any causal claim, because if you cannot show a dose–response at some level, you vitiate this type of argument using concentration data over a range. The only point of doing a range is to see a dose–response curve, and if you don't get one for sure, you then resist making conclusions until you do!

Martyn: These things are more easily arranged in the laboratory than in population-based research. The Canadian study did show a dose–response relationship, however.

Candy: Yes, but in the study of Neri & Hewitt (1991) there were too few cases to establish the baseline adequately. There were only 14 Alzheimer's disease patients and 17 matched controls in the <0.01 mg/l Al group, compared with a minimum of 425 cases in the other three Al concentration groups.

Martyn: That is less important if you have a stepwise increase, because then comparisons are not all made with reference to the baseline.

Candy: You can't be sure of the baseline and the stepwise increase is very small—that is my point!

Perl: You compared your rates to the lowest value and found no dose-response curve, Dr Martyn. You could just as easily, instead of looking at relative risk, conclude that living in an area with a very low Al content of water has prevented the development of dementia in some way that we don't understand. In other words, that's why the lower rate of dementia in that region is not seen in any other regions; and therefore you don't see a dose–response curve. This is another way of looking at your data.

Edwardson: It seems to me that there is a contradiction between what Bob Williams and Derek Birchall were saying. You said, Bob, that unless, like an enzymologist, you got a nice dose–response, you wouldn't accept the data. In contrast, Derek made the point that there was a critical threshold for Si at about the 100 μM level which, because of the non-stoichiometric nature of the interactions, resulted in the formation of a much more stable aluminosilicate compound. If that is true, then the Si hypothesis may be the most convincing explanation of these epidemiological data. But if it is true, would we expect a continuous dose–response curve, or would we expect a stepwise relationship?

Birchall: It would be stepwise, because I expect the effect on Al absorption to become strong at 100 μM Si, with much less effect below that.

Edwardson: What do you say to that?

Williams: It is very easy to reply to that! If you put in an inhibitor of a test system, you could get values all over the place, since the effects depend on the complexity of activator/inhibitor interactions. All that Birchall is saying is that he put an inhibitor into the system as well an activator, and if he does that in an arbitrary way, I have no idea what to expect.

Birchall: I am saying that there is no relationship between Al in water and Alzheimer's disease. There is not enough Al to have an effect (there is far more in food). What is present in water, that acts as an inhibitor, is silicon, but it is not a linear inhibition; it requies a threshold concentration of Si (about 100 μM). But the fact is that when you find Al in water to any sort of significant level, there will be very little Si; that's simply an accident of geochemistry. That is what I mean by a step function. In other words, you would have to go to, say, Cambridge (>200 μM) to observe the effect of water Si on dietary Al; it wouldn't happen in Glasgow (10 μM Si) or Manchester (25 μM Si) because, below the level of 100 μM, Si is not going to do much. Of course, the ingestion of Si from other dietary sources (e.g. cereals) will matter as well.

Williams: I am not denying your statements in any way! I was saying that if, in the absence of silica, there is an underlying dose–response curve, we are in business. If there is not, we are not. Silica *may* be confusing the results, but we don't know that; therefore, at the moment, you have to pass the judgement that we don't know that Al has anything to do with Alzheimer's disease or other parallel effects. Perhaps there is a muddled, complex situation. At this moment, we have no grounds for coming out of it on the Al side, because there is no dose–response.

Ward: In the water analysis used, Dr Martyn, were the water authorities using flame methods or colorimetric methods to measure Al? If colorimetric methods are used, do they detect the Al–Si complex? Because if not and you get a falsely low result, then your data stand.

Birchall: It depends on the method; if you go very acid to do it, then you destroy any Al–Si complex.

Martyn: I suspect that different methods were used by different places. I asked the Water Research Centre to collect the data on water Al because I wanted the collection of those data to be entirely independent of the estimation of rates of Alzheimer's disease. The WRC helped us plan the study; we chose areas of England and Wales where analytical methods were thought to be of a high standard.

Fawell: They were a combination of methods over time. We didn't take one analytical result; we took a mean of analytical data over about 10 years, which could include a range of methods. We tried to use areas of the UK where we would have the best available data at that time. If we were doing the study now, the analytical data would be considerably better than those used then; the standard of analysis has improved dramatically. It would be difficult to go back, so Christopher Martyn is doing the right thing by starting another study, to modern standards with modern analysis, and taking into account some of the other things we have learned from the first study. That seems wholly sensible.

Candy: You made a point in your talk, Dr Martyn, about other possible risk factors we should be looking at in water. Dr Flaten, in his thesis (Flaten 1986), found strong positive correlations between the Al concentration in drinking water and, for example, Mn, Fe, Si and Na. It would be more reassuring if some of the other epidemiological studies had not solely studied Al but had also looked at other trace elements in drinking water.

Williams: That might be intriguing. I remember the story that manganese miners suffer from parkinsonism (see Daniels et al 1979).

Martyn: An extrapyramidal syndrome, more dystonic than parkinsonian, is associated with chronic manganese exposure (Barbeau et al 1976).

Kerr: We clearly need some better marker for Al loading than serum or urine values, or hair or skin concentrations, and bone is unlikely to get past the ethics committee! Has anyone considered using desferrioxamine infusion and measuring Al excretion in the urine, 48 hours later? Is that an ethically acceptable alternative?

Martyn: It has been thought of. I don't know whether anyone has actually suggested it to an ethics committee. It might not be easy to do on a population scale. But, since we have no good measure of long-term exposure to aluminium, perhaps we should pursue the idea.

References

Barbeau A, Inoué N, Cloutier T 1976 Role of manganese in dystonia. Adv Neurol 14:339–352

Birchall JD, Chappell JS 1989 Aluminium, water chemistry and Alzheimer's disease. Lancet 1:953

Daniels AJ, Johnson LN, Williams RJP 1979 Uptake of manganese by chromaffin granules. J Neurochem 33:923–929

Flaten TP 1986 An investigation of the chemical composition of Norwegian drinking water and its possible relationships with the epidemiology of some diseases. Thesis No. 51. Institut for Uorganisk Kjemi, Norges Tekniske Hogskole, Trondheim

Martyn CN, Barker DJP, Osmond C, Harris EC, Edwardson JA, Lacey RF 1989 Geographical relation between Alzheimer's disease and aluminium in drinking water. Lancet 1:59–62

Neri LC, Hewitt D 1991 Aluminium, Alzheimer's disease, and drinking water. Lancet 338–390

Aluminium and the pathogenesis of Alzheimer's disease: a summary of evidence

D. R. McLachlan*, P. E. Fraser* and A. J. Dalton**

*Centre for Research in Neurodegenerative Diseases, University of Toronto, Tanz Neuroscience Building, Ontario, Canada M5S 1A8 and **New York State Institute for Basic Research in Developmental Disabilities, Staten Island, NY 10314-6399, USA*

Abstract. Known risk factors for Alzheimer's disease (AD) are few and insufficient knowledge is available to recommend steps to reduce AD in our ageing populations. Although not 'the cause', considerable evidence implicates human ingestion of aluminium as a possible risk factor for the expression of dementia of the Alzheimer type. A recent epidemiological study in Ontario relating the incidence of AD to aluminium in drinking water strongly supports this conclusion. To test further the hypothesis that aluminium may play a role in the pathogenesis of AD we conducted a clinical trial employing the trivalent metal ion binding compound, desferrioxamine. The design was a two-year randomized trial with behavioural assessments blinded to study assignment. Sixty-three patients with probable AD were selected who were living at home and were under 74 years. Forty-eight signed an informed consent and completed all initial testing. The main outcome measure was a video-recorded home-behavioural assessment of measures of skills of daily living. The principal outcome was that the mean slope for performance of the skills of daily living for the group without treatment was -1.72% maximum score/month, compared to -0.87% maximum score/month for the group treated with desferrioxamine ($P=0.038$). Considerable evidence supports the hypothesis that aluminium has an active role in the pathogenesis of AD.

1992 Aluminium in biology and medicine. Wiley, Chichester (Ciba Foundation Symposium 169) p 87–108

There is now a general agreement that aluminium is both highly neurotoxic in experimental models and present in toxic concentrations in affected brain regions in Alzheimer's disease (AD). In brain biopsies containing neurofibrillary tangles from early AD, and AD-affected tissue from advanced terminal disease at post mortem, elevated concentrations of aluminium are found in bulk neocortical grey matter (Crapper et al 1973, 1976, Trapp et al 1978, Yoshimasu et al 1981, Ward & Mason 1986). Aluminium has been reported to be elevated in five AD-affected cell or tissue compartments: in the nucleus (Crapper et al 1980), on dinucleosomes enriched in repressed (untranscribed) genes (Lukiw et al 1991); in neurofibrillary tangles (Perl & Brody 1980, Perl & Pendlebury 1984); in the cores of senile plaques (Duckett & Galle 1976, Masters et al 1985, Candy et al 1986, Edwardson & Candy 1990); in ferritin (Fleming & Joshi 1987); and in pial blood vessels (Crapper et al 1976). Two laboratories failed to find elevated concentrations of aluminium in AD-affected tissue (McDermott et al 1979, Markesbery et al 1981). In the former study, failure to detect increased amounts of Al may have been related to the sample size and the criteria for case selection for both controls and AD subjects (McLachlan et al 1980). In the latter report (Markesbery et al 1981), the failure to detect Al may be related to the difficulty in using certain nuclear reactors for neutron activation analysis applied to the measurement of aluminium (Krishnan et al 1987). Instrumental neutron activation is based on the thermal neutron reaction ^{27}Al(nγ)^{28}Al. Without pre-irradiation separation, the assay of aluminium in tissue is difficult and often inaccurate because of the interfering isotopes: ^{31}P (n,α)^{28}Al reaction from fast neutrons generated in most of the conventional reactors generally used for biological assays. In plain language, phosphorus in most reactors decays to ^{28}Al. Since the concentration of phosphorus is about 12 000 times higher than that of aluminium in biological specimens, it is often difficult to measure low concentrations of naturally occurring Al accurately by this method.

While most workers agree that elevated aluminium concentrations are present in AD-affected brain, the precise role of aluminium in the degenerative process of AD is controversial. Aluminum by itself does not induce AD-type neurofibrillary tangles composed of paired helical filaments (PHFs), or neuritic plaques with amyloid cores. Therefore, Al does not fulfil all of Koch's postulates, which are used to evaluate the role of a bacterium in the aetiology of a disease. However, aluminium is a 'dementing ion', capable of inducing changes in learning and memory and possibly contributing to the dementia of AD (McLachlan et al 1991a). Aluminium does induce some biochemical changes which may contribute to the production of AD-type tangles and amyloid accumulation. Reports in the past 18 months strengthen this possibility. In evaluating the role of this environmental agent in AD, Flaten & Garruto (1991) have argued that, taking all the evidence into account, Al neurotoxicity does fulfil the criteria put forward by Sir Austin Bradford Hill (1965) to evaluate the cause-and-effect relationships between a disease (AD) and a potential causative factor (Al). Hill suggested eight criteria:

(1) *Strength of the association.* A statistically significant association between the concentrations of aluminium in drinking water and the number of cases of AD, evaluated by epidemiological techniques, has been shown in five countries and in eight studies (this volume: Martyn 1992). The relative risk of AD in geographical regions characterized by high aluminium concentrations ranges

between 1.5 and 4, when compared to regions with low aluminium concentrations in drinking water.

A recent survey conducted in Ontario by Neri & Hewitt (1991) is perhaps the most convincing, for several reasons. (i) A significant association was found for processed water but not raw intake water. (ii) The diagnosis of probable AD was taken as the hospital discharge diagnosis (2344 AD patients who were matched for age, sex and place of residence to 2232 randomly selected non-psychiatric controls). Ontario has universal health care and all residents have equal access to the appropriate diagnostic facilities and laboratory tests required for the accurate diagnosis of AD. Second opinions by specialists are frequent and the diagnostic criteria for AD proposed by McKhann et al (1984) are widely used. Thus the accuracy of the clinical diagnosis of AD is likely to be high. (iii) The study was case controlled. (iv) The aluminium concentration in water was taken as the average for a 12-month period as measured and published by the Ontario Ministry of the Environment. This procedure greatly reduced the uncertainty concerning annual variation in Al concentration in the finished drinking water. (v) The postal code used to give the place of residence in this study also gives precise information about the source of drinking water, thereby reducing the uncertainty about the water source. (vi) There is a clear dose–response effect for aluminium in this study.

Importantly, an epidemiological study from Zurich failed to find a relation between cognitive impairment and the aluminium content of drinking water (Wettstein et al 1991). The reason for this apparent absence of a relation has not been investigated, but it could be related to the presence of naturally occurring protective factors. For instance, in a recent study on a population of 77-year-old males in Ontario by Forbes et al (1991), those persons who show no indication of impaired mental functioning are more likely to have resided in areas where the aluminium concentrations in drinking water are relatively low and where fluoride concentrations in drinking water are relatively high. This study suggests that fluoride may protect against a toxic effect of aluminium. Silicon may also protect against Al neurotoxicity and may be a confounding factor in epidemiological studies that would reduce the estimated risk (this volume: Birchall 1992).

Drinking water contributes only a small fraction of the total aluminium ingested. It is possible that the strength of the relation would be higher if both the total Al ingested from all sources and the precise species of Al were accurately measured. Nevertheless, the relative risk of AD from high Al levels in drinking water of 1.5 to 4 is the range assessed for the risk of cancer of the lung from second-hand smoke.

(2) *Consistency*. Although analytical difficulties remain, elevated levels of Al in brain tissue from AD patients have been reported by at least 10 laboratories employing six different analytical techniques on tissues obtained from various

geographical localities in four continents (review, McLachlan 1986), and this appears to be a consistent finding.

(3) *Specificity*. Elevated brain aluminium content occurs after prolonged renal failure requiring dialysis treatment. However, aluminium does not accumulate in brain diseases other than AD as a result of non-specific tissue damage (Traub et al 1981). Epidemiological studies indicate that several other neurological disorders, such as epilepsy, parkinsonism, amyotrophic lateral sclerosis and medical diseases predisposing to multi-infarct dementia, are not associated with increased concentrations of Al in drinking water.

(4) *Temporality*. Humans are exposed to aluminium over a lifetime and there is no known method for detecting when brain Al levels become elevated in living people. For aluminium to become a risk factor for AD, one would suppose that either the blood–brain barrier to aluminium becomes defective very early in the natural history of this disease, or aluminium is a contributing factor to the release of the aetiological 'causal' factors responsible for the disease. This laboratory favours the former hypothesis for several reasons. In cell nuclei from temporal lobe tissue, the mean aluminium content of dinucleosomes extracted from control brains free of neurofibrillary degeneration was 518 µg Al/g DNA (mean age 79 years, $n = 7$), compared with 2850 µg Al/g DNA in cell nuclei from temporal lobes of AD patients (16 AD brains, mean age 79.8 years) ($P = 0.0001$ on an analysis of variance) (Lukiw et al 1991). Elderly control subjects without AD do not accumulate aluminium in this intranuclear compartment with age. Thus, we have no evidence that the accumulation of Al on DNA precedes the AD process. Further, we submit that the conclusion of McDermott et al (1979) suggesting that aluminium accumulates in brain with ageing is erroneous, because their ageing 'control' samples included brains with AD-type neurofibrillary tangles and probably represented preclinical cases of AD (McLachlan et al 1980). Nevertheless, the observation that aluminium-treated human neurons in tissue culture develop epitopes characteristic of AD-type neurofibrillary tangles suggests that aluminium acts early in the development of AD neurofibrillary tangles (Guy et al 1991).

(5) *Biological gradient*. A clear dose–response gradient has been observed in epidemiological studies from Norway (Vogt 1986, Flaten 1990) and Canada (Neri & Hewitt 1991).

(6) *Plausibility*. Alzheimer's disease occurs in all populations throughout the world who survive to the age of risk. It is therefore plausible that a common environmental toxic agent might contribute to the AD degenerative process. At the chemical level, Al affects several enzymes and biomolecules relevant to AD; it is toxic to many species by several routes of exposure; and it is present

in several subcellular structures. Al also alters protein kinase activity and results in altered phosphorylation of cytoskeletal proteins which may be important in the formation of AD-type neurofibrillary degeneration (this volume: Jope & Johnson 1992).

(7) *Coherence*. The progression of neurological signs after a single intracranial injection of an Al dose lethal to 50% of animals does not seriously conflict with the generally known facts about the natural history of AD, although the time course is much slower. Indeed, the clinical course of the experimental model more closely reproduces AD than the encephalopathy associated with dialysis in renal failure. After Al injection, cats and rabbits remain asymptomatic by behavioural and electrophysiological criteria for seven to 15 days. Cats first exhibit selective deficits in performance of tasks measuring short-term memory and impairment in motor control after jumping. Several days later, there is evidence of defects in the motor control required for retrieving food and the progressive development of an increased 'lead pipe' type of tone in limb and truncal muscles. These animals may exhibit myoclonic jerks. Seizures occur but can be suppressed with anti-convulsant medications. Animals may survive for over three years with chronic neurological damage associated with the impaired acquisition of tasks measuring new learning. More recently, Strong et al (1991) have developed an aluminium-induced chronic neurodegenerative condition in rabbits which is not accompanied by seizures. Thus the aluminium-induced neurobehavioural and motor control deficits are similar in character and sequence of progression to those found in AD, although the time course is much shorter.

(8) *Experimental intervention*. A treatment trial employing a trivalent metal ion chelating agent, desferrioxamine, indicates that lowering the body and brain Al content retards the progression of Alzheimer's disease (McLachlan et al 1991b). This is the strongest observation in support of causative relationships, and the trial will be described in some detail.

A randomized two-year prospective, single-blind clinical trial was conducted to determine whether the sustained use of low-dose desferrioxamine would slow the progression of the dementia (McLachlan et al 1991b). A total of 48 people living at home in the early stages of the illness were randomly assigned to three treatment groups: desferrioxamine (125 mg injected intramuscularly twice daily, five days per week, for two years), lecithin (oral dose of 1.0 g/day), and no treatment. A structured performance test measuring daily living skills was videotaped in the home and was taken as a reliable measure of outcome over the two-year initial observation period. The tapes were analysed at random by trained behaviour raters who were unaware of the purpose and protocol of the study. There was no statistical difference in the average rate of decline in performance between the group receiving lecithin and the untreated group. However, the average two-year decline in the desferrioxamine-injected group was both practically and statistically significantly slower than that observed for the combined untreated and lecithin-treated groups, designated as the 'no-treatment' group.

Two outcome measures indicated therapeutic benefit from desferrioxamine treatment: death rate, and decline in skills of daily living. During the initial 24 months of observation, there were five deaths in the combined no-treatment group and no deaths in the treated group. During the total period of observation of 58 months, representing 185 patient-years of observation, there were nine deaths in the no-treatment group, with one death in the desferrioxamine-treated group. On the most conservative outcome measures derived from the videotaped record of skills of daily living home-behavioural assessment, the average two-year decline (over the first 24 months) in the desferrioxiamine-treated group was 25% of the maximum score, as compared with 57% in the no-treatment groups. This represents an average monthly rate of decline for the desferrioxamine-treated group of -0.87% of maximum and for the no-treatment group of -1.72%.

We also used a single statistic to evaluate drug efficacy which takes into account the deaths during the two-year period of observation and calculates the slopes of the performance scores measured five times in 24 months of observation on 25 items of structured behaviours such as brushing teeth, putting on gloves, etc. (Part A of the Video-Recorded Home-Behaviour Measures) in which scores were assigned a value ranging between 4 and 0. The slope of the changes in scores for each patient was calculated by dividing the score obtained at each test interval by the days between testing and assigning zero at the time of death. The average rate of deterioration (slope) for the no-treatment group was -2.491% of maximum score (i.e. 100%) per month ($n = 22$) and -0.745% per month ($n = 25$) for the desferrioxamine-treated group. This single statistic indicates a 3.3-fold faster rate of deterioration in the no-treatment group with the probability that the outcome is due to chance alone of $P = 0.0002$ on an analysis of variance. The pre-treatment y intercept of the slopes for the two groups is 97.023% of maximum score for the desferrioxamine-treated group and 96.903% of maximum score for the no-treatment group, indicating that the two groups were matched in performance before treatment began and were within the sensitive range of the test. The effect appears to be due to aluminium, because an analysis of the trace metal content of brains of patients from a previous study who died with advanced AD while receiving desferrioxamine has demonstrated that aluminium is the only trivalent metal that has been removed from the brain after prolonged treatment. This clinical trial strongly supports the conclusion that aluminium is indeed an important factor in the complex series of events associated with AD.

Mechanisms of neurotoxicity

Several molecular mechanisms may be targets for aluminium toxicity. Aluminium has multiple toxic actions, including the induction of the anomalous phosphorylation of proteins, disruption of calcium homeostasis, and the non-covalent cross-linking of proteins. Each of these toxic effects of aluminium may be necessary but not sufficient to induce the formation of PHFs.

The important histopathological marker for AD, neurofibrillary degeneration, is composed of intraneuronal accumulations of PHFs. The cores of PHFs are assembled from the microtubule-binding domain of inappropriately phosphorylated tau, a microtubule-associated protein (Crowther 1991). Tau exists as a family of four or five related proteins that display developmental and species variability. This family of proteins plays an important role in neurite extension and growth cone plasticity. Antibody recognition and biochemical studies have demonstrated that intraneuronal tau regulation is achieved by specific phosphorylation, which results in a direct reduction in microtubule affinity (Papasozomenos & Binder 1987). It appears that *in vivo* a functionally quiescent pool of phosphorylated tau protein is normally synthesized in the somatodendritic compartment of the neuron and then transported to the axon and activated by dephosphorylation.

In AD, new, possibly pathological, tau species have been identified (Lee et al 1991). The observation that normal tau characteristics can be restored by phosphatase treatment of PHF-derived tau indicates that one or more novel phosphorylation sites are involved in the conversion of tau to these new PHF-associated species. Epitope mapping has located a number of phosphate-dependent sites in both normal (tau-1, tau-2) and PHF-tau. The majority of antibodies recognize sites near the microtubule-binding repeats which constitute the structural core of PHF. Thus, phosphorylation at or near these sites may be the key step in the initiation of tau polymerization.

Alz 50 is a well characterized antibody with high affinity for PHF whose epitope could represent a crucial phosphorylation site (Wolozin & Davies 1987). Although some disagreement remains (Ksiezak-Reding et al 1990), a recent mapping study has located this epitope in the C-terminus of tau (Uéda et al 1990). This region contains at least three potential phosphorylation sequences: (1) the neurofilament Lys-Ser-Pro-Val repeat; (2) a Ser/Thr kinase site; and (3) a Ca^{2+}/calmodulin-dependent kinase site. The neurofilament repeat is known to be phosphorylated in PHF-tau, but most interesting is the correlation of electrophoretic mobility and the Ca^{2+}/calmodulin-dependent site. Using engineered tau constructs, Steiner et al (1990) showed that phosphorylation of a single site (Ser-405) is sufficient to produce shifts in electrophoretic mobility. Support for the idea that the Ca^{2+}/calmodulin-dependent kinase is important is the observation that elevated activity of this enzyme has been reported in AD-affected areas of brain (McKee et al 1990). However, the association of other protein kinase systems (review, Saitoh et al 1991) such as casein kinase, protein kinase C, Ser/Thr kinase and a novel tubulin-dependent kinase, with the increased phosphorylation of other neuronal proteins (e.g. synapsin and p60), may indicate a more general alteration in kinase activity.

The formation of PHFs in AD-type neurofibrillary degeneration probably represents a common neuronal response to injury, perhaps triggered by an impairment in calcium homeostasis. Increased intracellular calcium levels have been reported to induce the expression of epitopes found in AD tangles, indicating the induction of anomalous phosphorylation. De Boni & McLachlan (1985) obtained electron microscopic evidence of paired helical filament formation in human neuronal explants exposed to glutamic acid added to the culture media at concentrations of 2.2 mM or less. Importantly, glutamate-immunoreactive neurons are among those which develop neurofibrillary degeneration in AD. Epitopes related to hyperphosphorylated tau, characteristic of AD tangles, are also induced when human neuroblastoma cells are exposed to aluminium (Guy et al 1991). Both glutamate and Al manipulations raise intracellular calcium levels, which might also contribute to the activation of certain protein kinases. In AD, it is uncertain whether inappropriate phosphorylation or altered calcium homeostasis is the key event which initiates the formation of tangles. Perhaps different insults affect different components of the cascade, resulting in a common histopathology.

In addition to phosphorylation, aluminium could act as a non-covalent cross-linking agent. Unpublished electrothermal atomic absorption measurements from this laboratory on fractions of isolated paired helical filaments enriched in tau extracted from brain tissue of AD patients indicate a marked enrichment in aluminium content. The mean aluminium content in fractions from temporal lobe of seven patients was 173 µg Al per g protein, compared to 2 µg Al per g protein in bulk protein from neocortical grey matter. This represents approximately one aluminium atom for two molecules of tau, assuming all the protein in PHF is tau and has an average molecular mass of 60 kDa. The role of aluminium in the assembly of tau into PHFs requires further investigation.

Aluminium may also alter a number of phosphorylation processes important to learning and memory performance, thereby further contributing to the dementia of AD. *In vitro*, aluminium at low concentrations stimulates adenylate cyclase, possibly as AlF_4^- stimulation of the guanine nucleotide-binding regulatory component of the enzyme. Chronic oral administration of aluminium to rats results in significant increases in cyclic AMP levels and in the hyperphosphorylation of microtubule-associated protein 2 (MAP-2), of the 200 kDa neurofilament subunit, and of at least 10 other proteins (Johnson et al 1990). Aluminium also inhibits 3′,5′-cyclic nucleotide phosphodiesterase (Farnell et al 1985). Jope & Johnson (this volume 1992) have presented evidence that the G protein–InsP$_3$ second messenger system is also disturbed by aluminium. These results further support the notion that aluminium could contribute to the cascade of molecular events responsible for anomalous phosphorylation by altering second messenger systems affecting phosphorylation processes important in both the learning and memory mechanisms of the brain and the cytoskeletal systems.

Taking all the evidence into account from several independent lines of investigation, it is highly probable that the toxic consequences of Al accumulation are involved in the AD degenerative process.

Conclusion

(1) Desferrioxamine treatment appears to be a useful palliative treatment for selected AD patients early in the course of the illness. A double-blind, multicentre trial should be instituted to establish the efficacy of desferrioxamine in a wider clinical setting. Safe, effective oral aluminium-chelating agents should be developed.

(2) Human ingestion of and exposure to aluminium should be limited, because several segments of the ageing population may be unusually susceptible to aluminium neurotoxicity:

(a) The elderly may wish to reduce aluminium exposure, because the incidence of AD doubles about every 4.5 years after age 60 years (Jorm et al 1987). At age 85, about 25% to 50% of individuals exhibit impaired cognitive function.

(b) Down's syndrome (trisomy of chromosome 21) is the only established cause of AD. The early onset of AD histopathology and the presence of progressive cognitive deficits which are detectable even in the presence of mental handicap (Dalton et al 1974) indicate that a controlled study of the consequences of lowering lifetime exposure to and ingestion of aluminium for the dementia of AD should be considered for Down's syndrome patients. Since Down's patients are not considered in many countries to be competent to provide informed written consent to enter anti-dementia drug trials, this population will have to wait until the use of an anti-dementia drug becomes part of standard medical practice. However, a controlled trial in which exposure to a known neurotoxin is reduced in air, food, water, pharmaceuticals, cosmetics, toothpastes, anti-perspirants and soaps might be considered ethically acceptable.

(c) Family members in which AD occurs as a dominant inheritance factor may wish to reduce their aluminium ingestion. Even in identical twins, concordance for the expression of AD is reported to be only 44% (McLachlan et al 1991c, Table 6-1), which suggests that environmental factors may contribute to the clinically obvious component of the illness, dementia. The consequences for aluminium metabolism and the blood–brain barrier of the recently described mutations in the amyloid gene require further investigation.

(3) If a concerned individual is to be able to make an informed decision concerning aluminium ingestion, the aluminium content must be listed on the packages of all substances marketed for human ingestion and contact, including

processed foods, drinking fluids, cosmetics, toothpastes and pharmaceuticals. Excluding pharmaceuticals, the usual dietary intake ranges between 10 and 100 mg per day. A scientifically rigorous goal for human daily ingestion of aluminium should be established. We estimate that the daily intake from natural sources is about 3 mg.

(4) The risk of neurological damage associated with aluminium exposure in the work place should be fully investigated (Rifat et al 1990).

(5) Municipal processed drinking water should be regulated so that the concentration of aluminium, wherever possible, is less than 50 µg/l; a long-term goal should be less than 10 µg/l.

(6) Further research should be conducted to enable us to understand fully the health risks of aluminium.

Acknowledgements

This work was supported by grants from the Ontario Mental Health Foundation, the Medical Research Council of Canada, the Department of National Health and Welfare, the Scottish Rite Charitable Foundation and the Alzheimer Association of Ontario.

References

Birchall JD 1992 The interrelationship between silicon and aluminium in the biological effects of aluminium. In: Aluminium in biology and medicine. Wiley, Chichester (Ciba Found Symp 169) p 50–68

Candy JM, Oakley AE, Klinowski J et al 1986 Aluminosilicates and senile plaque formation in Alzheimer's disease. Lancet 1:354–357

Crapper DR, Krishnan SS, Dalton AJ 1973 Brain aluminum distribution in Alzheimer's disease and experimental neurofibrillary degeneration. Science (Wash DC) 180:511–513

Crapper DR, Krishnan SS, Quittkat S 1976 Aluminum, neurofibrillary degeneration and Alzheimer disease. Brain 99:67–79

Crapper DR, Quittkat S, Krishnan SS, Dalton AJ, De Boni U 1980 Intranuclear aluminum content in Alzheimer's disease, dialysis encephalopathy, and experimental aluminum encephalopathy. Acta Neuropathol 50:19–24

Crowther RA 1991 Review: structural aspects of pathology in Alzheimer's disease. Biochim Biophys Acta 1096:1–9

Dalton AJ, Crapper DR, Schlotterer GR 1974 Alzheimer's disease in Down's syndrome: visual retention deficits. Cortex 10:366–377

De Boni U, McLachlan DRC 1985 Controlled induction of paired helical filaments of the Alzheimer type in cultured human neurons, by glutamate and aspartate. J Neurol 68:105–118

Duckett S, Galle P 1976 Mise en évidence de l'aluminium dans les plaques de la maladie d'Alzheimer: étudié à la microsonde de Castaing. CR Hebd Seances Acad Sci Ser D Sci Nat 282:393–395

Edwardson JA, Candy JM 1990 Aluminum and the etiopathogenesis of Alzheimer's disease. Neurobiol Aging 11:314 (abstr 255)

Farnell BJ, Crapper McLachlan DR, Baimbridge K, De Boni U, Wong L, Wood PL 1985 Calcium metabolism in aluminum encephalopathy. Exp Neurol 88:68–83

Flaten TP 1990 Geographical associations between aluminium in drinking water and death rates with dementia (including Alzheimer's disease), Parkinson's disease and amyotrophic lateral sclerosis in Norway. Environ Geochem Health 12:152–167

Flaten TP, Garruto RM 1991 Desferrioxamine for Alzheimer's disease. Lancet 338:324–325

Fleming J, Joshi JG 1987 Ferritin: isolation of aluminum–ferritin complex from brain. Proc Natl Acad Sci USA 84:7866–7870

Forbes WF, Hayward LM, Agwani N 1991 Dementia, aluminium, and fluoride. Lancet 338:1593–1594

Guy SP, Jones D, Mann DMA, Itzhaki RF 1991 Human neuroblastoma cells treated with aluminum express an epitope associated with Alzheimer's disease neurofibrillary tangles. Neurosci Lett 121:166–168

Hill AB 1965 The environment and disease: association or causation? Proc R Soc Med 58:295–300

Johnson GVW, Cogdill KW, Jope RS 1990 Orally administered aluminum alters in vitro protein phosphorylation and protein kinase activities in rat brain. Neurobiol Aging 11:209–216

Jope RS, Johnson GVW 1992 Neurotoxic effects of dietary aluminium. In: Aluminium in biology and medicine. Wiley, Chichester (Ciba Found Symp 169) p. 254–267

Jorm AF, Korten AE, Henderson AS 1987 The prevalence of dementia: a quantitative integration of the literature. Acta Psychiatr Scand 76:465–479

Krishnan SS, Harrison JE, McLachlan DRC 1987 Origin and resolution of the aluminum controversy concerning Alzheimer neurofibrillary degeneration. Biol Trace Elem Res 13:35–42

Ksiezak-Reding H, Chien C-H, Lee M-Y, Yen S-H 1990 Mapping of the Alz 50 epitope in microtubule-associated proteins tau. J Neurosci Res 25:412–419

Lee VM-Y, Balin BJ, Otvos L Jr, Trojanowski JQ 1991 A68: a major subunit of paired helical filaments and derivatized forms of normal tau. Science (Wash DC) 251:675–678

Lukiw WJ, Krishnan B, Wong L, Kruck TPA, Bergeron C, McLachlan DRC 1991 Nuclear compartmentalization of aluminum in Alzheimer's disease. Neurobiol Aging 13:115–121

Markesbery WR, Ehmann WD, Hossain TIM, Alauddin M, Goodin DT 1981 Instrumental neutron activation analysis of brain aluminium in Alzheimer disease and aging. Ann Neurol 10:511–616

Martyn CN 1992 The epidemiology of Alzheimer's disease in relation to aluminium. In: Aluminium in biology and medicine. Wiley, Chichester (Ciba Found Symp 169) p 69–86

Masters CL, Multhaup G, Simms G, Pottgiesser J, Martins RW, Beyreuther K 1985 Neuronal origin of a cerebral amyloid: neurofibrillary tangles of Alzheimer's disease contain the same protein as the amyloid of plaque cores and blood vessels. EMBO (Eur Mol Biol Organ) J 4:2757–2763

McDermott JR, Smith AI, Iqbal K, Wisniewski HM 1979 Brain aluminum in aging and Alzheimer disease. Neurology 29:809–814

McKee AC, Kosik KS, Kennedy MB, Kowall NW 1990 Hippocampal neurons predisposed to neurofibrillary tangle formation are enriched in type II calcium/calmodulin-dependent protein kinase. J Neuropathol & Exp Neurol 49:49–63

McKhann G, Drachman D, Folstein M et al 1984 Clinical diagnosis of Alzheimer's disease: report of the NINCDS-ADRDA Work Group under the auspices of Department of Health and Human Services Task Force on Alzheimer's Disease. Neurology 34:939–944

McLachlan DRC 1986 Aluminum and Alzheimer's disease. Neurobiol Aging 7:525–532

McLachlan DRC, Krishnan SS, Quittkat S, De Boni U 1980 Brain aluminum in Alzheimer's disease: influence of sample size and case selection. Neurotoxicology 1(4):25–32

McLachlan DRC, Kruck TPA, Lukiw WJ, Krishnan SS 1991a Would decreased aluminum ingestion reduce the incidence of Alzheimer's disease? Can Med Assoc J 145:793–804

McLachlan DRC, Dalton AJ, Kruck TPA et al 1991b Intramuscular desferrioxamine in patients with Alzheimer's disease. Lancet 337:1304–1308

McLachlan DRC, Rupert JL, Kung Sutherland M, Grima EA 1991c Etiology of Alzheimer's disease. In: Berg JM, Karlinsky H, Lowy F (eds) Alzheimer's disease research: ethical and legal issues. Carswell Publishing, Toronto, p 93–119

Neri LC, Hewitt D 1991 Aluminium, Alzheimer's disease, and drinking water. Lancet 338:390

Papasozomenos SCh, Binder LI 1987 Phosphorylation determines two distinct species of tau in the central nervous system. Cell Motil Cytoskeleton 8:210–226

Perl DP, Brody AR 1980 Alzheimer's disease: x-ray spectrometric evidence of aluminum accumulation in neurofibrillary tangle-bearing neurons. Science (Wash DC) 208:297–299

Perl DP, Pendlebury WW 1984 Aluminum accumulation in neurofibrillary tangle-bearing neurons of senile dementia, Alzheimer's type: detection by intraneuronal X-ray spectrometry studies of unstained tissue sections. J Neuropathol & Exp Neurol 43:349

Rifat SL, Eastwood MR, McLachlan DRC, Corey PN 1990 Effect of exposure of miners to aluminium powder. Lancet 336:1162–1165

Saitoh T, Masliah E, Jin L-W, Cole GM, Wieloch T, Shapiro IP 1991 Protein kinases and phosphorylation in neurologic disorders and cell death. Lab Invest 64:596–616

Steiner B, Mandelkow EM, Biernat J et al 1990 Phosphorylation of microtubule-associated protein tau: identification of the site for Ca^{2+}-calmodulin dependent kinase and relationship with tau phosphorylation in Alzheimer tangles. EMBO (Eur Mol Biol Organ) J 9:3539–3544

Strong MJ, Wolff AV, Wakayama I, Garruto RM 1991 Aluminum-induced chronic myelopathy in rabbits. Neurotoxicology 12:9–22

Trapp GA, Miner GD, Zimmerman RL, Master AR, Heston LL 1978 Aluminium levels in brain in Alzheimer's disease. Biol Psychiatry 13:709–718

Traub RD, Rains TC, Garruto RM, Gajdusek DC, Gibbs CJ Jr 1981 Brain destruction alone does not elevate brain aluminum. Neurology 31:986–990

Uéda K, Masliah E, Saitoh T, Bakalis SL, Scoble H, Kosik KS 1990 Alz-50 recognizes a phosphorylated epitope of tau protein. J Neurosci 10:3295–3304

Vogt T 1986 Water quality and health—study of a possible relationship between aluminum in drinking water and dementia. Central Bureau of Statistics of Norway, Oslo (Abstr Sosiale og okonomiske studier 61)

Ward NI, Mason JA 1986 Neutron activation analysis techniques for identifying elemental status in AD. Modern trends in activation analysis. Proceedings of seventh conference (Copenhagen), p 925–934

Wettstein A, Aeppli J, Gautschi K, Peters M 1991 Failure to find a relationship between mnestic skills of octogenarians and aluminum in drinking water. Int Arch Occup Environ Health 63:97–103

Wolozin B, Davies P 1987 Alzheimer-related neuronal protein A68: specificity and distribution. Ann Neurol 22:521–526

Yoshimasu F, Yasui M, Yoshida H et al 1981 Aluminum in Alzheimer's disease in Japan and Parkinsonism dementia in Guam. XII World Congress of Neurology (Kyoto, Japan). Abstracts of papers presented. Excerpta Medica, Amsterdam (abstr 15.07.02)

DISCUSSION

Kerr: Professor McLachlan, in your controlled trial of desferrioxamine, it seems that a major benefit of this drug is to prolong the life of people with Alzheimer's disease, with nine deaths in your untreated group compared to one in the DFO-treated group, after 58 months. Can you tell us what these nine died of, that was prevented by giving them DFO?

McLachlan: Like most patients with Alzheimer's disease, eight out of nine died of pneumonia after becoming bedridden. One died suddenly, perhaps of myocardial infarction, but an autopsy was not performed and we have inadequate records of the agonal process.

There were a number of patients in the treated group who, within six months of the end of the 24-month trial, and before the code was open, I encouraged to stay on the drug. Some of those people continue to this day to have a reasonable quality of life; they are able to travel by air, some are able to go for walks by themselves. However, no treated patients improved. The drug treatment can only be considered as palliative and appears both to slow the deterioration in the quality of life and to increase life expectancy. I believe that all the patients will eventually die of Alzheimer's disease.

Martyn: You spoke about Bradford Hill's criteria for causality. I wonder if you misinterpret his criterion of temporality? Surely he was only saying that if there is a causal association, the cause must precede the effect. He did not insist that it has to be any particular length of time before the effect?

McLachlan: I agree with your interpretation of Hill's criterion of temporality. However, I wish to emphasize that we do not know when, in the course of this disorder, abnormal concentrations of Al first appear in neurotopic target sites. Does increased Al concentration occur as a very early event, or a late event? We do not have an independent way of assessing this at the present time. We wish to accumulate experimental evidence to address the hypothesis that Al accumulates only in damaged cells.

Zatta: Dr McLachlan, why did you decide to inject the desferrioxamine intramuscularly, rather than intravenously?

McLachlan: We have attempted to use slow infusion of desferrioxamine by pump in AD, but found this unsatisfactory for demented patients. Also, DFO in our study was given by the spouse, in the home, on a continuing basis for a two-year period. It is not legal in Canada for a non-trained individual to give intravenous injections.

Wisniewski: On the question of the paired helical filaments (PHF) which make up the NFT, and the presence of Al there, we know now that different populations of PHF exist: soluble and insoluble. The forms which chemists are isolating are the 'petrified' or insoluble PHF. Immunohistochemical studies of PHF suggest various stages of tangle maturation. Therefore, using the methods of Dr Perl and Dr Edwardson, we should look at tangle-bearing neurons at various stages of NFT maturation. Without this, it will be difficult to assign any meaning to the biochemically isolated PHF in relation to their Al content.

Dr McLachlan also referred to the presence of Al associated with nuclear DNA, comparing demented and non-demented groups of patients. Unfortunately, Alzheimer's disease is not such a 'black-and-white' condition. It depends on a quantitative difference in terms of the neuropathology. We know that there are people whom we consider to have preclinical Alzheimer's disease, because at a given moment they already have extensive AD neuropathology, but they can still 'tolerate' it; they don't show clinical symptoms of dementia (Evans et al 1989). Therefore, in Dr McLachlan's studies I would expect an overlap between the control and the AD patient.

As to treatment, desferrioxamine is a powerful chelating agent, so why should it chelate only Al, but not iron and many other elements which it can remove? I doubt that Al is the only element removed from persons who have been treated.

On the presence of Al and pathology, we know that when Al is given to certain animal species, they develop neurofibrillary tangles of the 10 nm type, but Al is not present there. So Al can induce a dramatic pathology, yet it is not found in the tangle caused by Al injection.

Blair: It is not possible to produce a model for Alzheimer's disease in its behavioural aspects in an experimental animal. It is however possible to produce biochemical models. We demonstrated that in man in Alzheimer's disease there is a marked disturbance in tetrahydrobiopterin metabolism, confirmed by others since. This deficiency is not unique to Alzheimer's disease (it occurs also in depression). We gave rats large doses of Al compounds in drinking water over three months and reproduced the biochemical parameters of choline acetyltransferase and tetrahydrobiopterin deficiency (Hamon et al 1987). We also reproduced other biochemical features which are known to be altered in senile dementia of Alzheimer type: for example, in relation to Dr Jope's findings, we have confirmed that cyclic AMP production is increased by Al. So you can produce a biochemical model in the rat in which you can reproduce many of the biochemical stigmata that have been observed in Alzheimer's disease.

In these models you can also reverse some of the biochemical changes by treating the tissues with a powerful chelator, transferrin, using it merely as a chemical and not making any implications about its role in the disease process.

So I would support Professor McLachlan; there is an excellent argument for Al being implicated in Alzheimer's disease. But we should remember that it is not a sharply defined disease; it is a heterogeneous disease with a spectrum of responses in patients.

One aspect, not mentioned yet, is of how many of these patients show Parkinson's disease. There is argument among the specialists in Parkinson's disease about co-morbidity with Alzheimer's disease, or an enhanced level of Alzheimer's disease (van Duijn & Hofman 1991).

Strong: In relation to Dr Wisniewski's comments, he in fact was one of the first investigators to study chronic aluminium exposure experimentally (Wisniewski et al 1980, 1982). Our attempts to develop chronic models of Al neurotoxicity were based on this work. It is now clear that if you inoculate New Zealand White rabbits with a low dose (100 µg) of Al chloride intracisternally, once monthly for 10 months, a chronic disease state is induced (characterized by a progressive spastic myelopathy) in which encephalopathic features, or seizures, are not seen and in which only the minority of animals (fewer than 10%) develop neocortical neurofibrillary pathology (Strong et al 1991). We originally published this on the basis of five rabbits; we have now seen the same picture in a further 40 rabbits. It's a chronic, slowly progressive disease in which, I agree, 10 nm filaments are deposited; but those neurofilamentous structures appear to be phosphorylated in an abnormal state.

We have begun to study this in an acute rabbit model, utilizing high dose (1000 µg) aluminium chloride to see if there is aberrant, or excessive, phosphorylation of neurofilament subunit proteins. We can't tell you the results yet, but we find that it takes much longer to enzymically dephosphorylate the high and intermediate weight neurofilament subunit proteins after Al treatment than in the control state. So although Al is not binding to the neurofilament subunit protein itself, we think it is altering the neurofilament phosphorylation state acutely, at least at higher doses of Al.

McLachlan: In human neurons in tissue culture, two laboratories have shown that low dose (100 µM) exposure to Al for 6–16 days results in the neurons exhibiting epitopes detected by antibodies which characteristically recognize neurofibrillary tangles (Guy et al 1991, Mesco et al 1991). These epitopes represent anomalous phosphorylation of tau protein. This argues that Al may act earlier in the time course of Alzheimer's disease, or at least in the formation of the paired helical filaments, than we have had evidence for previously. That is, aluminium 'turns on' the phosphorylation, rather than merely accumulates on the phosphorylated site on tau.

Strong: An additional point is that there are neuron-specific thresholds of aluminium toxicity (Strong & Garruto 1991). The early acute studies on Al toxicity were excellent from a morphological point of view, but they showed a way to kill an experimental animal quickly with Al and to induce massive neurofibrillary degeneration. Now, with chronic *in vivo* experiments, we are beginning to study mechanisms which may be more relevant to the chronic human disease state.

Wisniewski: I think there is evidence that proteins with 'slowed down' turnover are abnormally phosphorylated.

Strong: Within 48 hours of giving Al chloride, I can excessively phosphorylate neurofilament subunit proteins; that's not a long time.

Wisniewski: Can you induce abnormal phosphorylation of proteins with drugs?

Strong: You see abnormal *localization* of a phosphorylated neurofilament; but that is not necessarily abnormal phosphorylation.

Perl: We have evidence of different levels of Al accumulation in neurofibrillary tangles according to their localization in the brain, which seems to correlate with when the tangles were formed. These data would suggest that neurofibrillary tangles from areas of the brain that are affected earlier in Alzheimer's disease have the higher concentrations of Al.

Williams: From the analytical ratio of phosphorus to Al in the neurofibrillary tangles, and knowing the molecular weight of tau protein, we should be able to see whether there is a sensible relationship between the numbers of Al atoms that are claimed to be in the tangle, compared with the number of phosphates.

McLachlan: We find a molar ratio of about one Al atom to two molecules of tau, based on a 60 kDa molecular mass for tau. After exhaustive dialysis with EDTA, about one aluminium atom per two tau molecules remains, which suggests that aluminium is integral to the PHF. We cannot exclude the possibility that Al is required for the assembly of PHFs from anomalously phosphorylated tau. In chromatin there are, on average, 23 Al atoms per 200 DNA base pairs, based on isolated AD dinucleosomes. Control brain dinucleosomes prepared in identical solutions have about three Al atoms per 200 DNA base pairs (Lukiw et al 1991).

Williams: I don't know how many phosphates there are on tau; perhaps three per tau?

Wisniewski: There are many more than one in the phosphorylated tau.

Williams: Perhaps something like one Al for 8–10 phosphates in tau? And the analytical result for DNA is in the order of one Al per 20 bases? This is again a high number, when you think that for any one base there is only one negative charge, which is a phosphate diester. It's not like phosphate on a protein, which has two negative charges. We do not expect diester phosphates to bind aluminium. So it's a big number of Al ions, which may be explained by phosphate on proteins in the nucleosomes?

Zatta: Professor McLachlan, you previously published (Crapper et al 1980) the amount of Al bound to heterochromatin in AD- and dialysis encephalopathy-affected patients. In contrast to Alzheimer's disease, the brains of patients affected by progressive encephalopathy associated with renal failure and dialysis treatment showed aluminium levels in the grey matter lower than in the controls. However, the whole-tissue concentrations were elevated 10–15 times above the control concentrations. Can you confirm these findings?

McLachlan: Yes; in brains of patients who have undergone dialysis, we do not find elevated Al levels in nuclear compartments as we find in Alzheimer's disease. AD is the only neurological condition in which we find both excessive heterochromatization and an accumulation of Al on all chromatin fractions, particularly on dinucleosomes enriched in repressed genes (HNF-L). In renal dialysis we do not find a change in chromatin conformation or an increase in Al on dinucleosomes (Crapper et al 1980, Lukiw et al 1992).

Zatta: In my paper (p 186) I shall present evidence that different molecular species of Al produce very different biological effects on various cell types. We have also some preliminary evidence that in the rat, which is considered to be a species very resistant to Al toxicity, it is possible to produce tangles, but it depends on the Al species and on giving long-term Al treatment; so it is not just a case of 'resistance' to Al in this animal, but an Al speciation effect related to long-term treatment.

Wisniewski: The parkinsonism-dementia complex of Guam is, in my view, the closest approach to a condition in which Al may play a role in neurofibrillary pathology. Why should one not treat these patients with DFO, as Dr McLachlan has treated Alzheimer's disease patients?

McLachlan: I agree that the Guam PD/ALS condition might be a candidate for a desferrioxamine trial. Unfortunately, Guam is remote from where we have an analytical system with which to measure the side effects of DFO. In the past, we have monitored DFO metabolites in all our patients. We use HPLC techniques to look for toxic metabolites of desferrioxamine. Before embarking on such a trial, we have to provide for patient safety, and we are looking at ways to accomplish this at present.

Petersen: You spoke about a possible protective role of fluoride against Al toxicity. Is that because aluminium fluoride is not absorbed from the gut?

McLachlan: That is the most probable interpretation of the epidemiological studies such as that by Forbes et al (1991). However, CNS effects of fluoride cannot be excluded.

Petersen: Fluoride binds Al very strongly. Most of the stimulatory effects of fluoride on, for example, adenylate cyclase *in vitro* are actually effects of Al fluoride, because fluoride is said to etch Al off the glassware used. These effects are normally counteracted by desferrioxamine. So if aluminium fluoride is absorbed, one might expect it to have dramatic effects in the organism. If that is not the case, one might assume that AlF_3 is *not* absorbed.

Blair: If gallium is accepted as a marker for Al, gallium in the presence of fluoride is reduced in its absorption in the rat, both in the fed and in the fasting state. It is unusual for that kind of effect to occur in both those conditions (Farrar et al 1988).

Williams: What does the fluoride do to the levels of iron?

Blair: We haven't looked at iron absorption; we have just looked at gallium, as a marker for Al.

Williams: Iron should show the same type of behaviour. Note that excess fluoride causes iron deposits in teeth.

Blair: Yes; I would expect iron absorption to be limited in the presence of fluoride. So there is evidence here for reduced intestinal absorption of aluminium.

Williams: Bruce Martin has been examining the relationship between Al and F at various pH levels. We are particularly interested, in the present context, in the possibility of Al and fluoride coming together within the physiological range of observations, in terms of pH, $\log[F^-]$, etc.

Martin: We have been investigating the stability of Al–fluoride complexes, their ternary complexes with hydroxide, and their mixed complexes with nucleoside phosphates. The composition of the ternary complexes with hydroxide depends upon the pH and the analogous $pF = -\log[F^-]$. Figure 1 shows some of the results in the ternary system with hydroxide and is not as complicated as might appear at first sight.

The figure shows distribution curves for mole fraction on a total Al(III) basis versus pF, which refers to the free, ambient fluoride molar concentration. First we look at the dashed curves for pH 4. As the ambient fluoride concentration increases (pF decreases) we pass from the AlF^{2+} (labelled F_1) complex near the left to the AlF_2^+ complex (labelled F_2) that peaks at $pF = 4.5$ (0.03 mM F^-) at 0.65 mole fraction total Al(III). Addition of increasing amounts of fluoride yields successively the AlF_3, AlF_4^- and AlF_5^{2-} complexes (labelled F_3, F_4 and F_5, respectively), all of which peak close to 0.65 mole fraction.

This simple, classic picture of Al–fluoride complexes is valid only over a limited pH range. At pH <4 the weak acid HF begins to form ($pK_a = 3.1$), and at pH >5, Al(III) hydrolysis and ternary complexes with hydroxide start to appear. Many investigators have added Al–fluoride mixtures to systems at pH 7.5.

The solid curves in Fig. 1 show the distribution curves at pH 7.5, and it is immediately evident that there is a greater multiplicity of species present (Martin 1988). The main starting species on the left at high pF (low ambient F^-) is the tetrahedral complex $Al(OH)_4^-$ (labelled h_4, where the small h represents a hydroxide group). Addition of fluoride results in the successive replacement of hydroxide groups to give as the predominant species $(HO)_3AlF^-$, $(HO)_2AlF_2^-$, $(HO)AlF_3^-$ and AlF_4^-, which is no longer tetrahedral. (The respective labels in the figure are h_3F_1, h_2F_2, hF_3 and F_4.) Further addition of fluoride results in the substitution of coordinated water molecules about hexacoordinate Al(III) to yield AlF_5^{2-} and finally AlF_6^{3-}, just as at lower pH. Owing to intervention by bound hydroxide, at the higher pH, a greater fluoride concentration is required to obtain the same number of bound fluorides.

Investigators have added Al(III) and excess fluoride to protein systems such as G proteins (GTP-binding regulatory proteins) and have observed a peak effect at about 5 mM added fluoride. Inspection of the classic dashed curves in the figure reveals AlF_4^- as the dominant species at the corresponding $pF = 2.3$, and the results have been interpreted as showing this species serving as a tetrahedral pseudophosphate. However, in aqueous solution the AlF_4^- complex should be hexacoordinate with two bound waters. Moreover, most such experiments were

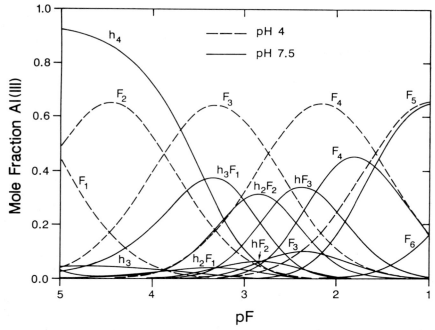

FIG. 1 (*Martin*)   Mole fraction of total Al(III) versus $pF = -\log [F^-]$, where $[F^-]$ is the ambient fluoride molar concentration, for fluoride complexes of aluminium at two pH values—dashed curves for pH 4 and solid curves for pH 7.5. Symbols on curves designate the number of fluoride (F) or hydroxy groups (h) bound to Al(III). Thus h_4 represents $Al(OH)_4^-$; F_4, AlF_4^-; and hF_3, $(HO)AlF_3^-$. Curves were constructed from Al^{3+} hydrolysis constants given in my paper, binary fluoride stability constants from my non-linear least-squares reanalysis of literature data that yields constants close to those reported (Brosset & Orring 1943), and an assumed statistical occurrence of ternary complexes (Martin 1988).

performed at pH 7.5, where the match with the solid curves in Fig. 1 suggests the predominant species to be $(OH)AlF_3^-$ with three fluorides.

In experiments with G proteins, the presumed tetrahedral pseudophosphate AlF_4^- is postulated to reside adjacent to GDP on the G protein (Chabre 1990). We have recently used multinuclear NMR spectroscopy to study the ternary system Al(III), F^- and nucleoside diphosphates (NDP) in aqueous solutions without protein (Nelson & Martin 1991). Under a wide variety of conditions, species with compositions $(NDP)AlF_x$ with $x = 0$ to 3 have been characterized; no species with $x = 4$ was detected. All of these complexes should have hexacoordinate Al(III). Under the conditions of the G protein experiments at pH 7.5 with up to 0.1 mM GDP and a maximum effect at about 3 mM F^-, most Al(III) occurs as $(HO)AlF_3^-$; the lesser amount of nucleoside diphosphate is not competitive. Thus, under these conditions, the distribution curves in the figure apply even to solutions with added nucleotides. This result is confirmed by the high pAl values for the F^- solutions in Table 1 of my paper (p 9) (this volume: Martin 1992). Therefore, the proposed G protein-bound ternary complex $(NDP)AlF_4$ does not occur to a significant extent in aqueous solution, where the presumed tetrahedral AlF_4^- is hexacoordinate with two bound water molecules. Evidently the G protein provides a very special binding site.

You asked me about physiological conditions, but what are they? That is the real question.

Williams: The fluoride concentration in cytoplasm will be very low, say less than $10\,\mu M$.

Martin: At $10\,\mu M$ there would be a pF of 5, so, according to Fig. 1, at pH 7.5 no fluoride complex forms.

Williams: So fluoride is not likely to interfere with uptake problems.

Blair: Dr Martin is discussing thermodynamic equilibria. In the physiological situation you must also consider the kinetic equilibrium; how could one Al species convert into another? If we had a physiological system which drew away AlF_4^-, how quickly would the other Al forms pass into that form?

Martin: That depends upon the other form. Probably, fairly promptly for mononuclear Al complexes, but much more slowly for polynuclear complexes and insoluble materials.

Petersen: It's often said that when you put fluoride into solution, you are etching Al off the glassware. Nobody has suggested that ATP can do that. This suggests that the affinity of fluoride for Al must be relatively high.

Martin: But HF attacks glass by forming covalent Si-F bonds, and releasing any Al^{3+} in the process.

Birchall: The answer perhaps is that there's enough Al in the chemicals used anyway, not to need to invoke getting it off the glass. I doubt the leaching of Al from borosilicate glass anyway.

Flaten: While we are discussing the Al–fluoride interaction in relation to the epidemiological studies of drinking water, a crucial question is the uptake of the different forms of Al in the gut. People have tended to think that fluoride is protective, preventing Al uptake. But in the USA and other places where fluoride is added to the drinking water, typical fluoride concentrations are around 0.05 mmol/l (pF around 4.3), resulting in the formation of considerable amounts of the neutral Al trifluoride, according to the diagram presented by Professor Martin. I would suspect that this small, neutral molecule is more easily taken up through the gut than most charged Al species; therefore, the added fluoride may actually help Al to get into the body.

Fawell: You have to bear pH in mind. The pH of drinking water would normally be adjusted to neutral or above; in many cases, we adjust the pH to 8 and above, in order to reduce the aggressiveness to distribution and plumbing materials. It depends on the local circumstances.

Flaten: The acidity of the stomach would in any case favour the formation of the neutral Al trifluoride.

Wisniewski: As mentioned before, gastric juice is acid; however, the mucus covering the mucosal epithelium is not.

Day: In water of pH 6.0, and at a pF of 4.0, the dominant Al species is somewhere between AlF_2^+ and AlF_3. If you raise the pH of the water, surely all you will do is substitute OH^- for F^-. You will still end up with roughly the same proportion of neutral species. So the fact that the pH of the medium might be 7 rather than 6 will not greatly affect the uptake.

Martin: It is not quite so simple; upon substitution there is some shifting in the predominant species at constant pF. If you look at my Fig. 1 (p 105), and pick pF = 3.3, you see F_3 or AlF_3^0 as the main complex at pH 4 and h_3F_1 or $(OH)_3AlF^-$ as the main complex at pH 7.5. At pF = 4.0 and pH 7.5 the dominant species is h_4 or $Al(OH)_4^-$ instead of AlF_2^+ and AlF_3^0 at pH 4. Supporting your point, at pF = 4.0, the species hF_2 or $HOAlF_2^0$ rises to a maximum 0.21 mole fraction at pH 6.4 and the species $(HO)_2AlF^0$ to a maximum 0.13 mole fraction at pH 6.7. Thus, all three neutral species, AlF_3, $HOAlF_2$ and $(HO)_2AlF$, appear in fluoridated drinking waters.

Day: Would the ionic strength of physiological fluids have a big effect on these proportions?

Martin: Because Al^{3+} is a multiply charged ion, there are appreciable ionic strength effects that will alter the quantitative mole fractions, but should not change the qualitative conclusions. I used the best Al^{3+}-F_x stability constants available and they were obtained in 0.5 M KNO_3. The effect of reducing the ionic strength to the lower values in physiological fluids or to still lower values found in drinking waters will be to heighten the formation of all neutral complexes to greater mole fractions than indicated above.

Blair: To my understanding, there is very little absorption of metals through the stomach, so the discussion of stomach acidity is not directly relevant to questions of absorption of Al. The main site of absorption of aluminium lies in the small intestine. The pH in that fluid is 6.5–7.00. At the surface of the jejunum it is much more acid, down to 4 or 5. So whatever species of aluminium you are dealing with has to pass through this acidic layer. This change from pH 6 to pH 3 must alter the position of the chemical equilibrium.

Martin: It would not change it very much, because hydroxyl complexes are just beginning to appear at pH 6. It would decrease the affinity, because you will be making some HF ($pK_a = 3.0$), but it wouldn't change the distribution appreciably.

Blair: In the rat experiments, we operated at 10^{-6} M gallium and 10^{-4} M fluoride and we found a reduction in short-term transport of gallium over two hours into the major tissues of the body by a factor of 10 (Farrar et al 1988).

References

Brosset C, Orring J 1943 Studies on the consecutive formation of aluminium fluoride complexes. Svensk Kemisk Tidskr 55:101–116

Chabre M 1990 Aluminofluoride and beryllofluoride complexes: new phosphate analogs in enzymology. Trends Biochem Sci 15:6–10

Crapper DR, Quittkat S, Krishnan SS, Dalton AJ, De Boni U 1980 Intranuclear aluminum content in Alzheimer's disease, dialysis encephalopathy, and experimental aluminum encephalopathy. Acta Neuropathol 50:19–24

Evans DA, Funkenstein H, Albert MS et al 1989 Prevalence of Alzheimer's disease in a community population of older persons. Higher than previously reported. JAMA (J Am Med Assoc) 262:2551–2556

Farrar G, Morton AP, Blair JA 1988 The intestinal speciation of gallium: possible models to describe the bioavailability of aluminium. In: Brätter P, Schramel P (eds) Trace element analytical chemistry in medicine and biology. Walter de Gruyter, Berlin, vol 5:342–347

Forbes WF, Hayward LM, Agwani N 1991 Dementia, aluminium, and fluoride. Lancet 338:1593–1594

Guy SP, Jones D, Mann DMA, Itzhaki RF 1991 Human neuroblastoma cells treated with aluminum express an epitope associated with Alzheimer's disease neurofibrillary tangles. Neurosci Lett 121:166–168

Hamon CGB, Edwards P, Jones SA et al 1987 Tetrahydrobiopterin metabolism in senile dementia. In: Pfleiderer W, Wachter H, Blair JA (eds) Biochemical and clinical aspects of pteridines. Walter de Gruyter, Berlin, vol 5:107–118

Lukiw WJ, Krishnan B, Wong L, Kruck TPA, Bergeron C, McLachlan DRC 1991 Nuclear compartmentalization of aluminum in Alzheimer's disease (AD). Neurobiol Aging 13:115–121

Martin RB 1988 Ternary hydroxide complexes in neutral solutions of Al^{3+} and F^{-}. Biochem Biophys Res Commun 155:1194–1200

Martin RB 1992 Aluminium speciation in biology. In: Aluminium in biology and medicine. Wiley, Chichester (Ciba Found Symp 169) p 5–25

Mesco ER, Kachen C, Timiras PS 1991 Effects of aluminum on tau proteins in human neuroblastoma cells. Mol Chem Neuropathol 14:199–212

Nelson DJ, Martin RB 1991 Speciation in systems containing aluminum, nucleoside diphosphates, and fluoride. J Inorg Biochem 43:37–43

Strong MJ, Garruto RM 1991 Neuron-specific thresholds of aluminum toxicity in vitro. Lab Invest 65:243–249

Strong MJ, Wolff AV, Wakayama I, Garruto RM 1991 Aluminum-induced chronic myelopathy in rabbits. Neurotoxicology 12:9–22

van Duijn CM, Hofman A 1991 Relation between nicotine intake and Alzheimer's disease. Br Med J 302:1491–1494

Wisniewski HM, Sturman JA, Shek JW 1980 Aluminum chloride induced neurofibrillary changes in the developing rabbit: a chronic animal model. Ann Neurol 8:479–490

Wisniewski HM, Sturman JA, Shek JW 1982 Chronic model of neurofibrillary changes induced in mature rabbits by metallic aluminum. Neurobiol Aging 3:11–22

Intestinal absorption of aluminium

Gijsbert B. van der Voet

Toxicology Laboratory, University Hospital, PO Box 9600, 2300 RC Leiden, The Netherlands

Abstract. The intestinal absorption of aluminium can contribute significantly to systemic exposure to this element. Aluminium can be absorbed not only from oral pharmaceuticals but also from solid food and drinking water. The absorption process is not restricted to patients with kidney disorders; other groups of patients and healthy subjects are not excluded. Details of the absorptive mechanism are mainly obtained from *in vitro* (everted gut sac) and animal studies (intestinal perfusion) rather than from controlled human studies and case reports. The process of absorption depends on the intraluminal speciation, the intraluminal quantity, the presence of competing (iron, calcium) or complexing (citrate) substances and the intraluminal pH. The condition of the exposed organism with respect to the gut also determines intestinal absorption (iron status, calcium [vitamin D, parathyroid hormone] status, age and kidney function). Various absorption sites and passage routes, both transcellular and paracellular, have been reported, each apparently related to a different aluminium species (hydrated ionic species, aluminium citrate complex etc.). No uniform mechanistic model allowing extrapolation to the clinical situation has yet emerged.

1992 Aluminium in biology and medicine. Wiley, Chichester (Ciba Foundation Symposium 169) p 109–122

The intestinal absorption of aluminium may be defined as the appearance of aluminium in the blood—in fact in the portal blood—after intraluminal exposure. It took the medical community quite a while to realize that aluminium could actually be absorbed after oral intake and could consequently reveal its toxicity. This delay took place in spite of reported warnings by Berlyne and his colleagues dealing with this absorption problem in the early 1970s, at the very beginning of the aluminium era (1970, 1972). This is even more frustrating when we recall that the phenomenon had already been discussed in a book by Ernest Ellsworth Smith in 1928. Some data seemed to exist even before that time. We now realize that aluminium is absorbed not only from oral pharmaceuticals but also from solid food and drinking water. It is also realized that the absorption of aluminium is not restricted to patients with renal disorder; other groups of patients and healthy subjects are not excluded. The relative contribution of oral aluminium to the systemic body burden can be as significant as (or even more significant than) any form of parenteral exposure. The evaluation of oral aluminium as a risk factor in the development of osteotoxicity and neurotoxicity is recognized as an extremely urgent problem for the renal patient. Oral aluminium is also under serious discussion as a potential risk factor in the patient with Alzheimer's disease. In this respect it should be stressed that these questions of the clinical relevance of oral aluminium as a risk factor seem to have overshadowed the study of the mechanistic aspects of the absorptive process itself.

Methodology

Most of the reported studies are either retrospectively analysed or deliberately designed to allow parameters of oral exposure such as daily dose or total dose, on the one hand, to be correlated to elevated blood and tissue levels or to parameters of toxicity, on the other. Essentially, information is required by which to establish a molecular model describing the routes that aluminium takes between its mucosal uptake and portal appearance. It is clear that not every experimental set-up provides suitable information for this purpose. Data on the intestinal absorption process are made available from case reports from adult and paediatric patients with renal deficiency. Oral aluminium-containing medication is the source of aluminium in most cases reported. A minority of the reports deal with infant formula, contaminated with aluminium, orally administered to very young children. Data also become available from controlled trials in patients and healthy volunteers. However, these human studies do not sufficiently clarify to what extent the intestinal absorption is affected by the actual intraluminal availability of aluminium, the exposure period, and the exact site of exposure and absorption. Moreover, the condition of the absorbing human—age, parathyroid hormone (PTH) levels, the state of calcium and phosphate metabolism, and so on—is largely beyond experimental control. Moreover, the characteristics of the molecular mechanism have not been established. Therefore, *in vitro* and animal experiments are performed to gain better experimental control and to achieve a more detailed mechanistic rationale. *In situ* perfusion of discrete parts of the intestine—leaving the blood circulation intact—can be combined with serial sample collection from perfusion media and blood. These perfusion systems provide information on both the mucosal uptake of aluminium and its appearance in the blood. An everted gut sac system prepared from a piece of jejunum or ileum and incubated in an aluminium-containing medium also provides information on the mucosal uptake process of aluminium and its passage, although the system is hampered by the absence of an intact blood circulation. The *in vitro* incubation of intestinal slices has recently regained more interest in the field of aluminium absorption. Current views on mechanisms have mainly been provided by studies performed with these experimental systems.

Characteristics of aluminium absorption

Today's research on the mechanism of the intestinal absorption of aluminium tends to concentrate on three working hypotheses: (1) the intraluminal chemical speciation of aluminium determines the actual availability of aluminium for a particular intestinal passage route; (2) the absorbing organism may have a certain predisposition for aluminium absorption with respect to the intestinal wall; and (3) because of its supposed lack of any physiological function, aluminium may use existing pathways for other—in this case essential—substances. These three aspects will be examined in turn.

Intraluminal speciation and availability

The aluminium compound absorbed is usually not the compound ingested. In fact, the problem lies not so much in what compound is actually ingested but in what is made of it at the exposure site, in what quantity, and how (bio)available it is. Given this information, a number of phenomena may be explained.

Intestinal absorption of aluminium depends on the intraluminal chemical species. For instance, aluminium chloride and aluminium citrate may produce intraluminally (given a low pH and comparable experimental conditions) equal amounts of two completely different species—either the aluminium ion or a neutral aluminium citrate complex. However, the aluminium citrate complex is much more (bio)available than the aluminium ion.

Intestinal absorption of aluminium depends on the intraluminal quantity. With respect to quantity, it is obvious that one and the same aluminium species may be absorbed in a concentration-dependent fashion within the restrictions of a particular route; at a higher concentration of the same species, more will be absorbed. But some interpretations show clearly that concentration dependence and species dependence may be confused. For instance, some people observe a difference in absorption at low pH when equal amounts of aluminium chloride and aluminium hydroxide are given. Ionic species are made available for absorption from both compounds. Therefore an absorption difference is explained only by the quantity of the (bio)available species, rather than by different species derived from these compounds. This confusion is also seen in the older studies when different pharmaceutical formulations containing aluminium hydroxide were examined for their absorptive potential. The absorption differences observed were often explained on the basis of different species, but should be explained on a quantitative base.

Intestinal absorption depends on the intraluminal presence of substances with structural similarity, like essential metals such as calcium and iron, or with a binding potential like that of citrate. The presence of other intraluminal substances may affect the speciation and quantity of available aluminium species. For instance, intraluminal phosphate will probably act as a protecting agent against intestinal absorption because of the ability to withdraw aluminium from the bioavailable pool. Calcium and iron are subject to a speciation process of their own and may produce species with great similarity in charge, mass and complexation characteristics which may as such compete with aluminium species for complexation or for an intestinal transport route. On the other hand, compounds are present which complex with aluminium and take it along for the ride, over the intestinal wall—for example, aluminium citrate. The actual availability of a certain aluminium species is the result of the relative strengths of a network of competing and complexing substances.

Intestinal absorption of aluminium depends on the intraluminal pH. The observation that the intestinal absorption of aluminium from aluminium chloride or aluminium hydroxide increases at a lower pH is not surprising. The pH is a very fundamental variable which affects all the others—the intraluminal species, the quantity, and the binding strength.

Condition of the absorbing organism

The absorbing human or animal may have a certain predisposition for aluminium absorption. A certain effect may be exerted on the intestinal wall.

Intestinal absorption of aluminium depends on the iron status of the absorbing organism. Aluminium absorption in iron-deficient rats leads to an enhancement of intestinal absorption of aluminium as compared to rats of normal iron status.

Intestinal absorption depends on the calcium (vitamin D, PTH) status of the absorbing organism. Reports on calcium could be taken to indicate that aluminium is partly subject to calcium-regulatory metabolism, especially to vitamin D metabolites and PTH. The data however seem largely confusing. It may be stated that high PTH levels in the presence of normal calcium leads to enhanced intestinal absorption of aluminium and that vitamin D deficiency may lead to reduced aluminium absorption.

Intestinal absorption of aluminium depends on the age of the absorbing organism. Paediatric patients are said to be more sensitive to oral aluminium loading than adults. The higher sensitivity of children is claimed for a number of reasons. Because most of human brain development should occur in the first year of life, the immature blood–brain barrier (and the poorly developed gastrointestinal tract) are expected to be more permeable to aluminium; also, the immature formation and function of the kidneys may lead to a higher body load; and so on. An age effect is very plausible; however, in spite of the reports, the distinction between age-dependent and dose-dependent absorption or toxicity is as yet very difficult to make. Younger children simply receive higher dosages.

Intestinal absorption depends on kidney function.   Kidney disorder is a very indirect variable. The claim that 'renal dysfunction is a risk factor for aluminium toxicity' is very easily made, but it fails to provide any firm data which would explain enhanced intestinal absorption. A renal condition does not mean elimination insufficiency alone; it implies disturbed calcium and phosphate metabolism, for instance. But there are some residual, unknown factors that surely enhance intestinal absorption.

Absorption sites and routes

The physiological process of intestinal absorption for any substance may include paracellular passage routes (along the enterocyte through the tight junctions) and transcellular passage routes (through the enterocyte, involving passage through the brush border membrane, through the cell cytoplasm, and passage through the basolateral membrane). Finally, the substance may appear in the portal and peripheral blood. It is a dynamic process which can work both ways. In more kinetic terms, the processes of intestinal absorption include on the one hand energy-*in*dependent diffusion processes—passive and facilitated—and on the other hand energy-dependent active transport. The paracellular route may be passive in nature. The transcellular process involves passive, facilitated and active routes. For instance, calcium transport takes place as passive diffusion through the paracellular route, while facilitated diffusion may take place over the brush border membrane and active transport over the basolateral membrane. For each essential group of substance—sugars, organic acids, water, metals—a route or even a set of routes can be described. It should be realized that different parts of the intestine have a potential for the intestinal absorption of a substance—for instance, calcium absorption, iron absorption and also perhaps citrate absorption take place mainly in the duodenum. On the other hand, other parts of the intestine—jejunum and ileum—also possess a potential for intestinal absorption of these substances, although less powerful than that of the duodenum. Physiological conditions, in fact, determine whether these regions are also used for absorption. The essential problem in almost all studies reported so far is the lack of knowledge of the aluminium species created; therefore comparison between most of the studies cannot be made and extrapolation to the human situation is hampered.

Adler & Berlyne in 1985 studied the intestinal absorption of aluminium chloride at pH 2.0 with an *in vivo* duodenal perfusion technique in vitamin D-deficient and vitamin D-sufficient rats. Their results suggest that aluminium is absorbed in the duodenum by both a non-saturable mechanism and a vitamin D-dependent saturable mechanism for which it may compete with calcium.

Cochran and colleagues in 1990 perfused the duodenum with aluminium chloride at pH 8.5. Their data suggest that energy-dependent transport plays an important role in aluminium's uptake. Calcium channels may provide an additional entry site.

Feinroth et al (1982) studied the intestinal absorption of aluminium given as the chloride at pH 7.4 in rat everted jejunal sac. Aluminium absorption appeared to involve an energy-dependent, carrier-mediated mechanism. Provan & Yokel (1988a) used the jejunal slice to determine if aluminium interacts with the gastrointestinal calcium-transporting system. Their results, obtained using various calcium channel blockers and activators, suggested that aluminium indeed interacts with this calcium-transporting system. Using an *in situ* jejunal perfusion system at pH 7.4 the same investigators reported in the same year (Proven & Yokel 1988b) that gastrointestinal uptake of aluminium occurs by an energy-dependent, sodium-dependent, paracellular pathway-mediated process.

In 1989, Froment and colleagues reported an extensive study on the site and mechanism of enhanced gastrointestinal absorption of aluminium citrate using oral gavage, everted duodenal and jejunal sacs, isolated duodenal loops and a Ussing chamber. The data suggested that enhanced aluminium absorption following the administration of aluminium citrate occurs in the proximal bowel, via the paracellular pathway, as a result of the opening of cellular tight junctions.

Most investigators have been guided by physiological or clinical conditions. For instance, if aluminium interacts with calcium absorption and if calcium is mainly absorbed in the duodenum then they study the duodenum, or if low pH affects aluminium solubility and enhances intestinal absorption, they look at the most acidic intestinal area, like the lower stomach and the proximal duodenum, and so on.

So far, various mechanisms have been postulated by each of these investigators. They have shown an absorbtive potential for aluminium for different parts of the intestine. There is not a complete lack of uniformity, but the differences observed are probably mainly due to the difference in intraluminal aluminium speciation. Nobody has produced the same species. A uniform mechanistic model for aluminium absorption is therefore unlikely to be achieved at present. Most probably, each aluminium species will prove to have a distinct absorption mechanism. The study of intestinal absorption needs more information on intraluminal aluminium speciation. Only then can a reliable mechanistic foundation be laid for the intestinal absorption of aluminium under various clinical conditions.

Aluminium citrate: transcellular or paracellular transport?

A standardized system of perfusing the rat small intestine was developed by us (Van der Voet & De Wolff 1984, Van der Voet et al 1989). This system allows the investigator to select a certain part of the intestine for perfusion during a certain time period. Moreover, the perfusion medium can be sampled serially in time; the disappearance of aluminium from this medium is a measure of mucosal uptake. A cannula may also be introduced into the portal vein, allowing one to make serial measurements of the portal appearance of aluminium, and also into the carotid artery, allowing one to monitor its peripheral appearance. Realizing the potential effect of species differences, we introduced aluminium as the chloride and as the citrate while the pH was kept at 4.0, expecting to produce the aluminium ion and the aluminium citrate complex, respectively. Moreover, to investigate whether the absorption of aluminium was energy dependent, we added the metabolic uncoupler dinitrophenol (DNP) to the perfusion medium. Experimentally, the small intestine of the rat was perfused with 20 mmol aluminium/l combined with 5, 20 and 80 mmol citric acid/l. Samples were collected serially during a 60-min period from the perfusion medium and from the portal and systemic blood, and were analysed for aluminium.

The appearance of aluminium (given as the ion) in the systemic blood correlated positively with (1) the amount of aluminium retained by the mucosa and negatively with (2) the aluminium concentration already present in the blood. In other words, the more aluminium was retained by the mucosa, the more appeared in the blood; and the more aluminium was already present in the blood, the less appeared there. These processes were not affected by DNP. The results may suggest that no metabolic energy is required for the transport of ionic aluminium and that a diffusion process is involved. Citric acid, in all the concentrations used, stimulated both mucosal uptake and the portal and systemic appearance of aluminium. Concentration dependence was noted, although the processes seemed to level out at the higher citric acid concentrations used. The level of aluminium (given as the citrate) in systemic blood was completely independent of the blood concentration before perfusion, in contrast to ionic aluminium. DNP partly inhibited the citric acid-enhanced absorption. These data may imply that intestinal absorption of an aluminium citrate complex may require metabolic energy and be active in nature. This phenomenon might be compatible with a transcellular passage route for aluminium. In our group we are now studying the effect of 2,4,6-triaminopyrimidine (TAP)—a substance claimed to inhibit paracellular transport (Moreno 1975)—on aluminium absorption. Preliminary results of this study (Fig. 1) showed again an enhancement of the aluminium absorption by citrate. TAP, however, inhibited the citrate-stimulated absorption, implying that aluminium citrate also might use a paracellular transport route. It should be stressed that the interpretation

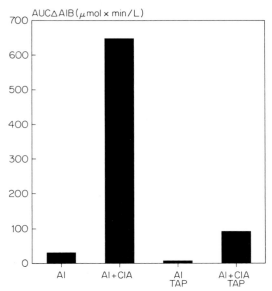

FIG. 1. Effect of triaminopyrimidine (TAP) (10 mmol/l) on the appearance of aluminium in the peripheral blood after perfusion of the jejunum and ileum with the ion (Al) (20 mmol/l) and as aluminium citrate (Al + CIA) (20 mmol/l) for 60 min at pH 4.0. Blood concentrations are corrected for the zero time value (ΔAlB in μmol/l) and effects are expressed as the area under the blood concentration curve (AUC in μmol × min/l) in 60 min. *Preliminary results giving mean values for only three rats per group.*

of any work with these paracellular or transcellular blockers requires great caution.

Inference

There is no room for very clear and concise conclusions yet in the field of the intestinal absorption of aluminium. It should be stated that investigations into mechanisms should not be solely evoked by the clinical problems, but more the other way around—that mechanistic toxicological research should be pursued more vigorously to provide enough information to transpose to, and ultimately explain, clinical phenomena.

Further details can be found in the reviews by De Wolff & Van der Voet (1990) and Van der Voet (1992).

References

Adler AJ, Berlyne GM 1985 Duodenal aluminum absorption in the rat: effect of vitamin D. Am J Physiol 249:G209–G213

Berlyne GM, Ben Ari J, Pest D et al 1970 Hyperaluminaemia from aluminum resins in chronic renal failure. Lancet 2:494–496

Berlyne GM, Yagil R, Ben-Ari J, Weinberger G, Knopf E, Danovitch GM 1972 Aluminium toxicity in rats. Lancet 1:564–568

Cochran M, Goddard G, Ludwigson N 1990 Aluminum absorption by rat duodenum: further evidence of energy-dependent uptake. Tox Lett 51:287–294

De Wolff FA, Van der Voet GB 1990 Intestinal absorption of aluminum. In: De Broe ME, Coburn JW (eds) Aluminum and renal failure. Kluwer Academic Publishers, Dordrecht, Boston & London, p 41–56

Feinroth M, Feinroth MV, Berlyne GF 1982 Aluminum absorption in the rat everted gut sac. Miner Electrol Metab 8:29–35

Froment DPH, Molitoris BA, Buddington B, Miller N, Alfrey AC 1989 Site and mechanism of enhanced gastrointestinal absorption of aluminum by citrate. Kidney Int 36:978–984

Moreno JH 1975 Blockage of gallbladder tight junction cation-selective channels by 2,4,6-triaminopyrimidinum (TAP). J Gen Physiol 66:97–155

Provan SD, Yokel RA 1988a Influence of calcium on aluminum accumulation by the rat jejunal slice. Res Commun Chem Pathol Pharmacol 59:79–92

Provan SD, Yokel RA 1988b Aluminum uptake by the *in situ* rat gut preparation. J Pharmacol Exp Ther 245:928–931

Smith EE 1928 Aluminum compounds in food. Paul B. Hoeber, New York, p 213–236

Van der Voet GB 1992 Intestinal absorption of aluminum: relation to neurotoxicity. In: Isaacson RL, Jensen KF (eds) The vulnerable brain and environmental risks. Vol 2: Toxins in food. Plenum Publishing, New York, in press

Van der Voet GB, De Wolff FA 1984 A method of studying the intestinal absorption of aluminium in the rat. Arch Toxicol 55:168–172

Van der Voet GB, Van Ginkel MF, De Wolff FA 1989 Intestinal absorption of aluminum in rats: stimulation by citric acid and inhibition by dinitrophenol. Toxicol Appl Pharmacol 99:90–97

DISCUSSION

Petersen: Your result with dinitrophenol was interesting. It seemed to have quite a dramatic effect in enhancing the absorption of Al, after the poisoning. An advantage of using DNP is that it can be washed out very easily. Was this enhancing effect reversible?

Van der Voet: We did not test the reversibility of the effect, but DNP is washed out easily. When we perfuse the intestine for one hour, the perfusion solution is yellow initially and then turns white in the course of 60 min. We lose the DNP, so DNP is subject to an absorption process of its own. I think this DNP effect as observed in our experiments is not only an effect of the metabolic inhibition; there is something more to it.

Petersen: For a toxic substance, the intestine should be a barrier. My way of looking at your result would be that if you poison the intestine, the barrier doesn't function so well any more.

Williams: Dinitrophenol is an uncoupler of oxidative phosphorylation and a proton transporter, so it would uncouple the citric acid cycle from ATP generation. The citric acid cycle would just produce protons which would not produce ATP. This being so, I don't think you could tie the observation on DNP to Al. DNP wrecks the whole of oxidative metabolism.

Van der Voet: I agree.

Blair: Triaminopyrimidine is a blocker of the junction between the cells of the intestinal mucosa. You say that the results show that you were looking at a paracellular route of transport. We found that TAP doesn't work as a blocker of the paracellular channel (Coogan 1982). Others have found this also.

Van der Voet: We used TAP because it was used as a paracellular blocker in the earlier literature on Al absorption. We tested different concentrations of TAP and used the concentration where we obtained an effect on Al absorption. We added it to the perfusion fluid. I don't know how we can explain the observation; perhaps we have to explain it in a completely different manner than blocking the paracellular route of Al absorption. But it is very difficult to work with these blockers and activators because they have a certain potential of their own, and all these absorption studies, using jejunal sacs and so on, may be interpreted in a completely different way.

Farrar: Can I ask you what proportion of mucosally bound aluminium was actually transported to the basolateral compartment?

Van der Voet: It was a very low proportion. We had mucosal uptake, measured in millimoles, and the portal and systemic appearance of Al, measured in micromoles, so there is roughly a hundred-fold to a thousand-fold difference between the concentrations. You can calculate a balance on these data, but you have to make rough estimates.

Farrar: We did everted sac experiments with gallium and about 1% of what was mucosally bound was transported (Farrar et al 1987).

Williams: How do you get a millimolar Al solution without citrate, Dr van der Voet?

Van der Voet: We use Al chloride and usually make it isotonic with, say, choline chloride. We try to buffer it, using a large range of buffers which (according to the literature) do not bind to any metal, and establish a pH of 4 in this study. We tried to interpret Bruce Martin's results and tried to establish a certain definite pH range; at a pH level of 4, we expect to have the ionic Al hexahydrate; for Al citrate, at this pH it is probably a neutral Al citrate complex. That is what we chose to work with.

Farrar: You were perfusing the small intestine at pH 4; did you do any viability studies?

Van der Voet: Yes; after each experiment parts of the intestine were checked for morphological damage. We also measured physiological parameters like calcium and sodium absorption, and their absorption potential over the intestine. We used these as an internal control of the viability of the system.

Blair: There is a certain amount of transferrin that appears in bile. What effect do you think that has on Al uptake?

Van der Voet: This is very complicated. We did some studies, comparing the intestinal absorption of trivalent and bivalent iron; they had different effects on aluminium absorption. In spite of much speculation we did not come to any reliable conclusion about the role of intestinal transferrin.

Blair: As a more general comment, many people would say that if you are studying the uptake of Al at 10^{-3} M and pH 4.0, the mechanisms you might reveal would have very little relevance to intestinal absorption in the gut, where the Al concentration level might be of the order of 10^{-6} M, and therefore alternative processes might be involved, and where the pH in health is about 7. The chemical forms of Al that you are studying are changed dramatically at pH 7; you are using a hydroxy Al species, but at a pH value of 4, you have principally an Al cation, with different charges and different sizes. The whole situation is very different from the *in vivo* one.

Van der Voet: We deliberately chose not to use physiological conditions at this stage. We want to approach the problem from a completely mechanistic point of view. So I am not interested, at this point, in what physiological pH exists in the intestine, and what species are to be expected there physiologically. I wanted only to assess the *potential* of certain parts of the intestine for the absorption of Al. When we have these details, we hope to be able to transfer this to the clinical situation. This is the reason that we chose these conditions, establishing a pH of 4.0 and comparing the absorption of ionic Al with the Al citrate complex. We hope to have the species under control.

Day: You found that the uptake of Al ions was inversely correlated to the concentration of Al already present in the blood. Was the blood concentration of Al established as an experimental variable, or was it simply the result of feeding the animal with low levels of Al over a long period?

Van der Voet: This was the natural background level in the rat. We only discovered this correlation because there were a few rats with rather high Al background levels. In other rat populations we did not pick up this correlation, because the range of concentrations that we were working with was very small. We now find this negative correlation in almost every experiment that we do. This is therefore a very significant finding: this correlation exists when you perfuse with Al chloride at pH 4.0; it does not exist on perfusion with the aluminium as the citrate. Using Al citrate, we find a completely horizontal relation, independent of the initial concentration in the blood.

Day: We can assume that Al in the blood would be present in its preferred equilibrium state, bound to transferrin, or whatever. It would not have been present as free Al^{3+} ions, so there could be no question of a transmembrane equilibrium. So is there some more complex feedback mechanism which is operating?

Van der Voet: I think so. So far we have just made the observation, but we can't come up with a plausible model which would explain the phenomenon.

Blair: We have recently looked at the NMR spectrum of Al citrate, using [27]Al (unpublished observations with Dr D. Williamson & Dr G. Moore at the University of East Anglia in Norwich). My impression is that at least three Al citrate species can exist, given appropriate pH changes and concentration ratios.

Martin: Yes, there are three main mononuclear citrate species over the pH range 3 to 9, where tetrahedral $Al(OH)_4^-$ takes over (Martin 1988). However, polynuclear species form over a period of time.

Blair: Yes; we did 24-hour scans.

Martin: The group in Umeå has published the definitive study on the Al^{3+} – citrate system, and they consider formation of polynuclear complexes over a period of time (Ohman 1988).

Day: [27]Al has a quadrupole; therefore in practice you can only see the species with high symmetry, because of line broadening. You would not necessarily observe *all* the Al complex species that were present in the citrate solutions.

Blair: We saw three separate peaks, so there are a minimum of three.

Klinowski: In solution it should not matter, because the quadrupole interaction is averaged by the thermal motion of molecules.

Day: Our observations with ^{27}Al NMR were not on the citrate system, but on aqueous perchlorate systems. At pH < 3, we observed the fully symmetrical, octahedral aqueous species $[Al(H_2O)_6]^{3+}$. As we raised the pH to above 4, this disappeared. We saw no further discrete species until we raised the pH to about 8, when the tetrahedral $[Al(OH)_4]^-$ species was observed (Tonge 1989).

Klinowski: Dr Akitt did this in Leeds.

Day: Yes, I am aware of his work, and I think we saw very much the same as he did (Akitt & Elders 1985). Taking all that into account, I think my comments about the citrate system would be valid.

Blair: The Al spectra are subject to very rigid exclusions. Nevertheless we could see the Al cation, and the hydroxy Al species, and we saw three peaks from the Al citrate system, with a large excess of citrate.

Williams: Dr Van der Voet mentioned calcium. In the conditions which he is studying, which would refer to the outside of a cell, on a mucosa or something like that, calcium–Al competition will be absolutely critical, especially at pH 4. I did an analysis of this competition which showed that if you lower the pH below 5, Al comes very much into its own, outside of the cell. As Bruce Martin showed, Al hydroxide formation is a cooperative phenomenon, so that over a very narrow range of pH lowering, you will suddenly see a large increase in free Al concentration, while the free ion concentration of Ca remains unchanged. Despite the fact that at this pH there is competition, say from protons for carboxylate groups, the power dependence (i.e. dependence on $[H^+]$) is only as the power of one, whereas the dependence on the Al concentration is of the power 4 in $[H^+]$. So the free Al level which is competing for carboxylate surfaces suddenly races up, if you drop the pH, and therefore Al starts to be able to cross-link the surface—for example, gill surfaces. This is the problem with the fish's gill! Your intestinal system is more like the fish gill than the inside of the cell, so the Ca level is absolutely critical for protection against Al.

It's interesting that you see vitamin D dependence in your results. The curious thing about vitamin D and Ca is that indirectly it is this vitamin which manipulates calcium uptake. We know which proteins control the regulatory genes for calcium uptake, via calcium carrier proteins. Calbindin is the Ca transport protein across the mucosal cells and vitamin D regulates its production via a zinc-finger DNA-binding protein. Vitamin D also controls osteocalcin, the protein acting on bones, inside the osteoclasts, so the concentration of calcium is changed if you alter the vitamin D levels in many ways.

So if you alter vitamin D, you play with a whole slew of Ca-binding proteins, both for carriage across cells and in the blood.

Birchall: We observe a dependence on Ca of the fish gill's response to aluminium: high Ca levels tend to reduce the response, presumably by competing with Al for binding.

McLachlan: We find by *in situ* hybridization in Alzheimer's disease brain tissue that CA1 and CA2 hippocampal neurons have reduced messenger RNA pool sizes for both calbindin-28K and vitamin D receptor. There's a positive correlation in the CA1 hippocampal neuron between how much vitamin D receptor there is, and calbindin. Do we have any data on calcium transport systems in non-nervous system tissues in Alzheimer's disease?

Edwardson: We performed absorption studies with ^{40}Ca in carefully diagnosed Alzheimer's disease patients and showed impaired gastrointestinal uptake of labelled Ca, in the presence of normal levels of vitamin D and parathyroid hormone (Ferrier et al 1990). We were quite excited about this at first, but we went on to look at patients with multi-infarct dementia and found similar changes in that group; so we concluded it was a non-specific change. The interesting thing is that these Alzheimer patients are similar to those in whom we have studied the absorption of Al citrate, where, as I mentioned earlier (p 46–47), there seems to be a significant increase in Al absorption, in contrast to the reduced absorption of radiolabelled Ca.

Kerr: What would Bob Williams predict would be the effect of administering vitamin D on Al absorption?

Williams: I can't get that far, as a chemist! Vitamin D is the main hormone controlling the levels of certain calcium proteins. The major ones known are osteocalcin, which controls bone resorption, and calbindin, which is the Ca-transporting protein inside nearly all cells. It was formerly called the intestinal binding protein, because it was first found in the gut, but subsequently it was found in many cells, including brain cells. You can think of it as buffering the Ca inside the cell as well as transporting it across. Its structure is like that of half of calmodulin, but calbindin doesn't trigger intracellular events in the same way (calmodulin will bind aluminium). So there is no doubt that if you can get the free Al high enough in cells, all manner of things will be Al dependent, but how high can you get the Al level at the sites where these proteins are? There's the problem.

Kerr: My reading of the scant clinical studies is that if vitamin D has an effect on gastrointestinal Al absorption in renal failure, it is a small one. Many patients in renal failure before or on dialysis are given vitamin D analogues and we don't see an upsurge of Al absorption, as judged by plasma Al levels.

References

Akitt JW, Elders JM 1985 Aluminium-27 nuclear magnetic resonance studies of the hydrolysis of aluminium (III). J Chem Soc Faraday Trans 81:1923–1930
Coogan MJ 1982 Analysis of a model to describe lead transport by the small intestine. PhD thesis, Aston University, Birmingham, UK
Farrar G, Morton AP, Blair JA 1987 The intestinal absorption and tissue distribution of aluminium, gallium and scandium: a comparative study. Biochem Soc Trans 15:1164–1165
Ferrier IN, Leake A, Taylor GA et al 1990 Reduced gastrointestinal absorption of calcium in dementia. Age Ageing 19:368–375
Martin RB 1988 Bioinorganic chemistry of aluminum. Metal Ions Biol Syst 24:1–57
Ohman L 1988 Stable and metastable complexes in the system H^+-Al^{3+}-citric acid. Inorg Chem 27:2565–2570
Tonge MD 1989 The aqueous and analytical chemistry of aluminium in relation to renal dialysis. PhD Thesis, University of Manchester (Chapter 6: Al-27 NMR studies of aluminium speciation in aqueous solution)

Aluminium intoxication in renal disease

David N. S. Kerr*, Michael K. Ward**, Hewett A. Ellis***, William Simpson† and
Ian S. Parkinson††

*Department of Medicine, Royal Postgraduate Medical School, Du Cane Road, London
W12 0NN, **Department of Medicine, University of Newcastle upon Tyne, Newcastle
upon Tyne NE2 4HH, ***Department of Pathology, University of Newcastle upon Tyne,
Royal Victoria Infirmary, Newcastle upon Tyne NE1 4LP, †Department of Radiology,
Newcastle General Hospital, Newcastle upon Tyne NE4 6BE and ††Serono Diagnostics
Limited, 21 Woking Business Park, Halbert Drive, Woking, Surrey GU21 5JY, UK

Abstract. Aluminium intoxication in renal failure occurred over weeks or months
when dialysis fluid or parenteral solutions were heavily contaminated and over
many years when the main source was oral administration of aluminium-containing
phosphate binders. Encephalopathy was common during subacute intoxication
but in slow aluminium poisoning the main brunt was borne by the bones. However,
in both tempos of intoxication several organs or systems were involved.
Encephalopathy was usually accompanied by bone disease, bone disease by
parathyroid suppression and both by anaemia. The heart and the lymphocytes
are probably damaged by aluminium overload. Among the many questions left
unanswered 15 years after the incrimination of aluminium as the cause of this
multi-system illness are: (1) does low level aluminium overload in renal failure
cause gradual deterioration in cerebral function? And, if so, (2) does it resemble
Alzheimer's disease or a slow-onset version of dialysis encephalopathy? The
evidence we review suggests that the answer to (1) is 'yes' and to (2) 'probably
the latter'.

*1992 Aluminium in biology and medicine. Wiley, Chichester (Ciba Foundation
Symposium 169) p 123–141*

Aluminium intoxication was presaged in the early years of the twentieth century
and recognized after industrial exposure, but its clinical manifestations were
first fully explored in patients with chronic renal failure on replacement therapy
in the 1970s (Kerr & Ward 1986, 1988). Four factors were responsible: (1)
exposure to large volumes of contaminated fluid during haemodialysis,
peritoneal dialysis, haemofiltration and, occasionally, intravenous therapy; (2)
ingestion of grams of aluminium daily as a phosphate binder; (3) loss of the
renal excretory pathway for aluminium; and (4) increased aluminium absorption
from the gut in uraemia. The early 'epidemics' of aluminium intoxication were
largely due to heavy contamination of haemodialysis fluid made from tap water
with inadequate purification and were arrested by the introduction of water
treatment with combinations of filtration, softening, carbon adsorption, reverse

osmosis (RO) and de-ionization, tailored to the individual water supply (Parkinson et al 1981; Kerr et al 1986). However, the epidemics still occur in countries where stark economic choices inhibit proper water treatment (Chan et al 1990).

The clinical and pathological features of the syndromes attributed to aluminium intoxication and the evidence that aluminium is their cause have been well documented in several monographs (Taylor 1986, De Broe & Coburn 1989, Sigel & Sigel 1988), special issues of renal journals (*Clinical Nephrology* vol 24, Supplement 1, 1985; *American Journal of Kidney Diseases* vol 6, No. 5, 1985; *Kidney International* vol 29, Supplement 18, 1986; *Trace Elements in Medicine* vol 8, Supplement 1, 1991) and over 1000 original articles. In this chapter we describe the syndromes largely from our own experience and draw attention to a number of still unanswered questions. We emphasize that aluminium intoxication in renal failure affects many organs and systems.

Acute aluminium intoxication

This syndrome was described during a brief episode in which one manufacturer's supply of peritoneal dialysis (PD) fluid became heavily contaminated with aluminium from a filter; aluminium concentrations in the affected batches ranged from 23 to 54 μmol/l. Fortunately, three of the centres supplied (Paris, Glasgow and Edinburgh) were monitoring serum aluminium and recognized the cause of the previously undescribed syndrome before irreparable harm was done. Over the course of a month plasma aluminium rose to peaks of 10–54 μmol/l. The symptoms of anorexia, nausea, vomiting, cramp-like abdominal pains, weight loss and general malaise appeared within three days of first using the faulty PD fluid. These symptoms were reversed rapidly when the plasma aluminium level was lowered by haemodialysis against almost aluminium-free dialysate (Cumming et al 1982). This is a unique paradigm of acute aluminium intoxication and one major question has been left unanswered. These patients were subjected, for periods of up to a month, to very high serum aluminium concentrations, many times higher than informed guesses at those encountered at Camelford, Cornwall, where 20 tons of aluminium sulphate were accidentally discharged into the treated water in 1988 (Lowermoor Incident Health Advisory Group 1991). Did the PD patients suffer any long-term ill effects? Eleven years later we would expect about half of them to be alive, some of them after successful renal transplantation and therefore without much further aluminium exposure. A follow-up report would be most informative; prolonged normal cerebral function in this group would allay some of the fears aroused at Camelford.

Subacute aluminium intoxication from haemodialysis fluid

During the 'epidemics' of disease from this source, some dialysis fluid was made from tap water with concentrations of aluminium over 30 μmol/l (Kerr & Ward

1988). Aluminium was taken up by the patient against a strong concentration gradient because of the protein binding of aluminium. Plasma concentrations rose rapidly across a single dialysis; in Newcastle, with only moderately elevated tap water Al levels, serum Al concentrations rose from about 50 to over 200 µg/l (c. 1.9–7.4 µmol/l) (Parkinson et al 1981), so it is probable that much higher post-dialysis serum Al levels were attained in centres with high tap water levels of Al. Possibly, acute aluminium intoxication syndrome occurred before 1981, but was not recognized: our aluminium-intoxicated patients were non-specifically ill but nausea, vomiting and malaise were then such common symptoms during haemodialysis that they were attributed to other causes. Consequently, in centres with very high tap water levels of aluminium, such as Stobhill in Glasgow (Elliott et al 1978) and Eindhoven (Flendrig et al 1976), the first features of intoxication to be recognized were those of dialysis encephalopathy.

Dialysis encephalopathy

The original description of this disease by Alfrey and colleagues (1972) has never been bettered. Our experience of it in the Newcastle outbreak, which is

TABLE 1   Clinical features at presentation of dialysis encephalopathy and at any time in the illness in 27 Newcastle patients

Clinical feature	Presentation	Any time
Speech disturbance	19	27
Myoclonus	16	27
Seizures	2	10
Memory loss	6	23
Dementia	—[a]	23
Dyspraxia	6	27
Dysphagia	0	7
Fatigue	2	27

[a]Not formally tested. Deterioration in personality and ability to cope with home haemodialysis were common in the weeks before presentation with the symptoms listed above.

TABLE 2   Investigations in 27 patients with dialysis encephalopathy at Newcastle

Investigation	Result
Abnormal EEG	24/24
CSF pressure normal	21/21
CSF examination normal	21/21
CT scan normal	6/6
Serum Al (mean ± SD)	239 ± 114 µg/l (contemporary normal < 11)
Brain Al (mean ± SD)	24 ± 18 µg/g (contemporary normal < 4)

summarized in Tables 1 and 2 and described in detail elsewhere (Ward et al 1976, Bates et al 1985, Kerr et al 1985), was virtually identical. The syndrome differed in many respects from Alzheimer's disease. The commonest presenting symptom was a striking abnormality of speech which often began as sudden interruption in mid-sentence. An equally dramatic resumption of the conversation followed an intravenous injection of diazepam, suggesting that this was an epileptiform phenomenon. Myoclonic jerking and grand mal seizures were also common. The disabling apraxia was initially obvious as a deterioration in handwriting which could be used to chart recovery after removal of the source of aluminium (Platts & Anastassiades 1981), renal transplantation or treatment with the trivalent ion chelator, desferrioxamine. The electroencephalographic (EEG) abnormalities (Figs. 1 and 2) were universal.

Eight of our 27 patients received renal transplants, six of them early in the disease, and five of these six made a slow recovery. Four patients transferred to a purer water supply had arrest or reversal of the syndrome, and more striking recovery has been described by others who rigorously excluded all sources of aluminium (Pierides et al 1980). There is often an initial deterioration in the early post-transplantation period, despite the observation that the serum aluminium level falls steadily, though fairly slowly, from the onset of transplant function. A similar deterioration is reported in the first few days of

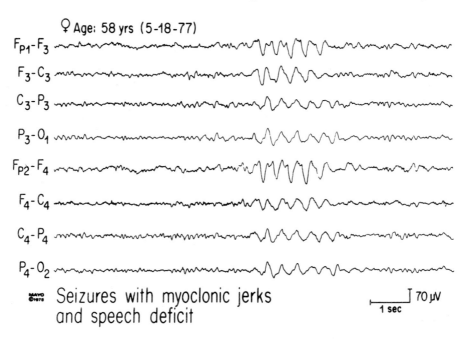

FIG. 1   Abnormal EEG of a Newcastle patient during dialysis dementia.

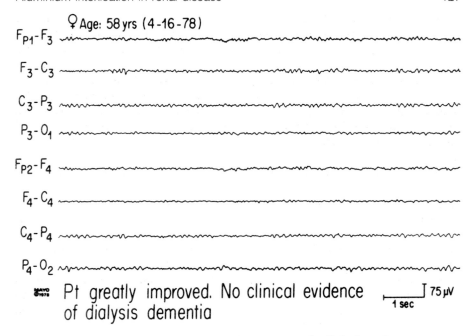

FIG. 2 EEG of same patient as in Fig. 1 after treatment by dialysis against pure water and infusions of desferrioxamine, showing return to normal.

desferrioxamine treatment (Sherrard et al 1988), possibly due to the fact that the Al–desferrioxamine complex (of molecular mass about 600 Da) enters cerebrospinal fluid much more readily than transferrin–Al (Ellenberg et al 1990). The complication can probably be avoided by using low initial doses of desferrioxamine (e.g. 10 mg/kg) and infusing it a few hours before a haemodialysis (Sherrard et al 1988).

The clinical, EEG and other features encountered at Newcastle were closely mirrored in other outbreaks, making clinical diagnosis easy in the context of an endemic disease, but the neuropathological findings were slight in comparison with the clinical picture, inconsistent, and largely non-specific (Bugiani & Ghetti 1989, Boyce 1989). In our series (Ward et al 1976, Bates et al 1985), prominent features at autopsy were astrocytosis of cortex and brainstem and loss of Purkinje cells in the cerebellum; but the latter must be a late manifestation, because substantial recovery from apraxia can occur if the disease is treated early. Despite several hundred known deaths from this disease the literature on light microscopy, electron microscopy and aluminium localization studies of the brain are scant and inconclusive. Since the disease, if recognized, is now treatable, the opportunity for these studies has largely passed. Any opportunity for autopsy study that does arise should be eagerly grasped.

The second unanswered question about dialysis dementia concerns its pathophysiology. Postulated mechanisms of action include: interference in glucose metabolism by displacement of Mg from the MgATP complex by aluminium; inhibition of adrenergic transmitter synthesis through reduced tetrahydrobiopterin production, again through displacement of Mg; interruption of axonal transport by binding with calmodulin and retardation of the polymerization of tubulins; increased blood–brain permeability; and neurofibrillary degeneration (Savazzi 1991). However, direct evidence for the importance of these and other known neurotoxic effects of aluminium in dialysis dementia is lacking.

To a third question about dialysis dementia we have only partial answers: 'if massive aluminium overload causes lethal encephalopathy, does modest aluminium exposure produce modest cerebral damage?' During our own epidemic in Newcastle, cerebral function was studied in the exposed but not clinically encephalopathic subjects. Overall IQ was normal but the patients had significant impairment in three tests of performance IQ: digit symbol, block design and picture arrangement (English et al 1978). In Manchester a similar patient population showed defects in speech performance (Ackrill et al 1979). These patients were heavily loaded with aluminium; four developed fatal encephalopathy in each of the two studies. In an area of lower aluminium exposure, Gilli et al (1980) found progressive deterioration of verbal and performance IQ with time on dialysis which did not correlate with serum aluminium. However, the actual serum aluminium levels were not quoted, so it is difficult to judge whether Gilli et al were observing chronic mild aluminium intoxication from oral intake or the effects of some other aspect of prolonged uraemia or dialysis. Three more recent reports have come from centres where aluminium overload has resulted from the use of aluminium hydroxide as a phosphate binder. Rovelli et al (1988) found impaired performance in the word span test in patients with serum aluminium levels above 100 μg/l (about 4 μmol/l) but only in those with an abnormal EEG suggestive of early dialysis encephalopathy. Altmann et al (1989) studied 27 patients with lower aluminium exposure (mean serum aluminium 59 μg/l; c. 2.2 μmol/l) and found defects in several tests of psychomotor function, including symbol digit coding; some improvement in psychomotor function occurred during treatment with desferrioxamine. This group of patients had substantial increase in aluminium stores, judged by the rise in serum aluminium levels during desferrioxamine treatment. The overload was probably much greater than any that exists in Alzheimer's disease, but the desferrioxamine test as used in renal failure would not be informative in Alzheimer patients with nearly normal renal function. Sprague et al (1988) found memory impairment, but no other cognitive defects, and an increased incidence of myoclonus, asterixis and muscle weakness in patients with a positive desferrioxamine test.

The interpretation of all such studies is confounded by our uncertainty whether prolonged chronic renal failure or some aspect of haemodialysis other than aluminium intoxication causes cognitive impairment. Several studies have shown such defects, particularly in children (Fennell et al 1990) and in adults who have spent several years on haemodialysis (Ventura et al 1990). However, these reports, surprisingly, give no information on aluminium intake or aluminium concentrations in serum or tissues and do not even mention aluminium in the discussion.

Aluminium-induced anaemia

A progressive fall in haemoglobin levels in the year before the onset of dialysis dementia was first reported in the Stobhill outbreak (Elliott et al 1978), but the definitive studies were carried out in Edinburgh, where Short et al (1980) observed the progressive development of microcytic anaemia as aluminium load increased and the reversal of the process over a few months when effective water treatment was introduced. This sequence of events has been well documented subsequently but it remains uncertain how far minor degrees of aluminium overload contribute to the anaemia of renal failure. There are two persuasive arguments for such an effect. (1) There is a rise in haemoglobin level which accompanies desferrioxamine treatment in patients with only modest aluminium overload and anaemia (Altmann et al 1988, Bia et al 1989) and even occurs in some with normocytosis and normal Hb (Ihle et al 1986). The decline of Hb

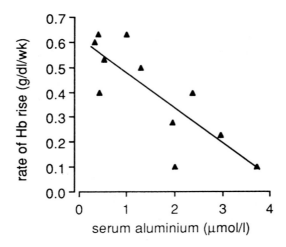

FIG. 3   Relationship between serum Al and rate of rise of Hb level after erythropoietin injections (Stevens et al 1991). (Reproduced by permission of the author and the Editor of *Nephrology, Dialysis and Transplantation.*)

to its previous low level within a few months of the discontinuation of desferrioxamine (Altmann et al 1988) is a puzzling feature which has engendered speculation that desferrioxamine improves erythropoiesis by some other mechanism, such as making iron more available to red cell precursors. However, it may well be due to inadequate length of treatment, since desferrioxamine was given for only three months in Altmann's study. (2) Many recent studies at the Hammersmith Hospital (Fig. 3; Stevens et al 1991) and elsewhere have shown that the response to erythropoietin is inhibited by quite modest elevations of serum aluminium (Grützmacher et al 1990, Rosenlöf et al 1990). This 'erythropoietin test' can be likened to an exercise test for coronary artery disease; it brings to light minor degrees of inhibition of erythropoiesis which are not apparent in the resting state. It is therefore possible that systemic low-grade aluminium intoxication in other diseases (e.g. Alzheimer's) would be manifest not by anaemia but by a slower than normal rise in Hb after stresses such as haemorrhage or haemolysis.

Bone disease and parathyroid function

In our Newcastle patients, heavily overloaded with aluminium from the dialysis fluid, bone disease developed over a few years. Bone pain, myopathy and pathological fractures affected more than half the patients by the fourth year of haemodialysis. These symptoms were accompanied by radiological changes, including partial and complete non-healing fractures and osteopenia which accompanied the development of osteomalacia and a reduction in calcified bone area (Simpson et al 1976). This epidemic of crippling bone disease ceased with the introduction of water purification (Kerr et al 1986).

The study of 610 bone biopsies at the start of and during dialysis showed that hyperparathyroidism remitted as osteomalacia progressed (Ellis et al 1977). Serum parathyroid hormone (PTH) and alkaline phosphatase levels were less elevated in patients with osteomalacia or aplastic bone disease than in those with still active hyperparathyroidism, in striking contrast to the situation before dialysis, when osteomalacia was accompanied by very high serum alkaline phosphatase and PTH levels, showing that we were dealing with a new form of osteomalacia and suggesting that some factor during dialysis suppressed parathyroid function. One possible explanation is the tendency to spontaneous hypercalcaemia in patients with dialysis osteomalacia (Cochran et al 1981), but this was not a common phenomenon in our patients and would not explain the observations of Piraino et al (1986). Direct effects of aluminium on the parathyroids—inhibition of PTH release and cellular damage—have been shown *in vitro* at concentrations commonly encountered in patients on dialysis (Bordeau et al 1987) and seem a more plausible explanation.

Chronic aluminium overload from phosphate binders

Although dramatic epidemics of encephalopathy and bone disease ceased with proper water treatment, aluminium toxicity has remained a serious worldwide problem as nephrologists have continued to prescribe aluminium hydroxide as the most effective phosphate binder. Until the recognition by Alfrey and his team that aluminium was the cause of dialysis encephalopathy and that aluminium was absorbed after the ingestion of aluminium hydroxide (Kaehny et al 1977), aluminium-containing compounds were regarded as non-toxic and were prescribed for the great majority of patients in renal failure to control plasma phosphate levels. The practice dies hard because the only alternative phosphate binders currently on the market are salts of calcium and magnesium which have their own side effects, particularly when the patient also receives alfacalcidol or calcitriol to compensate for the loss of the renal α-hydroxylation, which produces the active metabolite of vitamin D. However, there is now plentiful evidence that oral Al from this source can be reduced almost to zero by careful use of alternatives (Salusky et al 1991).

There is wide variation in Al absorption from aluminium hydroxide, as judged by the rise in serum Al during a course of treatment. The role of spontaneous gastric acidity in this phenomenon is probably small (Beynon & Cassidy 1990), though complete suppression of gastric acid by ranitidine does reduce Al excretion after an oral dose (Rodger et al 1991). Ingestion of citrate (Molitoris et al 1989) and possibly of other organic acids (Ittel et al 1991) has a more dramatic effect on Al absorption that can generate Al levels in the encephalopathic range. It is interesting to recall that in one of the worst outbreaks of encephalopathy due to contaminated water supply (Flendrig et al 1976), the patients also received a large dose of aluminium hydroxide and calcium citrate, the importance of which was then unrecognized.

The time scale and the pattern of disease from oral absorption of Al differ from those seen in the presence of contaminated dialysis fluid. Encephalopathy from oral aluminium intake has occurred sporadically and even in a few epidemics, and it particularly affected small children, who receive disproportionately high doses for their size and absorb aluminium more readily than adults (Andreoli et al 1984), before the hazard was recognized. However, the main manifestation now is bone disease of gradual onset. In the early stages the admixture of hyperparathyroidism and aluminium overload has puzzled even the experts (Malluche & Faugere 1989), but we believe that most of the data can now be explained on a simple schema. (1) Initially, PTH stimulates aluminium absorption and movement into bone, so that patients with active hyperparathyroidism have as high a bone burden as those already developing typical aluminium osteodystrophy. (2) However, as long as osteitis fibrosa remains active, the results of aluminium overload are not manifest in the bone histology, radiology or symptoms. (3) As body burden increases the parathyroids are suppressed and bone histology changes from osteitis fibrosa to osteomalacia or inert bone. (4) A speeded-up sequence of these events is seen after parathyroidectomy, which precipitates the onset of osteomalacia (de Francisco et al 1987).

Severe aluminium osteodystrophy is now common in patients in their second decade of dialysis in centres that have used aluminium hydroxide throughout. In Seattle, 15/23 patients undergoing bone biopsy after more than eight (mean 13) years of dialysis had aluminium staining over more than 25% of the osteoid surface, and 10 had inert bone disease (Andress et al 1986).

Other manifestations of aluminium intoxication in renal failure

During our epidemic of aluminium intoxication in Newcastle the patients were 'generally sick'. Aspects of this systemic illness are still being described. Aluminium-loaded patients tolerate renal transplants well (Nordal et al 1988) and infections badly (Davenport et al 1990); they have defects of lymphocyte function which may explain both phenomena (McGregor et al 1992). A cardio-myopathy has been attributed to aluminium overload (London et al 1989).

Implications of the syndromes of aluminium overload in renal failure for other disease states

Three main lessons may be learnt from this story. (1) Aluminium overload causes widespread damage to the structure and function of many organs and tissues. Those who seek an explanation for Alzheimer's disease or Guam amyotrophic lateral sclerosis/parkinsonian dementia in aluminium overload should look carefully outside the central nervous system, particularly at the bones, blood, immune system and cardiac muscle. (2) The clinical manifestations of aluminium overload vary considerably with the rate of intoxication. We should be prepared for surprises when studying possible aluminium intoxication in diseases with incubation periods an order of magnitude greater than that of dialysis encephalopathy. (3) Despite this caveat, we should not underestimate the differences between dialysis encephalopathy and Alzheimer's disease, amyotrophic lateral sclerosis and parkinsonian dementia. Even in prolonged moderate aluminium overload in renal failure, none of the studies to date has revealed a neurological syndrome resembling these.

References

Ackrill P, Barron J, Whiteley S, Horn AC, Ralston AJ 1979 A new approach to the early detection of dialysis encephalopathy. Proc Eur Dial Transpl Assoc 16:659–660
Alfrey AC, Mishell JM, Burks J et al 1972 Syndrome of dyspraxia and multifocal seizures associated with chronic hemodialysis. Trans Am Soc Artif Intern Organs 18:257–261

Altmann P, Plowman D, Marsh F, Cunningham J 1988 Aluminium chelation therapy in dialysis patients: evidence for inhibition of haemoglobin synthesis by low levels of aluminium. Lancet 1:1012–1015

Altmann P, Dhanesha U, Hamon C, Cunningham J, Blair J, Marsh F 1989 Disturbance of cerebral function by aluminium in haemodialysis patients without overt aluminium toxicity. Lancet 2:7–12

Andreoli SP, Bergstein JM, Sherrard DJ 1984 Aluminum intoxication from aluminum-containing phosphate binders in children with azotemia not undergoing dialysis. N Engl J Med 310:1079–1084

Andress DL, Endres DB, Ott SM, Sherrard DJ 1986 Parathyroid hormone in aluminium bone disease: a comparison of parathyroid assays. Kidney Int 29 (suppl 18):S87–S90

Bates D, Parkinson IMS, Ward MK, Kerr DNS 1985 Aluminium encephalopathy. In: D'Amico G, Colasanti G (eds) Advances in nephrology and dialysis. (Contributions to nephrology vol 45) Karger, Basel, p 29–41

Beynon H, Cassidy MJD 1990 Gastrointestinal absorption of aluminium. Nephron 55:235–236

Bia MJ, Cooper K, Schnall S et al 1989 Aluminum induced anemia: pathogenesis and treatment in patients on chronic hemodialysis. Kidney Int 36:852–858

Bordeau AM, Plachot J-J, Cournot-Witner G, Pointillart A, Balsan S, Sachs C 1987 Parathyroid response to aluminum in vitro: ultrastructural changes and PTH release. Kidney Int 31:15–24

Boyce BF 1989 Cellular and subcellular localization of aluminium. In: De Broe ME, Coburn JW (eds) Aluminum and renal failure. Kluwer, Dordrecht, p 167–177

Bugiani O, Ghetti B 1989 Aluminum encephalopathy: experimental vs human. In: De Broe ME, Coburn JW (eds) Aluminum and renal failure. Kluwer, Dordrecht, p 109–125

Chan MK, Varghese Z, Li MK, Wong WS, Li CS 1990 Newcastle bone disease in Hong Kong: a study of aluminium associated osteomalacia. Int J Artif Organs 13:162–168

Cochran M, Platts MM, Moorhead PJ, Buxton A 1981 Spontaneous hypercalcaemia in maintenance dialysis patients: an association with atypical osteomalacia and fractures. Mineral Elect Metab 5:280–286

Cumming AD, Simpson G, Bell D, Cowie J, Winney RJ 1982 Acute aluminium intoxication in patients on continuous ambulatory peritoneal dialysis. Lancet 1:103–104

Davenport A, Davison AM, Newton KE, Will EJ, Giles GR, Toothill C 1990 Aluminium accumulation increases the susceptibility to sepsis following renal transplantation. Nephrol Dial Transplant 4:306 (abstr)

De Broe ME, Coburn JW (eds) 1989 Aluminum and renal failure. Kluwer, Dordrecht

de Francisco AM, Ellis HA, Ward MK, Kerr DNS 1987 Effect of parathyroidectomy (PTX) in bone aluminum accumulation in uremic patients. Kidney Int 32:620 (abstr)

Ellenberg R, King AL, Sica DA, Posner M, Savory J 1990 Cerebrospinal fluid aluminium levels following desferrioxamine. Am J Kidney Dis 16:157–159

Elliott HL, Dryburgh F, Fell GS, Sabet S, Macdougall AI 1978 Aluminium toxicity during regular haemodialysis. Br Med J 1:1101–1103

Ellis HA, Pierides AM, Feest TG, Ward MK, Kerr DNS 1977 Histopathology of renal osteodystrophy with particular reference to the effects of 1-α-hydroxyvitamin D_3 in patients treated by long term haemodialysis. Clin Endocrinol 7:31S–38S

English A, Savage RD, Britton PG, Ward MK, Kerr DNS 1978 Intellectual impairment in chronic renal failure. Br Med J 1:888–890

Fennell RS, Fennell EB, Carter RL, Mings EL, Klausner AB, Hurst JR 1990 A longitudinal study of the cognitive function of children with renal failure. Pediatr Nephrol 4:11–15

Flendrig JA, Kruis H, Da HA 1976 Aluminium intoxication: the cause of dialysis dementia? Proc Eur Dial Transpl Assoc 13:355–361

Gilli P, Bastiani P, Rosati G et al 1980 Impairment of the mental status of patients on regular dialysis treatment. Proc Eur Dial Transpl Assoc 17:306–311

Grützmacher P, Ehmer B, Limbach J, Messinger D, Kulbe KD, Scigalla P 1990 Treatment with recombinant erythropoietin in patients with aluminium overload and hyperparathyroidism. Blood Purif 8:279–284

Ihle BU, Becker GJ, Kincaid-Smith PS 1986 Clinical and biochemical features of aluminum-related bone disease. Kidney Int 29 (suppl 18):S80–S86

Ittel TH, Griessner A, Sieberth HG 1991 Effect of lactate on the absorption and retention of aluminium in the remnant kidney rat model. Nephron 57:332–339

Kaehny WD, Hegg AP, Alfrey AC 1977 Gastrointestinal absorption of aluminum from aluminum-containing antacids. N Engl J Med 296:1389–1390

Kerr DNS, Ward MK 1986 The history of aluminium related disease. In: Taylor A (ed) Aluminium and other trace elements in renal disease. Ballière Tindall, London, p 1–14

Kerr DNS, Ward MK 1988 Aluminum intoxication; history of its clinical recognition and management. In: Sigel H, Sigel A (eds) Metal ions in biological systems, vol 24: Aluminum and its role in biology. Marcel Dekker, New York, p 217–258

Kerr DNS, Ward MK, Parkinson IS 1985 Dialysis encephalopathy. Questions and answers. In: Cummings NB, Klahr S (eds) Chronic renal disease. Causes, complications and treatment. Plenum Medical, New York, p 185–200

Kerr DNS, Ward MK, Arze RS et al 1986 Aluminum-induced dialysis osteodystrophy: the demise of "Newcastle bone disease"? Kidney Int 29 (suppl 18):S58–S64

London GM, de Vernejoul M-C, Fabiani F et al 1989 Association between aluminum accumulation and cardiac hypertrophy in hemodialyzed patients. Am J Kidney Dis 13:75–83

Lowermoor Incident Health Advisory Group 1991 Water pollution at Lowermoor North Cornwall. HMSO, London

Malluche HH, Faugere M-C 1989 Renal osteodystrophy. N Engl J Med 321:317–319

McGregor FJ, Naves ML, Birly AK et al 1992 Interaction of aluminium and gallium with human lymphocytes: the role of transferrin. Biochim Biophys Acta, in press

Molitoris BA, Froment DH, Mackenzie TA, Huffer WH, Alfrey AC 1989 Citrate: a major factor in the toxicity of orally administered aluminum compounds. Kidney Int 36:949–953

Nordal KP, Dahl E, Albrechtsen D et al 1988 Aluminium accumulation and immunosuppressive effect in recipients of kidney transplants. Br Med J 297:1581–1582

Parkinson IS, Ward MK, Kerr DNS 1981 Dialysis encephalopathy, bone disease and anaemia; the aluminium intoxication syndrome during haemodialysis. J Clin Pathol 34:1285–1294

Pierides AM, Edwards WG, Cullum UX, McCall JT, Ellis HA 1980 Hemodialysis encephalopathy with osteomalacic fractures and muscle weakness. Kidney Int 18:115–124

Piraino BM, Rault R, Greenberg A et al 1986 Spontaneous hypercalcemia in patients undergoing dialysis. Etiologic and therapeutic considerations. Am J Med 80:607–615

Platts MM, Anastassiades E 1981 Dialysis encephalopathy: precipitating factors and improvement in prognosis. Clin Nephrol 15:223–228

Rodger RSC, Muralikrishna GS, Halls DJ et al 1991 Ranitidine suppresses aluminium absorption in man. Clin Sci 80:505–508

Rosenlöf K, Fyhrquist F, Tenhunen R 1990 Erythropoietin, aluminium and anaemia in patients on haemodialysis. Lancet 335:247–249

Rovelli E, Luciani L, Pagani C, Albonico C, Colleoni N, D'Amico GD 1988 Correlation between serum aluminum concentration and signs of encephalopathy in a large population of patients dialyzed with aluminum-free fluids. Clin Nephrol 29:294–298

Salusky IB, Foley J, Nelson P, Goodman WG 1991 Aluminum accumulation during treatment with aluminum hydroxide and dialysis in children and young adults with chronic renal disease. N Engl J Med 324:527–531

Savazzi GM 1991 Uremia and mechanisms of aluminium neurotoxicity; an overview. Int J Artif Organs 14:13–17

Sherrard DJ, Walker JV, Boykin JL 1988 Precipitation of dialysis dementia by desferrioxamine treatment of aluminum-related bone disease. Am J Kidney Dis 12:126–130

Short AIK, Winney RJ, Robson JS 1980 Reversible microcytic anaemia in dialysis patients due to aluminium intoxication. Proc Eur Dial Transpl Assoc 17:226–233

Sigel H, Sigel A (eds) 1988 Metal ions in biological systems, vol 24: Aluminum and its role in biology. Marcel Dekker, New York

Simpson W, Ellis HA, Kerr DNS, McElroy M, McNay RA, Peart KN 1976 Bone disease in long-term haemodialysis: the association of radiological with histological abnormalities. Br J Radiol 49:105–110

Sprague SM, Corwin HL, Tanner CM, Wilson RS, Green BJ, Goetz CG 1988 Relationship of aluminum to neurocognitive dysfunction in chronic dialysis patients. Arch Intern Med 148:2169–2177

Stevens JM, Auer J, Strong CA et al 1991 Stepwise correction of anaemia by subcutaneous administration of human recombinant erythropoietin in patients with chronic renal failure maintained by continuous ambulatory peritoneal dialysis. Nephrol Dial Transplant 6:487–494

Taylor A 1986 Aluminium and other trace elements in renal disease. Ballière Tindall, London

Ventura MC, Gonzalez R, Alarcon A et al 1990 Curve of perceptive intellectual deterioration in hemodialysis patients. Int J Artif Organs 13:87–92

Ward MK, Pierides AM, Fawcett P et al 1976 Dialysis encephalopathy syndrome. Proc Eur Dial Transpl Assoc 13:348–354

DISCUSSION

McLachlan: During the formation of mammalian erythrocytes and the extrusion of the nucleus, there is an arrest of all nuclear transcription processes accompanied by heterochromatization. Is there premature arrest of transcription and heterochromatization associated with this particular cause of microcytic anaemia?

Kerr: I don't know of any evidence of that. We haven't studied it ourselves.

McLachlan: Have you stained the erythroblast cells for Al and have you looked at nuclear conformation in these precursor cells in this condition?

Kerr: The bone marrow has been stained and does contain Al, but I don't know its localization. Kaye (1984) suggested that the macrophages of the reticuloendothelial system were the main site. However, Al bound to transferrin does inhibit haemoglobin synthesis in red cell precursors (Abreo et al 1990).

McLachlan: One of the histones involved in gene repression is histone H1°. This protein is homologous to the avian erythrocyte H5 which is responsible for the heterochromatization of the mature avian erythrocyte. We have found a linear correlation between the amount of Al and the amount of histone H1° in dinucleosomes prepared from temporal grey matter in Alzheimer's disease (Lukiw et al 1991). H1° is a transcription-repressing protein and is associated with inappropriate repression of genes such as HNF-L. It would be of interest to see if there is a relationship between H1°, aluminium and premature arrest of transcription in Al-induced microcytic anaemia of dialysis.

Kerr: The microcytic anaemia of Al intoxication looks very like iron-deficiency anaemia, but of course it doesn't respond to iron. People are now studying the mechanism by which Al-containing transferrin gets into the red cell, and are trying to look at the mechanism by which the Al interferes with iron uptake once it gets in.

Wisniewski: In experimental animals after systemic injections of Al, the white cell count is depressed. There are reports in the old literature about Al being used in the treatment of leukaemia. Did you look at the white cell count?

Kerr: The total leucocyte count is not depressed in Al intoxication but Dr F. J. McGregor and his colleagues in Glasgow have shown a number of defects in lymphocyte function during Al overload (McGregor et al 1992).

Birchall: Did you find Al in the osteocytes in osteomalacia?

Kerr: The two sites at which Al is easily detected histochemically are the ossification front and the cement lines. A number of studies (using methods such as LAMMA) show that Al and Si are co-located at those two points (Augsten & Stein 1988). The osteocytes do not appear to be a major site of localization. However, not all tissues stain equally with histochemical stains. For instance, the parathyroid glands are clearly suppressed during Al overload and Al can be demonstrated in them biochemically (Berland et al 1988), but it is difficult to demonstrate it histochemically. In other tissues where Al has been shown by LAMMA, histochemical staining has been negative (Verbueken et al 1985). So perhaps the osteocytes are affected more than we know.

What I find particularly interesting is that just when the anaemia starts to be corrected soon after you prevent Al overload, the bone starts to grow again, while there is still a lot of Al at the ossification front. This Al is buried deep while bone growth resumes and is left as an archaeological deposit showing when bone growth was once retarded.

Birchall: Have you any idea of the ratio of Al and Si in these 'co-locations' of Al and Si?

Kerr: The Al signal, by the X-ray technique, is sometimes higher than the Si signal, sometimes lower, but I don't know what that represents in molar ratios (Augsten & Stein 1988).

Williams: Why does bone fracture under such circumstances?

Kerr: In the really severe cases of osteomalacia, the bone is osteopenic on X-ray. However, that underestimates the damage. Histologically only wisps of irregular, moth-eaten calcified bone remain; the rest is soft osteoid. It is easy to see why it has little resistance to traumatic fracture. Osteomalacic bone also develops pseudofractures called 'Looser's zones' and some of the fractures start as pseudofractures that spread across the bone. They don't heal; serial X-rays taken over four or five years have shown fractures unhealed for that time (Simpson et al 1973).

Perl: We have done some LAMMA microprobe studies on aluminium-related bone disease (unpublished). A lot of Al is found at the growth lines. In many of these cases the Al was combined with iron. In that combination, the Aluminon stain was not very positive; iron seems to interfere with this histochemical reaction. We have not found Al in osteoblasts. We did find Al and Fe in the erythroid precursors in the bone marrow of the bone biopsies, in both nuclei and cytoplasm. Fe and Al go together very frequently, and I think that's another aspect of this whole story that has been forgotten; iron has its own toxic properties.

Kerr: Studies about eight years ago suggested that it was more damaging to have Fe and Al deposited together at the ossification front than to have Al alone (Pierides & Myli 1984, Ackrill et al 1988). There has been no confirmation in subsequent reports and it now seems that iron at the ossification front is not very important, though osteomalacia has occasionally been reported in pure iron overload (Vernejoul et al 1982).

Blair: You mentioned the decline in dihydropteridine reductase (DHPR) activity and reduced tetrahydrobiopterin synthesis with increasing levels of Al. This was measured in the erythrocyte, and of course the site of action of Al is the brain. However, in the rat we showed that erythrocyte DHPR levels (both with and without the addition of Al) correlated well with brain DHPR (Cutler & Blair 1987). So the background evidence is there for the rat.

Wischik: What is the function of DHPR in the brain?

Blair: This is a key enzyme in the maintenance of intracellular levels of tetrahydrobiopterin (BH_4), which is found in all cells of the body. Its principal function relevant to our discussion is the maintenance of the biosynthesis of the neurotransmitters dopamine, noradrenaline and serotonin, and control of the release of these compounds from the neuron into the synaptic cleft and their transmission to the postsynaptic neuron as a rate-controlling function. The enzyme DHPR is a salvage enzyme, which regenerates the cofactor tetrahydrobiopterin. Gross deficiency of BH_4 in the young results in severe neurological problems. BH_4 deficiency from genetic defects of DHPR is devastating. It can be reversed by certain therapies which do something to alleviate the condition. Dr Wisniewski has referred to childhood neurological disease being an early stage of dementia, so one could regard these disorders as early dementias, although my paediatrician colleagues prefer to call them amentias, because the children show *no* intellectual development.

Wisniewski: That is mental retardation.

Blair: They are hardly mentally retarded; the infants just lie in their cots waiting for death to overtake them. It is a question of total failure of brain function, in most cases.

Wischik: In a recent study of Wilson's disease (a copper storage disease), a significant correlation was found between personality change, sociopathic personality, difficulties with personality, and myoclonus and that kind of neurological disturbance, as opposed to a correlation with disturbance in handling abstract symbols (Dening & Berrios 1990). This is psychiatrically very interesting. The copper accumulates in the basal ganglia, in the caudate. Are there data on the region of Al accumulation in acute or chronic cases of Al intoxication? Which regions are vulnerable?

Edwardson: I shall be describing this extensively in my paper (see p 165).

Candy: In collaboration with Mike Ward, we have studied the focal accumulation of Al in the frontal cortex of 15 renal dialysis patients using imaging secondary ion mass spectrometry (SIMS), and also have measured bulk Al levels by atomic absorption spectrometry. Eleven out of the 15 patients were treated with $Al(OH)_3$, to control hyperphosphataemia, for periods of up to seven years. We found highly significant correlations between the mean serum Al concentrations measured over the period of dialysis treatment, and the period of dialysis, the duration of oral $Al(OH)_3$ administration, and the total amount of $Al(OH)_3$ administered. We have also found that the concentrations of Al in the frontal cortex correlated with the mean serum Al concentrations, the period of oral $Al(OH)_3$ administration, and the total amount of $Al(OH)_3$ prescribed. Thus, the mean serum Al concentration reflects the $Al(OH)_3$ intake, and brain Al concentration reflects both the mean serum Al concentration and the amount of $Al(OH)_3$ administered (Candy et al 1992). Is it therefore possible that in long-term dialysis patients we have a subpopulation who might, if dialysed and given $Al(OH)_3$ for longer, develop encephalopathy?

Ward: These are our current patients, dialysed on reverse osmosis-treated water and given $Al(OH)_3$, up to six capsules per day (475 mg), but usually given intermittently. We monitor them and no clinical evidence of dementia has occurred yet; but we have not done psychometric studies on these patients serially, to look for any change in cognitive function. In terms of the reduction in Al exposure, these patients are getting much less Al than the original group of dialysed patients.

Kerr: The answer to Dr Candy's question is 'yes'; a number of patients have developed this encephalopathy without dialysis, when taking aluminium hydroxide, some of them for quite short periods of time. Our first case was on Aludrox for only about three years before she became demented. In others—particularly infants (Rotundo et al 1982)—dementia has developed within a few months.

Candy: It might be possible to get an indication of the brain Al concentration from the mean serum Al concentration, so you could identify patients with a high concentration of Al in the brain who are at risk of developing encephalopathy, and reduce their Al intake. We had the brain from a patient from another dialysis centre who had been dialysed over an eight-year period and had received 6 g of Al daily. He had a brain Al concentration of 14 µg/g dry weight. The lowest brain concentration of Al found in encephalopathy patients was about 20–25 µg/g dry weight, so it is possible, over an extended period of oral $Al(OH)_3$, that you might start to see encephalopathy.

Ward: The serum level is nevertheless difficult to interpret, because if a patient is taking Al hydroxide, the serum level rises and then reaches a plateau; if you stop Aludrox, the Al comes down but may not return to the previous basal level. This presumably reflects mobilization from tissue deposits. So interpreting serum values in terms of a body burden of Al is quite difficult. Others have suggested that if you stop taking Aludrox and serum Al is measured, you can get a crude correlation with the amount of Al in bone.

Candy: We found, in this series of dialysis patients, no clear correlation between the presence of bone Al, determined histochemically, and the concentration of Al in the frontal cortex.

Day: Unlike Dr Candy, we have not found any relationship between mean monthly plasma Al levels and long-term exposure to Al hydroxide, in something like 400 patients (unpublished findings). Our conclusion has been that the serum or plasma Al level reflects short-term exposure in relation to the immediate clearance of Al, by whatever means. If the dialysis is more aggressive, or the kidney function is good, the balance between intake and excretion sets a lower plasma Al level, and this doesn't necessarily relate much to body stores of Al. A better long-term indicator is the erythrocyte Al content, which gives an integration over something like three months. The only good long-term indicator of the body burden of Al appears to be bone Al, as determined by biopsy.

In the first patient treated with desferrioxamine, in 1979 in Manchester, there was over 600 mg/kg of Al in the bone, an enormous load. After 10 months of treatment with desferrioxamine, on and off, the total removal of Al from the patient was about one gram. He recovered fairly well but had a recurrence of aluminium intoxication about two years later, and was treated with desferrioxamine again, with the removal of more Al.

He has had several recurrences and has been treated with desferrioxamine on each occasion. On the first occasion, at the beginning of the chelation treatment, his blood Al was 70–80 µg/l, or about 3 µM. But since then, at each recurrence of toxic symptoms, his plasma Al has been much lower, of the order of 20–40 µg/l. This is clearly no reflection of his body Al burden, which must still be considerable, despite our removing in total perhaps 4 g from him over the years. His plasma Al is very low but he still has a lot of Al in him, as measured by bone biopsy. Perhaps this is an extreme example, but in general we think that plasma Al is no more than a short-term reflection of the balance between intake and excretion and shouldn't be taken as a reliable indicator of long-term Al accumulation in all cases.

Martin: Do you find a correlation between erythrocyte and plasma Al?

Day: We do find a correlation in situations where a steady state could have been achieved. Thus, if we have a patient dialysing on low Al water, who has been like that for a while, and his Al intake is steady as well (he may be receiving Aludrox), then there is a clear relationship between erythrocyte and plasma Al.

Martin: What is that ratio? Are the concentrations about equal?

Day: Yes; the concentrations, expressed as mass per unit volume, are roughly equal. But if you then stop the Al intake, or treat the patient with desferrioxamine, the plasma Al level may change markedly, while the erythrocyte Al stays put in the short term, but will drop over the course of a few weeks. For example, if you treat with DFO, plasma Al may go up to four or even 10 times the ambient level; you will clear Al through the dialyser or by excretion, and then plasma Al drops back to rather less than it was before. The erythrocyte Al changes much more slowly. In two patients we studied, the level declined to almost zero, with a half life of about 40 days (Hewitt et al 1986).

Martin: You are suggesting that once Al gets in the red cells, it stays there?

Day: Yes. I don't think there's any dynamic equilibrium between the red cell and the plasma. Al gets in the red cell at its formation and stays there.

Edwardson: Your study with ^{26}Al showed that, didn't it?

Day: Yes; it showed that tracer ^{26}Al, entering the plasma, does not immediately appear in the red cell, although that is so far a limited study, with only one subject (me). After six hours, ^{26}Al was found in the plasma but none in the red cell (Day et al 1991).

Kerr: The course in the patient with recurrent episodes of encephalopathy is what you would predict, because after that length of time on dialysis, he will have lots of Al buried in cement lines, which will be released only as his bone remodels; so you would expect a gradual build-up to more mobile stores before the encephalopathy appears again. Then you treat him with desferrioxamine and remove his aluminium load, stop the treatment, and wait until another bunch of cement lines is remodelled over the next few years.

Day: From the serial bone biopsies, that looks very much like what happened in the patient I described.

Blair: We were talking about the relationship between observed parameters and Al load. We have shown that visual evoked response data correlate well with total accumulated Al in haemodialysis patients without obvious signs of aluminium poisoning (Altmann et al 1989). The data of course are an interesting indication of the steady onset of dementia in these patients.

Flaten: Professor Kerr, can you explain why Al-containing compounds are still used as phosphate binders in patients with total renal failure, when they can be replaced by Al-free compounds?

Kerr: I personally don't use them now, but many nephrologists still do. They say that patients don't like taking Ca carbonate because it causes indigestion or makes them sick; some become constipated, whereas others get diarrhoea. My view is that physicians have got into the habit of using Aludrox. There are plenty of studies showing that you can manage plasma phosphate with Ca carbonate, Ca acetate and other Ca salts. It gets more difficult if you are giving vitamin D or its analogues; then if you use Ca carbonate, you get more episodes of hypercalcaemia and must monitor serum calcium carefully. Magnesium hydroxide mixed with Ca carbonate will combat the constipation, but you then have to monitor serum Mg as well. Everybody is waiting for Ca alginate to come on the market. All the published data suggest that it is better than any other phosphate binder, but it is still stuck with the regulatory committees.

References

Abreo K, Glass J, Sell M 1990 Aluminum inhibits hemoglobin synthesis but enhances iron uptake in Friend erythroleukemia cells. Kidney Int 37:677–681

Ackrill P, Day JP, Ahmed R 1988 Aluminum and iron overload in chronic dialysis. Kidney Int 33 (suppl 24):S162–S167

Altmann P, Dhanesha U, Hamon CGB, Cunningham J, Blair JA, Marsh F 1989 Disturbance of cerebral function by aluminium in haemodialysis patients without overt aluminium toxicity. Lancet 2:7–12

Augsten K, Stein G 1988 Scanning electron microscopy and X-ray microanalysis investigation of aluminum deposition in samples from hemodialyzed patients. Trace Elem Med 5:55–59

Berland Y, Charbit M, Henry JF, Toga M, Cano JP, Olmer M 1988 Aluminium overload of parathyroid glands in haemodialysed patients with hyperparathyroidism: effect on bone remodelling. Nephrol Dial Transplant 3:417–422

Candy JM, McArthur FK, Oakley AE et al 1992 Aluminium accumulation in relation to senile plaque and neurofibrillary tangle formation in the brain of patients with renal failure. J Neurol Sci 107:210–218

Cutler P, Blair JA 1987 The effect of lead and aluminium on rat dihydropteridine reductase. Mechanisms and models in toxicology. Arch Toxicol Suppl 11:227–230

Day JP, Barker J, Evans LJA et al 1991 Aluminium absorption studied by ^{26}Al tracer. Lancet 337:1345

Dening TR, Berrios GE 1990 Wilson's disease: a longitudinal study of psychiatric symptoms. Biol Psychiatry 28:255–265

Hewitt CD, Day JP, Ackrill P 1986 The determination of aluminium in red blood cells. In: Taylor A (ed) Aluminium and other trace elements in renal disease. Baillière Tindall, London

Kaye M 1984 Bone marrow aluminium storage in end stage renal disease. Kidney Int 25:145 (abstr)

Lukiw WJ, Krishnan B, Wong L, Kruck TPA, Bergeron C, McLachlan DRC 1991 Nuclear compartmentalization of aluminum in Alzheimer's disease. Neurobiol Aging 13:115–121

McGregor FJ, Naves ML, Birly AK et al 1992 Interaction of aluminium and gallium with human lymphocytes: the role of transferrin. Biochim Biophys Acta, in press

Pierides AM, Myli MP 1984 Iron and aluminum osteomalacia in hemodialysis patients. N Engl J Med 310:323 (letter)

Rotundo A, Nevins TE, Lipton M, Lockman LA, Mauer SM, Michael AF 1982 Progressive encephalopathy in children with chronic renal insufficiency in infancy. Kidney Int 21:486–491

Simpson W, Kerr DNS, Hill AVL, Siddiqui JY 1973 Skeletal changes in patients on regular hemodialysis. Radiology 107:313–320

Verbueken AH, van de Vyver FL, Van Grieken RE, De Broe ME 1985 Microanalysis in biology and medicine: ultrastructural localization of aluminum. Clin Nephrol 24 (suppl 1):S57–S58

Vernejoul MC de, Girot R, Gueris J et al 1982 Calcium phosphate metabolism in patients with homozygous thalassaemia. J Endocrinol Metab 54:276–281

Aluminium and Alzheimer's disease

H. M. Wisniewski and G. Y. Wen

New York State Institute for Basic Research in Developmental Disabilities, 1050 Forest Hill Road, Staten Island, NY 10314, USA

Abstract. The hypothesis that aluminium (Al) is a cause of (or a risk factor in) the development of β-amyloid plaques and neurofibrillary tangles (NFT) and dementia in Alzheimer's disease (AD) is based on studies by Wisniewski et al, Klatzo et al and Terry & Peña in 1965 that showed that injection of experimental animals with Al compounds induces the formation of NFT. Other publications revealed that Al affects cognitive functions in experimental animals and humans undergoing dialysis for renal failure. Electron probe and laser microprobe mass analysis (LAMMA) studies have demonstrated the presence of Al in NFT and cores of amyloid stars and nuclei of neurons in AD patients. Other studies have indicated the association between amyotrophic lateral sclerosis/Guam parkinsonism-dementia complex and Al in the environment. A recent report suggests that the chelating agent desferrioxamine slows the rate of cognitive decline in AD patients. Extensive studies of the pathology of AD and Al-induced encephalopathy by our group and others indicate that Al does not cause Alzheimer's disease neuropathology. However, under certain conditions, cognition can be affected when Al enters the brain. Therefore, for individuals with renal failure or undergoing dialysis or individuals with a damaged blood–brain barrier, the intake of Al should be controlled.

1992 Aluminium in biology and medicine. Wiley, Chichester (Ciba Foundation Symposium 169) p 142–164

The neurotoxicity of aluminium in animals was first demonstrated in 1897, when the subcutaneous inoculation of Al solutions into the animals resulted in the development of bulbar paralysis, tremor, spasticity, rapid weight loss, and spinal cord changes, including the presence of small 'sickle-like' and 'glass-like' cells (Dollken 1897). Direct application of Al paste to the cortex of monkeys produced a chronic status of convulsion with recurrent seizures that simulated epilepsy in man (Kopeloff et al 1942). Aluminium attracted the further attention of neuropathologists when it was reported that its salts induced neurofibrillary changes in rabbit nerve cells (Klatzo et al 1965, Wisniewski et al 1965, Terry & Peña 1965). In addition to inducing neurofibrillary tangle (NFT) production, Al also induced cognitive deficits (Crapper & Dalton 1973, Petit et al 1980, Rabe et al 1982, King et al 1975), a reduction in the uptake of GABA and glycine neurotransmitters (Sturman et al 1983), death, and epileptic seizure, both in experimental animals (Wisniewski et al 1990a) and in humans as a result of the

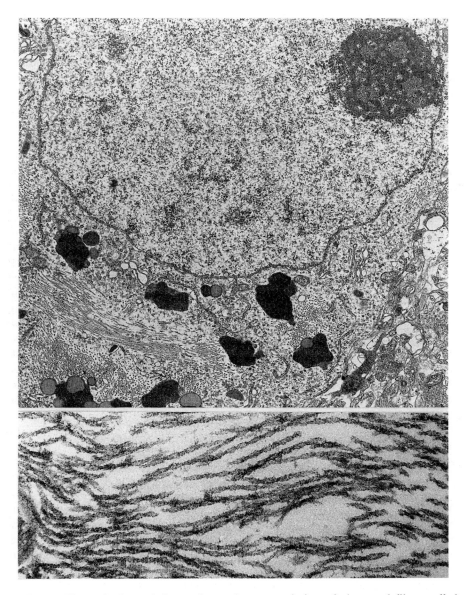

FIG. 1. The main (*upper*) figure shows the accumulation of abnormal fibres called paired helical filaments (PHF), which make up the neurofibrillary tangles, and lipofuscin (known as lipopigment or ageing pigment), in a neuron from an Alzheimer's disease patient. Magnification 12 320 ×. The *lower* portion shows a higher magnification of the PHF. Magnification 64 400 ×. (From Wisniewski & Sturman 1989 by permission of Marcel Dekker Inc.)

accidental implantation of metallic Al into the brain (Foncin & El Hachimi 1986). Al was also found to cause dialysis encephalopathy (Arieff 1990, Alfrey et al 1976) and was linked to Guam amyotrophic lateral sclerosis (ALS) and parkinsonism-dementia of Guam (Garruto et al 1990). Elevated amounts of Al were observed in brains of Alzheimer's disease patients (Crapper & Dalton 1973, Trapp et al 1978, Yoshimasu et al 1981). Al in drinking water was further linked to the incidence of Alzheimer's disease (AD) (Still & Kelley 1980, Vogt 1986, Leventhal 1986, Martyn et al 1989). With an electron probe and laser microprobe mass analysis (LAMMA), Al was found in the NFT and core of amyloid stars of AD patients (Perl et al 1986, Duckett & Galle 1976, Perl & Brody 1980, Crapper et al 1980, Masters et al 1985, Candy et al 1986). When Al was associated with ferritin in the brains of AD patients (Joshi et al 1985), it was implicated as a possible cause of the disease. However, other studies (Stern et al 1986, Mori et al 1988, Jacobs et al 1989, Moretz et al 1990) were unable to replicate these observations by electron probe and LAMMA. No correlation was found between the mean Al concentration and the degree of occurrence of NFT in AD brains (McDermott et al 1979). A recent thorough study has indicated that the Al content of drinking water is not related to AD pathogenesis (Wettstein et al 1991). The accidental implantation of metallic Al in to the human brain induced epilepsy but not neurofibrillary changes (Foncin & El Hachimi 1986). The paired helical filaments (PHF) characteristic of AD were not found in dialysis dementia patients' brains with a high content of Al (Arieff 1990, Alfrey et al 1976). A recent review emphasized that desferrioxamine, the trivalent ion-specific chelating (binding) agent, removed Al^{3+} from the body and thus slowed the progression of dementia, including AD (McLachlan et al 1991, Kalow & Andrews 1991). Desferrioxamine may also remove other trivalent ions (such as Fe^{3+} and P^{3+}) from the body and may not be the specific remover of Al^{3+}. Therefore, the implication that Al is the causal agent of Alzheimer's disease remains a speculative interpretation/hypothesis.

Comparison of AD neuropathology and Al encephalopathy

Alzheimer's disease is characterized by the presence of significant numbers of NFT (Fig. 1), β-amyloid plaques (diffuse, primitive and classical), amyloid angiopathy and neuronal loss, plus memory and other cognitive deficits that occur in middle or late life (Craun 1990). Lewy bodies, granulo-vacuolar degeneration and Hirano bodies are also present in the brains of AD patients (Probst et al 1991). The NFT of AD are made up of 20 to 24 nm PHF (Fig. 1), whereas the Al-induced NFT of experimental animals (Figs. 2, 3, 4 and 7) are composed of 10 nm neurofilaments (Fig. 2). Ultrastructural and immunohistochemical studies showed that the protofilaments making up the NFT of AD are larger than those of Al-induced NFT (32 ± 4 Å vs 20 ± 3 Å) (Wisniewski & Wen 1985), and the Al-induced NFT did not react with

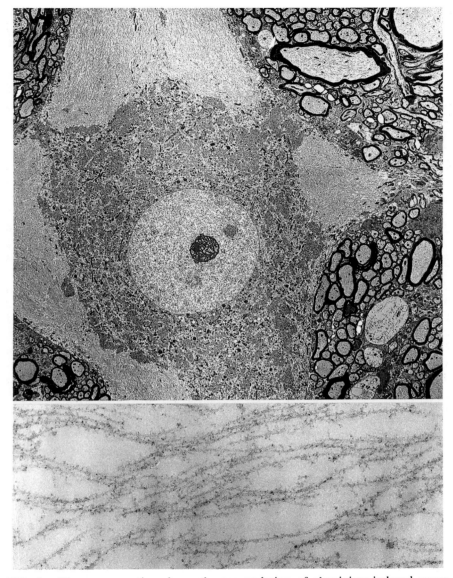

FIG. 2. The *upper* portion shows the accumulation of aluminium-induced neuro-fibrillary tangles (NFT) in a motor neuron of the rabbit spinal cord. Magnification 1736×. (Note the overall distribution of aluminium-induced NFT in the entire cross-section of the rabbit spinal cord shown in Fig. 5). The *lower* portion shows a higher magnification of aluminium-induced neurofilaments making up the NFT. Magnification 64 400×. Note that the diameter or size of the PHF shown in Fig. 1 is approximately twice that of a neurofilament. (From Wisniewski & Sturman 1989 by permission of Marcel Dekker Inc.)

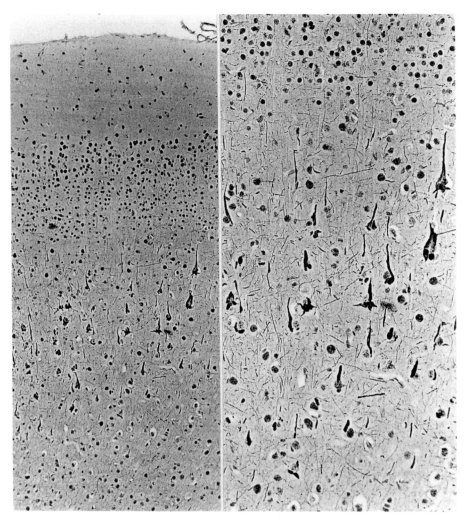

FIG. 3. Distribution of aluminium-induced neurofibrillary tangles (NFT) in the pyramidal neuron layer of the rabbit cerebral cortex (*left*). Magnification 165×. Higher magnification of NFT in pyramidal neuron layer (*right*). Magnification 430×.

anti-PHF monoclonal antibody 5–25 (Grundke-Iqbal et al 1985). It is noteworthy that both Pick body-like inclusions and flame-shaped NFT coexist in the presubiculum area of hippocampus of the Al-treated rabbit (Fig. 4). With silver impregnation stain (Figs. 3, 4 and 7) and toluidine blue stain (Fig. 5), AD NFT and Al-induced NFT are found to be morphologically similar to each other at the light microscopic level. With the thioflavin S stain, however, NFT of AD

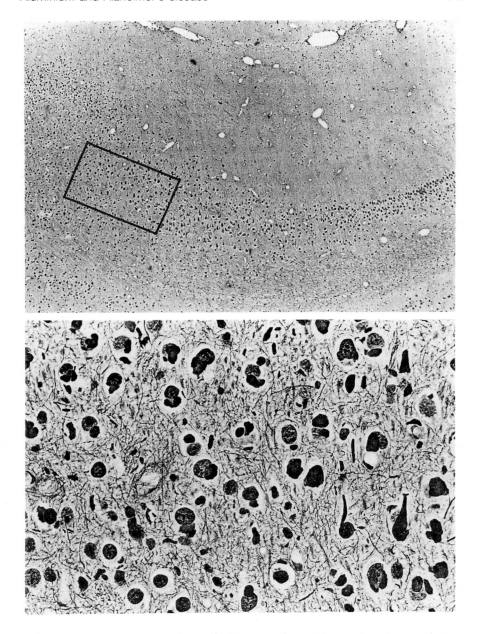

FIG. 4.   Coexistence of both Pick body-like inclusions and flame-shaped neurofibrillary tangles (NFT) in the presubiculum area (framed area) of hippocampus from an aluminium-treated rabbit. Magnifications 63× (*upper*), 630× (*lower*).

exhibited strong fluorescence, and Al-induced NFT did not (Fig. 6) (see colour plate). Furthermore, after Congo red staining, AD NFT revealed the birefringence of the β-pleated sheet, and the Al-induced NFT did not. The accidental implantation of metallic Al into the human brain induced epilepsy but no neurofibrillary changes (Foncin & El Hachimi 1986). In addition, Al caused the spongy type of dialysis dementia encephalopathy (Arieff 1990, Alfrey et al 1976) without a trace of PHF in the brain. The differences and similarities between AD NFT and Al-induced NFT are listed in Table 1. These differences rule out the possibility that aluminium is the causal agent producing NFT in Alzheimer's disease.

From the neuropathologist's point of view, even more important than the presence of NFT is the occurrence of many diffuse, primitive and classical

TABLE 1   A comparison of Al-induced and Alzheimer's disease neurofibrillary tangles

	Al-induced NFT	*Alzheimer's disease NFT*
Diseases associated with NFT	Al encephalopathy	AD, ALS, head trauma, DP, DS, GPD, HSD, LP, PiD, SSPE, TS, and in elderly people
Topography of NFT:		
In CNS	Cortex, hippocampus	Cortex, hippocampus
In PNS (spinal cord)	+	−
In cell	Perikaryon & proximal part of axon & dendrite, but not in terminal	Perikaryon, proximal & terminal of axon, & dendrite
Composition of NFT	10 nm NF	20 to 24 nm PHF
Peptide composition	68, 150 and 200 kDa (triplets of NF)	45 to 62 kDa (tau, ubiquitin)
Protofilament diameter	20 ± 3 Å	32 ± 4 Å
Immunoreactivity to:		
Antibody to 70 kDa NF	+	−
Anti-tau and anti-ubiquitin	−	+
Fluorescence after thioflavin S stain	−	+
Birefringence after Congo red (β-pleated sheet)	−	+
Silver impregnation stain	+	+
Toluidine blue stain	−	−

AD, Alzheimer's disease; ALS, amyotrophic lateral sclerosis; DP, dementia pugilistica; DS, Down's syndrome; GPD, Guam parkinsonism-dementia; HSD, Hallervorden-Spatz disease; LP, lipo-fuscinosis; NF, neurofilaments; NFT, neurofibrillary tangles; PHF, paired helical filaments; PiD, Pick's disease; SSPE, subacute sclerosing panencephalitis; TS, tuberous sclerosis, +, yes; −, no.

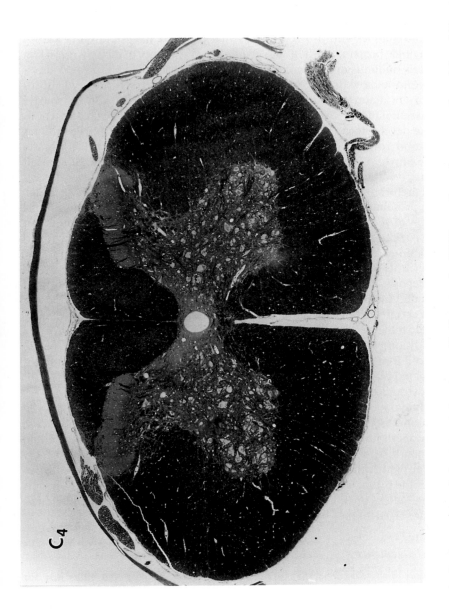

C₄

FIG. 5. Distribution pattern of aluminium-induced neurofibrillary tangles (NFT) over the entire cross-section of the rabbit spinal cord at the C₄ level (toluidine blue-stained 1 μm section). Magnification 28×.

amyloid plaques. A quantitative study of 100 brains from Down's syndrome patients revealed that amyloid plaques formed approximately 10–20 years earlier than NFT (Wisniewski et al 1987). Chemically, the major protein component of amyloid plaques and of cerebral vascular amyloid (angiopathy) is known as β-amyloid protein, with a molecular mass of approximately 4200 Da. The protein subunits are folded in a particular pattern called a β-pleated sheet. An amyloid plaque is a complex structure, made up of amyloid fibrils (7–10 nm in diameter) and abnormal neurites. Microglia and astrocytes are always an integral part of the β-amyloid plaque. Studies of the pathogenesis of plaque development indicate that the deposition of amyloid fibrils initiates the neuropil response and plaque formation. β-Amyloid protein originates from the amyloid precursor protein (APP), a ubiquitous protein. As a result of alternative splicing from a single gene on chromosome 21, several APP proteins are produced, including APP_{695}, APP_{751}, APP_{770} and APP_{714}. Which of these protein(s) is or are the source of the β-fragment responsible for the formation of the amyloid fibrils is unknown. A recent study also pointed out that microglial cells actively participate in the formation of amyloid fibrils (Wisniewski et al 1990b). β-Amyloid protein is the sole subunit protein of the amyloid fibrils in AD, but other proteins are closely associated with β-amyloid protein deposits. These other constituents of the deposits are β-amyloid-associated proteins and include α_1-antichymotrypsin; complement factors C1q, C3c and C3d; the serum amyloid P protein; heparan sulphate proteoglycans; and nucleoside diphosphatase (NDPase) (Wisniewski et al 1990b, Selkoe 1991a,b).

In addition, β-amyloid protein and its parent molecule, protease nexin-II/amyloid β-protein precursor (PN-2/APP), are present in the α-granules of human blood platelets, which are the major circulating repository for PN-II/APP (Van Nostrand et al 1990, 1991).

In summary, extensive studies of the pathology of AD and Al-induced encephalopathy by our group and others have revealed that Al does not cause AD neuropathology. However, the wide distribution, broad application and potent neurotoxicity of Al suggest a need for further investigation of the biological significance of Al in human health. Cognition can be affected when Al enters the brain. Therefore, the intake of Al should be controlled in people with renal failure or those who are undergoing dialysis, as well as in individuals with a damaged blood–brain barrier.

Acknowledgements

The authors wish to express their appreciation to Marj Agoglia and Janis Kay for their excellent secretarial assistance. Our appreciation is also extended to Maureen Stoddard Marlow for copy-editing the manuscript. This study was supported in part by funds from the New York State Office of Mental Retardation and Developmental Disabilities and the Fund for the Center for Trace Metal Studies and Environmental Neurotoxicology.

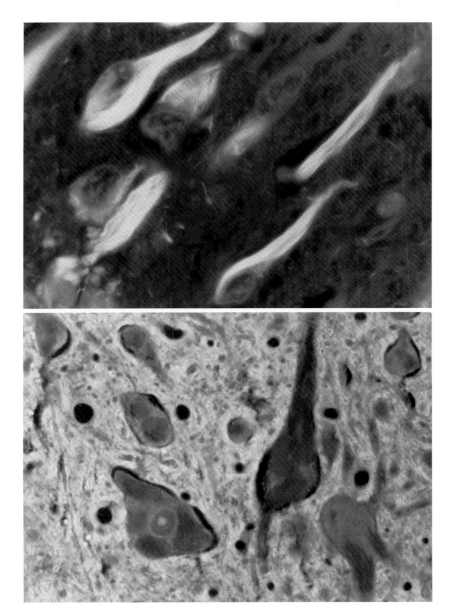

FIG. 6 *(Wisniewski & Wen)* Thioflavin S-stained sections showing the presence of fluorescence in the neurofibrillary tangle (NFT) of Alzheimer's disease *(upper)* and the absence of fluorescence in the aluminium-induced NFT of the rabbit *(lower)*. Magnifications 980 × *(upper)*, 800 × *(lower)*. (From Wisniewski & Sturman 1989, with permission of Marcel Dekker Inc.)

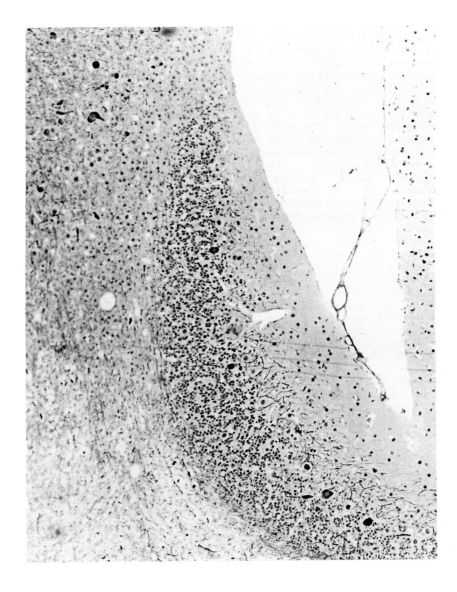

FIG. 7.   Presence of aluminium-induced neurofibrillary tangles (NFT) in the Purkinje
cells of the part of the rabbit cerebellum that is attached to the nucleus cochlearis. The
cerebellum is generally free of (or resistant to) NFT formation. Magnification 161 ×.

References

Alfrey AC, LeGendre GR, Kaehny WD 1976 The dialysis encephalopathy syndrome. N Engl J Med 294:184–188

Arieff AI 1990 Aluminum and the pathogenesis of dialysis dementia. Environ Geochem Health 12:89–93

Candy JM, Oakley AE, Klinowski J et al 1986 Aluminosilicates and senile plaque formation in Alzheimer's disease. Lancet 1:354–357

Crapper DR, Dalton AJ 1973 Alterations in short-term retention, conditioned avoidance response acquisition and motivation following aluminum-induced neurofibrillary degeneration. Physiol Behav 10:925–933

Crapper DR, Quittkat S, Krishnan SS, Dalton AJ, DeBoni U 1980 Intranuclear aluminum content in Alzheimer's disease, dialysis encephalopathy and experimental aluminum encephalopathy. Acta Neuropathol 50:19–24

Craun GR 1990 Review of epidemiologic studies of aluminum and neurological disorders. Environ Geochem Health 12:125–135

Dollken V 1897 Über die Wirkung des Aluminiums mit besonderer Berucksichtigung der durch das Aluminium verursachten Lasionen im Centralnervensystem. Arch Exp Pathol Pharmacol 98–120

Duckett S, Galle P 1976 Mise en évidence de l'aluminium dans les plaques de la maladie d'Alzheimer: étudié à la microsonde de Castaing. CR Hebd Seances Acad Sci Ser D Sci Nat 282:393–395

Foncin JF, El Hachimi KH 1986 Neurofibrillary degeneration in Alzheimer's disease: a discussion with a contribution to aluminum pathology in man. In: Bes A, Cahn J, Cahn R, Hoyer S, Marc-Vergnes JP, Wisniewski HM (eds) Current problems in senile dementias, No. 1. John Libbey Eurotext, London-Paris, p 202–210

Garruto RM, Yanagihara RT, Gajdusek DC 1990 Model of environmentally induced neurological disease: epidemiology and etiology of amyotrophic lateral sclerosis and Parkinsonism-dementia in the Western Pacific. Environ Geochem Health 12:137–151

Grundke-Iqbal I, Wang GP, Iqbal K, Wisniewski HM 1985 Alzheimer paired helical filaments: identification of polypeptides with monoclonal antibodies. Acta Neuropathol 68:279–283

Jacobs RW, Duong T, Jones RE, Trapp GA, Scheibel AB 1989 A re-examination of aluminium in Alzheimer's disease: analysis by energy dispersive X-ray microprobe and flameless atomic absorption spectrophotometry. Can J Neurol Sci 16:498–503

Joshi JG, Fleming J, Zimmerman A 1985 Ferritin aluminum binding. XIII World Congress of Neurology (Hamburg). Abstracts of papers presented. Excerpta Medica, Amsterdam (abstr. 05.07.03)

Kalow W, Andrews DF 1991 Intramuscular desferrioxamine in patients with Alzheimer's disease. Lancet 337:1304–1308

King GA, DeBoni U, Crapper DR 1975 Effect of aluminum upon conditioned avoidance response acquisition in the absence of neurofibrillary degeneration. Pharmacol Biochem Behav 3:1003–1009

Klatzo I, Wisniewski H, Streicher E 1965 Experiemental production of neurofibrillary degeneration. J Neuropathol & Exp Neurol 24:187–199

Kopeloff LM, Barrera SE, Kopeloff N 1942 Recurrent convulsion seizures in animals produced by immunologic and chemical means. Am J Psychiatry 98:881–902

Leventhal GH 1986 Alzheimer's disease and environmental aluminum in Maryville and Morristown, Tennessee. PhD thesis, University of Tennessee, TN, USA

Martyn CN, Barker DJP, Osmond C, Harris EC, Edwardson JA, Lacey RF 1989 Geographical relation between Alzheimer's disease and aluminium in drinking water. Lancet 1:59–62

Masters CL, Multhaup G, Simms G, Pottgiesser J, Martins RW, Beyreuther K 1985 Neuronal origin of a cerebral amyloid: neurofibrillary tangles of Alzheimer's disease contain the same protein as the amyloid of plaque cones and blood vessels. EMBO (Eur Mol Biol Organ) J 4:2757–2763

McDermott JR, Smith AI, Iqbal K, Wisniewski HM 1979 Brain aluminum in aging and Alzheimer disease. Neurology 29:809–814

McLachlan DRC, Kruck TP, Lukiw WJ, Krishnan SS 1991 Would decreased aluminum ingestion reduce the incidence of Alzheimer's disease? Can Med Assoc J 145:793–804

Moretz RC, Iqbal K, Wisniewski HM 1990 Microanalysis of Alzheimer disease NFT and plaques. Environ Geochem Health 12:15–16

Mori H, Swyt C, Smith QR, Atack JR, Rapoport SI 1988 In situ X-ray microanalysis of senile plaques in Alzheimer's disease. Neurology 38 (suppl 1):230

Perl DP, Brody AR 1980 Alzheimer's disease: x-ray spectrometric evidence of aluminum accumulation in neurofibrillary tangle-bearing neurons. Science (Wash DC) 208: 297–299

Perl DP, Muñoz-Garcia D, Good P, Pendlebury WW 1986 Laser microprobe mass analyzer (LAMMA)—a new approach to the study of the association of aluminum and neurofibrillary tangle formation. In: Fisher A, Hanin I, Lachman C (eds) Alzheimer's and Parkinson diseases: strategies for research and development. Plenum Press, New York, p 241–248

Petit TL, Biederman GB, McMullen PA 1980 Neurofibrillary degeneration, dendritic dying back, and learning-memory deficits after aluminum administration: implications for brain aging. Exp Neurol 67:152–162

Probst A, Langui D, Ulrich J 1991 Alzheimer's disease: a description of structural lesions. Brain Pathol 1:229–239

Rabe A, Lee MH, Shek JW, Wisniewski HM 1982 Learning deficit in immature rabbits with aluminum-induced neurofibrillary changes. Exp Neurol 76:441–446

Selkoe DJ 1991a The molecular pathology of Alzheimer's disease. Neuron 6:487–498

Selkoe DJ 1991b Amyloid protein and Alzheimer's disease. Sci Am 265:68–78

Stern AJ, Perl DP, Muñoz-Garcia D, Good PF, Abraham C, Selkoe DJ 1986 Investigation of silicon and aluminum content in isolated senile plaque cores by laser microprobe mass analysis (LAMMA). J Neuropathol & Exp Neurol 45:361

Still CN, Kelley P 1980 On the incidence of primary degenerative dementia vs. water fluoride content in South Carolina. Neurotoxicology 4:125–131

Sturman JA, Wisniewski HM, Shek JW 1983 High affinity uptake of GABA and glycine by rabbits with aluminum-induced neurofibrillary changes. Neurochem Res 8:1097–1109

Terry RD, Peña C 1965 Experimental production of neurofibrillary degeneration. J Neuropathol & Exp Neurol 24:200–210

Trapp GA, Miner GD, Zimmerman RL, Master AR, Heston LL 1978 Aluminum level in brain in Alzheimer's disease. Biol Psychiatry 13:708–718

Van Nostrand WE, Schmaier AH, Farrow JS, Cunningham DD 1990 Protease nexin II (amyloid β-protein precursor): a platelet α-granule protein. Science (Wash DC) 248:745–748

Van Nostrand WE, Schmaier AH, Farrow JS, Cunningham DD 1991 Platelet protease nexin-2/amyloid beta-protein precursor. Possible pathologic and physiological functions. Ann NY Acad Sci 640:140–144

Vogt T 1986 Water quality and health—study of a possible relationship between aluminum in drinking water and dementia. (Sosiale og okonomiske studier 61; English abstract) Central Bureau of Statistics of Norway, Oslo

Wettstein A, Aeppli J, Gautschi K, Peters M 1991 Failure to find a relationship between mnestic skills of octogenarians and aluminum in drinking water. Int Arch Occup Environ Health 63:97–103

Wisniewski HM, Wen GY 1985 Substructures of paired helical filaments from Alzheimer's disease neurofibrillary tangles. Acta Neuropathol 66:173–176

Wisniewski HM, Sturman JA 1989 Neurotoxicity of aluminum. In: Gitelman HJ (ed) Aluminum and health—a critical review. Marcel Dekker, New York & London, p 125–165

Wisniewski HM, Terry RD, Peña C, Streicher E, Klatzo I 1965 Experimental production of neurofibrillary degeneration. J Neuropathol & Exp Neurol 24:139 (abstr)

Wisniewski HM, Rabe A, Wen GY, Wisniewski KE 1987 Gene dose effect on early formation of neuritic plaques in Down syndrome (DS). J Neuropathol & Exp Neurol 46:334 (abstr)

Wisniewski HM, Moretz RC, Sturman JA, Wen GY, Shek JW 1990a Aluminum neurotoxicity in mammals. Environ Geochem Health 12:115–120

Wisniewski HM, Vorbrodt AW, Wegiel J, Morys J, Lossinsky AS 1990b Ultrastructure of the cell forming amyloid fibers in Alzheimer disease and scrapie. Am J Med Genet (suppl) 7:287–297

Yoshimasu F, Yasui M, Yoshida H et al 1981 Aluminum in Alzheimer's disease in Japan and Parkinsonism dementia in Guam. XII World Congress of Neurology (Kyoto, Japan). Abstracts of papers presented. Excerpta Medica, Amsterdam (abstr 15.07.02)

DISCUSSION

Blair: How common are Lewy bodies in Alzheimer's disease?

Wisniewski: Some people feel that they are quite common, occurring in around 30% of cases. There is a big debate on how to interpret the presence of Lewy bodies in AD; some think we are dealing with a Lewy body disease which is, in a way, separate from Alzheimer's disease. Others say that there is an Alzheimer's disease variant with Lewy body inclusions.

Blair: Am I right in thinking that Lewy bodies are often taken as diagnostic when they appear in the substantia nigra of Parkinson's disease? They are almost the 'plaque and tangle' of that disease?

Wisniewski: That is absolutely right, but in this situation (i.e. in AD) the Lewy bodies appear in the cortical neurons.

Strong: I have a few concerns about your concept of brain amyloidosis as the pathogenesis of Alzheimer's disease. The first relates to the deposition of amyloid within the cerebellum, yet the lack of clinical features involving cerebellum, or of degenerative phenomena within the cerebellum, at the microscopic level.

Wisniewski: The cerebellum is not as massively involved as the cerebral cortex, in terms of β-amyloid deposits, in Alzheimer's disease. The deposits in the cerebellum are mostly in the form of diffused plaques and there are not the large numbers of classical and primitive plaques seen in cortex. One can also have a lot of plaques without tangles in the neocortex without signs or symptoms of dementia. There is thus a substantial tolerance to β-amyloidosis. There is evidence that involvement of cerebellum is a late event in the occurrence of β-amyloid deposits in the course of the AD history.

Strong: Neuropathologically, do you find much cerebellar degeneration at autopsy? Is the cerebellum atrophic?

Wisniewski: No; but people have not done morphometric studies of cerebellum in AD.

Strong: There is a recent concept that Alzheimer's may not be so much a *degenerative* phenomenon as an abortive regenerative one. Unfortunately, the regenerative process, in the form of sprouting, produces misshaped and misdirected neuronal processes. Uchida and colleagues (Uchida & Tomonaga 1989, Uchida et al 1988, 1991) have suggested that the *absence* of a growth inhibitory factor is crucial to the induction of the neuritic dystrophy in Alzheimer's disease, and that this factor is dependent, in part, upon the presence of astrocytes. In Alzheimer's you see a lot of this, whereas you do not see it in other neurodegenerative diseases. How does that relate to amyloid?

Wisniewski: Amyloid and NFT pathology, in terms of their topographical distribution in brain, don't correlate. That is to say, in areas of extensive β-amyloid deposition, you do not necessarily find many neurons with NFT.

Strong: I am referring not to neurofibrillary tangles, but to the neuropil threads.

Wisniewski: 'Threads' are tangles within nerve cell processes. Neuritic 'threads' are the NFT, or bundles of PHF, in nerve cell processes and terminals, mostly unmyelinated nerve fibres. So the threads are part of the neurofibrillary pathology. It is important to note that NFT can occur in many and unrelated pathological conditions, like hydrocephalus or subacute sclerosing panencephalitis (SSPE), and in people with a history of head trauma.

Wischik: Dr Wisniewski, I agree with your broad judgement of the significant determining factors in AD as regards clinical phenomenology, and the view that the amyloid pathology and the PHF (neuritic) pathology are driven by different underlying mechanisms. There are points of detail with which I disagree. The first is the issue of old age as being preclinical Alzheimer's disease. In the Boston study (Evans et al 1989), 3600 cases of AD were examined. In a study recently completed in Cambridge in the UK (O'Connor et al 1989), 5000 cases were selected randomly from general practice registers. The Boston study assessed 460 cases in detail; in the Cambridge study, 520 cases were assessed in detail. CAMDEX, which was designed as a specific diagnostic instrument for Alzheimer's disease, as opposed to being a simple memory impairment instrument, was used. The US study claimed an incidence of 47%, but we feel this overstates the situation. In Cambridge, the overall incidence of probable Alzheimer's disease (in the sense of minimal or more severe impairment) in the general population over the age of 65 is 10.25%. In the group over 85, it reaches 28%, but not the 47% that the Boston survey finds. This lower figure is important, genetically, because it is clearly not the case that AD (adult onset) is a universal, autosomal dominant, genetic disease.

Wisniewski: But the Boston study shows age-associated increased prevalence of AD. My feeling is that we *are* indeed dealing with a preclinical AD, and the impact of that will have great significance when we improve our diagnostic tools.

Wischik: In fact, in the over-85 group in the Cambridge study we have a group known as the 'supernormals'. In that (85 +) group, the incidence of dementia decreases, compared to the 75–85 cohort.

Blair: We need to define what we mean by 'cognitive defect' very carefully. Numerous tests are applied, some more subtle than others, and some of the tests really don't imply that the person cannot function normally in the community—only that they could not, in front of a computer on that particular day, reach a particular score.

Williams: It is certainly important to specify the approaches and tests used, because they can vary.

Edwardson: We are discussing the prevalence of dementia in the context of environmental factors. The one study which I am aware of, where the investigators used the same instruments to compare the prevalence of dementia in New York and London, found a higher prevalence in New York (Copeland et al 1987). This suggests that there may be geographical differences in prevalence rates.

Wischik: On the point that β-amyloid deposition is of enormous diagnostic significance in Alzheimer's disease, it is true that you cannot make a neuropathological diagnosis of AD in the absence of β/A4 deposition. On the other hand, the presence of β/A4 deposition by itself does not give you a neuropathological diagnosis of Alzheimer's disease (Duyckaerts et al 1990, Crystal et al 1988).

Wisniewski: Alzheimer's disease means the presence of 'signs and symptoms'. The presence of pathology does not always lead to clinical expression. The best example of that is cancer: the moment I find a cancer cell, a patient has 'a cancer'; but if that person lacks the clinical expression of the disease in question, he does not have clinical cancer. You are bringing up the question of whether the presence of one plaque and one neurofibrillary tangle is already a lesion which will end up with a massive presence of plaques and tangles leading to the Alzheimer disease process. To overcome this problem, I suggest a compromise. As a pathologist, I do not make a diagnosis of AD; I say that the numbers (few, many, very large numbers) are or are not compatible with a diagnosis of AD. The clinician has to determine the presence or absence of signs and symptoms of dementia, and make the final diagnosis of AD. The brain is a complex structure; because Alzheimer's disease is a global encephalopathy, and we don't understand the compensatory mechanisms available, I cannot always predict, on the basis of pathology, whether an individual is demented or not.

Wischik: A simpler concept is the concept of necessary and sufficient cause: the presence of β/A4 pathology, even if it is very extensive, in many areas of the cortex, is possible without any cognitive deficit (Delaère et al 1990). So the β-amyloid deposition is a necessary precondition for AD. But what is both necessary and sufficient (I shall present data to support this in my paper: p 268–302) in cases with clinical dementia is the presence of extensive paired helical filament accumulations in some form, whether they are in cell bodies, or in dystrophic neurites. If so, β/A4 deposition is a precondition but not a diagnostic feature of the disease. And, in fact, Dutch cerebral angiopathy would then be an instance of extensive β/A4 deposition that does not move into the next step of neurofibrillary pathology (Levy et al 1990). In my view, one would not make a diagnosis of Alzheimer's disease in that case.

Edwardson: I think we are straining at a gnat! We now know of at least three genetic lesions at codon 717 in the gene for the β-amyloid precursor protein (APP) in some rare forms of early-onset, familial Alzheimer's disease. We also know that mutation at this codon, will, after 40, 50 or 60 years, depending on the pedigree, result in a brain full of plaques and tangles. We also know, from studies of Down's syndrome, that the first neuropathological changes are diffuse plaques; then there is development of mature plaques; ultimately, you get end-stage pathology. The fact that at different stages of development there are different patterns of pathology is, in some respects, irrelevant; what counts is what will happen in the course of time.

McLachlan: I disagree; all we know at present is that this mutation is *linked* to the expression of familial Alzheimer's disease in rare families. There is no proof yet, that the amyloid mutations cause AD.

Wisniewski: We don't know that everyone with a point mutation in that codon will develop AD. It may be only a risk factor.

Edwardson: In the New Brunswick pedigree, we don't know what the mutation is, but we know that over seven generations, approximately 50% of those in each generation develop Alzheimer's disease; it is an autosomal dominant gene, therefore, and the genetic lesion is clearly determining the disease (Nee et al 1983). Whatever happens there in terms of pathogenesis, we know that a genetic lesion, after approximately 50 years, leads to a brain with plaques and tangles. This is not merely a 'risk factor'; it has to be a *cause* of Alzheimer disease.

Garruto: We are discussing familial Alzheimer's disease. For clarification, what percentage of patients with Alzheimer's disease is familial and what percentage is 'sporadic', and are point mutations considered to be involved in the cause of sporadic Alzheimer's disease?

Wisniewski: Roughly 5% of AD cases are familial (FAD). Epidemiological surveys have suggested that first-degree relatives of patients with AD have an age-adjusted risk of developing AD which approximates the 50% risk ratio expected for an autosomal dominant trait.

Flaten: Dr Edwardson referred to a study of Alzheimer's disease which reported a higher prevalence in New York than in London. Without making any claims of causality, I would like to repeat my earlier comment that the average intake of Al in the USA is more than twice that in England.

Wischik: I have been struck by Henry Wisniewski's demonstration of β-amyloid fibril formation in the rough endoplasmic reticulum of the *microglia*, and by Yamaguchi's demonstration of diffuse β-amyloid deposits in the immediate vicinity of *astrocytes* (Yamaguchi et al 1991). Could you comment on this difference in cellular location?

Wisniewski: In diffuse plaques we also see microglia making amyloid fibrils. I consider all amyloid fibril formation in the brain to be cell-associated. We have evidence from studying vessels with amyloid angiopathy that the perivascular cells are making the amyloid fibrils. Indeed, astrocytes are also in close contact with amyloid deposits. However, our data suggest that astrocytes participate in the fragmentation and disintegration of amyloid deposits and not in their formation (Wisniewski & Wegiel 1991).

Candy: Ultrastructural morphology is often poor in post mortem material: how can you be sure that the presence of amyloid fibrils in microglial cells isn't due to phagocytosis, and that the microglial cell is the *processing cell* rather than the *producing cell*?

Wisniewski: In order to further distinguish between microglia/macrophage cells involved in fibril formation and those involved in the removal of amyloid deposits, we have recently examined the cells involved in β-amyloid phagocytosis after a stroke (Wisniewski et al 1991). Comparison of the morphology of the cells that produce and the cells that digest β-amyloid indicates that the newly formed and phagocytosed amyloid fibrils are located in different cellular compartments: in altered smooth endoplasmic reticulum (SER) and phagosomes, respectively.

Kerr: Most other forms of amyloid are being removed as well as being deposited. That certainly happens in amyloid secondary to sepsis; if you remove the septic limb, the amyloid is gradually removed. Is there any mechanism for getting rid of amyloid in the brain?

Wisniewski: As you point out, in systemic amyloidosis, the removal of amyloid is slow. However, under certain conditions it could be very fast. There are reports that in patients with lung amyloidosis who developed lobar pneumonia, and survived, both the pneumonia and the amyloid cleared. So, the release of white cell enzymes during the recovery from pneumonia also clears up the amyloid. As I mentioned before, in stroke, the macrophages can degrade plaque amyloid. We have evidence that ectoenzymes of astrocytes may be responsible for the dissolution and removal of amyloid fibrils. Our assumption is that in AD amyloid *can* be removed. If, therefore, we could stop the production of amyloid fibrils, we could start to talk about new approaches to the treatment of Alzheimer's disease. In other words, I would like to treat the cells making amyloid fibrils like neoplastic cells; if I can learn how to control or remove them, I think I can stop the progression of brain amyloidosis and AD.

Perl: You consider Alzheimer's disease to be a form of cerebral amyloidosis, yet you also say that in many elderly people there can be a substantial amount of amyloid accumulation in the brain in the absence of any symptomatology. It has been our observation too, and that of other neuropathologists, that elderly people can accumulate every bit as much amyloid in the brain as one sees in Alzheimer's disease and yet be entirely asymptomatic. It is only when the patient develops neurofibrillary tangles, neuritic threads, and dystrophic neurites that we begin to see the symptoms that we associate with Alzheimer's disease. But NFTs, neuropil threads and dystrophic neurites all contain paired helical filaments, so why do you emphasize cerebral amyloidosis as the disease, rather than the PHF accumulation, which really is correlated with the symptomatology—the development of the features by which the disease is recognized?

Wisniewski: In my opinion, in AD, β-amyloidosis, as in Down's syndrome, starts the cascade of AD neuropathology. Amyloid is defined by its physical and optical properties; PHF show all the properties of amyloid—they are birefringent after Congo red staining, and show green fluorescence after staining with thioflavin S. The tangles in Al-treated animals do not show these properties. So, by definition, both the proteins making NFT and β-amyloid fibrils show properties of amyloid; so I can call them both amyloid, by definition.

Williams: You said that the cells which may be causing the problem could be macrophages or microglia. If that is the case, do you know whether these cells contain Al? The macrophages (microglia) would presumably scavenge aluminium.

Wisniewski: This is not known. Under certain conditions, macrophages can phagocytose amyloid—for example, in brains containing numerous primitive and classical plaques in stroke (Wisniewski et al 1991). However, as I said earlier, phagocytosed amyloid is located in phagosomes. On the other hand, the newly formed amyloid fibrils are seen in altered smooth endoplasmic reticulum of the microglia cells. Antibodies raised to a synthetic peptide of the β-protein stain not only the amyloid deposits, but also neuronal lipofuscin. Western blotting of a lipofuscin-enriched fraction shows a β-protein-immunoreactive polypeptide migrating at approximately the 31 kDa position on SDS–polyacrylamide gel electrophoresis. These results suggest that a large non-fibrillar fragment of the amyloid precursor protein is associated with lipofuscin (Bancher et al 1989).

Flaten: You say that the microglia seem to be very important producers of amyloid. As far as I know, the microglia are also the most ferritin-rich cells in the brain. I do not know whether there is any connection between these two observations, but some people have suggested that the Al content of ferritin is elevated in Alzheimer's disease.

Wisniewski: Indeed, the best marker for identifying activated microglia is anti-ferritin antibody. However, its relevance to β-amyloid fibril formation is not clear.

Garruto: In dialysis-associated arthropathy, Netter and colleagues (1990, 1991) reported up to a 10-fold increase in Al in synovial tissue in dialysis patients ingesting gels containing Al than in patients with normal renal function. What was also interesting was the presence of β_2-microglobulin, a systemic amyloid, in the same synovial tissue of these patients.

Kerr: Certainly, Al has been found in the joints of patients on dialysis and has been proposed as a contributory factor in amyloid arthritis (Canavese et al 1990). β_2-Microglobulin levels are raised in the plasma of all dialysis patients, and in some of these it forms amyloid deposits in and around joints. If you look for Al in these affected joints you find it, but I am not sure that it is routinely raised in concentration, compared to unaffected joints (Netter 1991).

Ward: I don't think it's routinely raised, but then few analyses have been done.

Perl: We published a paper on the occurrence of β_2-microglobulin in amyloid deposits in the femoral head in a patient with accumulation around the joint and found prominent Al accumulation in the mineralization fronts (Heller et al 1989). We haven't studied other cases, but this one was very dramatic. I understand from my colleagues in renal medicine that these amyloid accumulations in bone are being seen with increasing frequency in renal dialysis patients. I don't know whether you are seeing this in Great Britain?

Kerr: In Charing Cross Hospital, where they have a population of dialysis patients in their second decade of haemodialysis with Cuprophan membranes, 97% of these have the clinical features of dialysis arthropathy, and in most the diagnosis of amyloidosis has been confirmed by histology or ^{125}I-SAP (serum amyloid-P) imaging (Sethi et al 1990 and personal communication).

Perl: So this appears to represent an instance of amyloid accumulation that is related to Al.

Ward: You don't know that it's Al-related, surely?

Perl: Aluminium was detected in the mineralized bone and in the calcification front.

Ward: You don't know that either, unless you have done bone histopathology.

Perl: In this particular patient, un-decalcified plastic-embedded sections were examined morphologically and by LAMMA (laser microprobe mass analysis).

Williams: And is silicon located there as well?

Perl: No; silicon was not detected by LAMMA in this bone sample.

Wisniewski: Patients with chronic osteomyelitis often develop systemic amyloidosis. Do your dialysis patients have recurrent infections?

Kerr: Not necessarily. Dialysis amyloid seems to be an inevitable consequence of being on dialysis for more than 10 years (with current dialysis membranes) and is associated with a very high level of β_2-microglobulin in the plasma. The amyloid contains β_2-microglobulin and, like any other amyloid, serum amyloid-P (SAP) and glycosaminoglycans. It does not seem to require recurrent infection to form, because we find it in otherwise well patients.

Wischik: In Alzheimer's disease, the amyloidosis, in the generic sense, whether intracellular or extracellular, appears to be related to defective proteolytic processing. In other words, you produce a proteolytic fragment in the course of the breakdown of a normal protein which has the unfortunate property of polymerization; that polymerization then inhibits further proteolysis and the product accumulates. Is that the case in these other amyloidoses? Is the actual molecular mechanism in long-term dialysis amyloidosis the same kind of defective proteolytic processing and the accumulation of a partial cleavage product?

Kerr: I know of no studies to show that. Dialysis amyloidosis has a very peculiar distribution; it is found in bones, as punched-out lesions; it is also found in the connective tissue surrounding joints. It occurs in the carpal tunnel, causing carpal tunnel syndrome. It is not present in liver, kidney or spleen, so there may be some specific cell in connective tissue that encourages the deposition of β_2-microglobulin.

Wischik: A unifying theme might be the presentation of a potentially amyloidogenic protein to a cell which doesn't have the specialized proteolytic apparatus for dealing with it, as for instance a non-specific scavenger system might. For most proteins, the cell's proteolytic repertoire is successful, but with these dangerous proteins, that repertoire is inadequate, so one gets amyloidosis. This could be a general mechanism.

Edwardson: It is important to stress that there is no general agreement that β-amyloid fibrils originate from microglial cells. Allsop et al (1989) have studied the earliest stages of plaque formation in Down's syndrome brains and shown that the first visible event is the deposition of amyloid adjacent to the cell body of morphologically normal neurons. These authors argue strongly for the neuronal origin of plaque amyloid. An alternative interpretation of Dr Wisniewski's electron photomicrographs is that microglial cells are trying to remove β-amyloid deposits.

Wisniewski: The pictures of microglia cells that I showed illustrate cells making and secreting amyloid fibrils (Wisniewski et al 1990). As discussed previously, on the basis of ultrastructural morphology we can differentiate the process of 'eating' amyloid fibrils from that of making it. However, I am not excluding the possibilities that APP or its peptides produced by the neurons or coming from the blood are picked up by the brain pericytes and microglia, which have difficulty with the processing of these amyloidogenic proteins. In other words, the microglia are not 'eating' amyloid fibrils but may take up amyloidogenic proteins and, as a result of defective 'recycling', start to make amyloid fibrils. In diffuse plaques we often see β-amyloid immunoreactivity around the neuronal cell body and apical dendrites. However, according to our data, the neuronal satellite microglia is the source of β-amyloid fibrils.

Wischik: It is a general feature that in cortex and hippocampus each pyramidal cell has a satellite cell nearby. J. Ulrich and colleagues are looking at levels of mRNA expression of APP by *in situ* hybridization specifically in these satellite cells (unpublished). If this amyloid deposit were intracellular, within satellite cells, it would be very interesting, because it would suggest that the pyramidal cells are not responsible for amyloidogenic processing.

Birchall: Given abnormal protein phosphorylation, and given that the deposits contain Al, it's not surprising that the structure wouldn't be easily proteolysed. Al has been used to tan leather for that very reason.

Strong: We did *in situ* hybridization in developing hippocampal neurons in culture (Strong et al 1990), and more recently have examined the patterns of expression of the various mRNA transcripts for the amyloid precursor protein (Strong & Jakowec 1991). The concept that the amyloid precursor protein is necessarily a disaster for a neuron may be wrong. It's clear that there is a developmentally time-linked pattern of expression of the amyloid precursor protein mRNA transcripts; it is probably linked to the acquisition of synaptic integrity, so that when a functional synapse is formed, the pattern of APP mRNA transcription within that neuron changes into a more mature developmental pattern. In fact, the serine protease inhibitor sequence-containing transcript (APP_{751}) becomes the predominant transcript.

If one then considers only the intraneuronal deposition of β/A_4 protein, which is a cleavage product of the amyloid precursor protein, Bauxbaum et al (1990) have suggested that an abnormal phosphorylation of the C-terminal segment of the APP protein, outside the transmembrane portion containing the β/A_4 protein, will yield the necessary cleavage of the protein to give rise to the β/A_4 protein. If we are talking about Al toxicity, and mechanisms of aberrant phosphorylation, including aberrant phosphorylation of tau protein, one way of looking at this is to postulate a fundamental abnormality in phosphorylation within the neuron itself. In the case of the β/A_4 protein, a normally expressed protein, aberrantly cleaved, yields a highly insoluble protein. Thus, intraneuronal amyloidosis may be a neuronal mistake; it's simply a side-phenomenon within the neuron that is a marker of the process of abnormal phosphorylation. That may bring us back to the Al mechanism, if it has a role intraneuronally.

References

Allsop D, Haga S-I, Haga C, Ikeda S-I, Mann DMA, Ishii T 1989 Early senile plaques in Down's syndrome brains show a close relationship with cell bodies of neurones. Neuropathol Appl Neurobiol 15:531–542

Bancher C, Grundke-Iqbal I, Iqbal K, Kim KS, Wisniewski HM 1989 Immunoreactivity of neuronal lipofuscin with monoclonal antibodies to the amyloid beta-protein. Neurobiol Aging 10:125–132

Bauxbaum J, Gandy E, Cicchetti P et al 1990 Processing of Alzheimer β/A_4 amyloid precursor protein: modulation by agents that regulate protein phosphorylation. Proc Natl Acad Sci USA 87:6003–6006

Canavese C, Pacitti A, Portigiatti M et al 1990 Aluminium and dialysis arthropathy. Nephron 56:455–456 (letter)

Copeland JRM, Gurland BJ, Dewey ME et al 1987 Is there more dementia, depression and neurosis in New York? A comparative community study of elderly in New York and London, using the computer diagnosis, AGECAT. Br J Psychiatry 151: 466–473

Crystal H, Dickson D, Fuld P et al 1988 Clinico-pathologic studies in dementia: non-demented subjects with pathologically confirmed Alzheimer's disease. Neurology 38:1682–1687

Delaère P, Duyckaerts C, Masters C, Beyreuther K, Plette F, Hauw J-J 1990 Large amounts of neocortical β/A4 deposits without neuritic plaques nor tangles in a psychometrically assessed, non-demented person. Neurosci Lett 166:87–93

Duyckaerts C, Delaère P, Hauw J-J et al 1990 Rating of the lesions in senile dementia of the Alzheimer type: concordance between laboratories. A European multicenter study under the auspices of EURAGE. J Neurol Sci 97:295–323

Evans DA, Funkenstein H, Albert MS et al 1989 Prevalence of Alzheimer's disease in a community population of older persons. Higher than previously reported. JAMA (J Am Med Assoc) 262:2551–2556

Heller DS, Klein MJ, Gordon RE, Good P, Perl D 1989 Intraosseous beta-2-microglobulin amyloidosis. Detection and measurement of osseous aluminum in a patient who was receiving long-term hemodialysis. J Bone Jt Surg 71:1083–1089

Levy E, Carman MD, Fernandez-Madrid IJ et al 1990 Mutation of the Alzheimer's disease amyloid gene in hereditary cerebral haemorrhage, Dutch type. Science (Wash DC) 248:1124–1126

Nee LE, Polinsky RJ, Eldridge R, Weingartner H, Smallberg S, Ebert M 1983 A family with histologically confirmed Alzheimer's disease. Arch Neurol 40:203–208

Netter P 1991 Aluminium and dialysis associated arthropathy. Nephron 59:699 (letter)

Netter P, Kessler M, Gaucher A, Bannwarth B 1990 Does aluminum have a pathogenic role in dialysis associated arthropathy? Ann Rheum Dis 49:573–575

Netter P, Fener P, Steinmetz et al 1991 Amorphous aluminosilicates in synovial fluid in dialysis-associated arthropathy? Lancet 337:554–555

O'Connor DW, Pollitt PA, Hyde JB et al 1989 The prevalence of dementia as measured by the Cambridge Mental Disorders of the Elderly examination. Acta Psychiatr Scand 79:190–198

Sethi D, Naunton Morgan TC, Brown EA et al 1990 Dialysis arthropathy: a clinical, biochemical, radiological and histological study of 36 patients. Q J Med 77:1061–1082

Strong MJ, Jakowec DM 1991 Differential expression of βA_4 precursor protein mRNA transcripts during hippocampal neuron development in vitro. Soc Neurosci Abstr 17(1):52

Strong MJ, Svedmyr A, Gajdusek DC, Garruto RM 1990 The temporal expression of amyloid precursor protein mRNA in vitro in dissociated hippocampal neuron cultures. Exp Neurol 109:171–179

Uchida Y, Tomonaga M 1989 Neurotrophic action of Alzheimer's disease brain extract is due to the loss of inhibitory factors for survival and neurites formation of cerebral cortical neurons. Brain Res 481:190–193

Uchida Y, Ihara Y, Tomonaga M 1988 Alzheimer's disease brain extract stimulates the survival of cerebral cortical neurons from neonatal rats. Biochem Biophys Res Commun 150:1263–1267

Uchida Y, Takio K, Titani K, Ihara Y, Tomonaga M 1991 The growth inhibitory factor that is deficient in the Alzheimer's disease brain is a 68 amino acid metallothionein-like protein. Neuron 7:337–347

Wischik CM, Harrington CR, Mukaetova-Ladinska EB, Novak M, Edwards PC, McArthur FK 1992 Molecular characterization and measurement of Alzheimer's disease pathology: implications for genetic and environmental aetiology. In: Aluminium in biology and medicine. Wiley, Chichester (Ciba Found Symp 169) p 268–302

Wisniewski HM, Wegiel J 1991 Spatial relationships between astrocytes and classical plaque components. Neurobiol Aging 12:593–600

Wisniewski HM, Vorbrodt AW, Wegiel J, Morys J, Lossinsky AS 1990 Ultrastructure of the cell forming amyloid fibers in Alzheimer disease and scrapie. Am J Med Genet (suppl) 7:287–297

Wisniewski HM, Barcikowska M, Kida E 1991 Phagocytosis of β/A4 amyloid fibrils of the neuritic neocortical plaques. Acta Neuropathol 81:588–590

Yamaguchi H, Nakazato Y, Yamazaki T, Shoji M, Kawarabayashi T, Hirai S 1991 Subpial β/A4 amyloid deposition occurs between astroglial processes in Alzheimer-type dementia. Neurosci Lett 223:217–220

Aluminium accumulation, β-amyloid deposition and neurofibrillary changes in the central nervous system

J. A. Edwardson, J. M. Candy, P. G. Ince, F. K. McArthur, C. M. Morris, A. E. Oakley, G. A. Taylor and E. Bjertness*

*MRC Neurochemical Pathology Unit, Newcastle General Hospital, Westgate Road, Newcastle upon Tyne, NE4 6BE, UK and *Department of Epidemiology, National Institute for Public Health, Oslo, Norway*

Abstract. Deposition of β-amyloid and the formation of neurofibrillary tangles (NFTs) are central to the aetiopathogenesis of Alzheimer's disease (AD). The possible effects of aluminium on these processes have been investigated in patients with renal failure who are exposed chronically to high blood levels of aluminium. Focal accumulation of aluminium was observed in neurons with high densities of transferrin receptors, indicating transferrin-mediated uptake, in regions such as cortex and hippocampus which are selectively vulnerable in AD. Increased staining for the β-amyloid precursor protein (APP) in cortical pyramidal neurons was evident in the majority of renal patients and immature senile plaques were present in 30% of cases, suggesting that aluminium may induce or accelerate β-amyloid deposition. The absence of neurofibrillary changes in this group of renal patients indicates that aluminium does not directly cause the formation of NFTs. The brain aluminium content was not raised in neuropathologically assessed cases of AD and we have been unable to confirm claims of defective transferrin binding in this disorder. If aluminium contributes to the development of sporadic AD, it must do so indirectly, perhaps via effects on the synthesis or metabolism of APP, or by contributing generally to the age-related attrition of neurons and thus reducing the threshold for deficits produced by more specific disease-related processes.

1992 Aluminium in biology and medicine. Wiley, Chichester (Ciba Foundation Symposium 169) p 165–185

Evidence for the neurotoxicity of aluminium in man has emerged from clinical situations where the gastrointestinal barrier to absorption is bypassed, or renal excretion impaired, or in rare cases of massive occupational exposure. In contrast, the suggestion that aluminium may have a role in the aetiopathogenesis of Alzheimer's disease (AD) is based upon circumstantial evidence, much of which is conflicting, and the 'aluminium hypothesis' remains highly controversial.

Alzheimer's disease is the commonest cause of senile dementia and is characterized neuropathologically by the development of senile plaques, neurofibrillary tangles and loss of neurons in selectively vulnerable regions of the brain such as the cortex, hippocampus and amygdala. Mature senile plaques consist of extracellular deposits of fibrillary aggregates of β-amyloid protein surrounded by dystrophic neurites and glial cell processes (Muller-Hill & Beyreuther 1989), while neurofibrillary tangles consist of intracellular accumulations of paired helical filaments in which the microtubule-associated protein, tau, is a major constituent (Lee et al 1991). Several familial forms of AD with early onset show linkage to chromosome 21, and three genetic mutations have been identified which involve the same codon in the gene for the β-amyloid precursor protein (APP) which is located on this chromosome (e.g. Chartier-Harlin et al 1991). These mutations give rise to different amino acids (Val→Ile/Phe/Gly) at a point just distal to the C-terminal sequence of the amyloidogenic β-amyloid protein, located in the transmembrane-spanning domain of APP. The data indicate that altered synthesis or disordered metabolism of APP is probably central to the pathogenetic processes which occur in AD, and may lead eventually to the full spectrum of neuropathological changes, including the formation of neurofibrillary tangles (Hardy & Allsop 1991). Studies on late-onset AD have failed to show linkage to chromosome 21 and, notwithstanding the problems inherent in investigating the genetic bases of disorders which present in old age, the consensus of opinion is that environmental factors may be important in this disorder (St George-Hyslop et al 1990).

Among the key issues which arise in relation to the possible role of aluminium in the neuropathological changes that occur in AD are: (i) the mechanisms of Al uptake and transport by the CNS and whether these are abnormal in AD, (ii) whether the brain content and/or distribution of Al are altered in AD, and (iii) whether the intraneuronal accumulation of Al directly leads to β-amyloid deposition or the formation of neurofibrillary tangles. These issues form the substance of this review.

Mechanisms of aluminium uptake and transport by the CNS

Transferrin-mediated transport is the principal mechanism by which Al enters the brain from the circulation. Serum transferrin is about 30% saturated with iron and apotransferrin binds aluminium with high affinity (Martin 1986). Studies using gallium-67 as a marker for aluminium have shown unidirectional uptake in rat brain and a regional distribution of ^{67}Ga which parallels the distribution of transferrin receptors (Pullen et al 1990). A postmortem study, using dynamic secondary ion mass spectrometry (SIMS) in an imaging mode, in patients with chronic renal failure exposed to elevated blood levels of Al as a consequence of impaired excretion and treatment with Al-containing phosphate binders, has shown focal, intracellular accumulation in regions such as cortex, hippocampus and amygdala (Fig. 1) which are rich in transferrin receptors

(Morris et al 1989). Transferrin receptors in these regions are located mainly on the surface of pyramidal and other large diameter neurons (Fig. 2) which have a high requirement for iron for the synthesis of respiratory chain enzymes (Morris et al 1990). Thus, aluminium appears to accumulate within brain regions and neuronal populations that are among the most vulnerable in both AD and dialysis encephalopathy, largely as a consequence of transferrin-mediated uptake. A recent study by Cannata et al (1991) on rats dosed intraperitoneally with $Al(OH)_3$ has shown that the brain content of Al is significantly increased in iron-depleted animals and significantly decreased in iron-overloaded animals, indicating that iron-deficient individuals may be at increased risk from the accumulation of Al by the nervous system.

Aluminium accumulation by the brain in chronic renal failure reflects exposure to aluminium-containing phosphate binders (Candy et al 1992). Mean serum levels of Al measured during life and the brain content measured post mortem correlated with both the duration and the total amount of $Al(OH)_3$ which had been taken. No correlation was observed between these indices and the bone Al content, indicating that longitudinal monitoring of serum Al may provide a more reliable index of brain exposure than bone biopsy in dialysis patients. Furthermore, imaging SIMS revealed significant focal accumulation of Al in the brains of some dialysis patients where the bulk content was within the normal range. Thus, there may be an increase in the Al content of pyramidal and other neurons with high densities of transferrin receptors without any detectable change in total brain content—a finding which may be significant in relation to the changes that occur in AD.

Transferrin-mediated transport of aluminium in Alzheimer's disease

While the brain content of transferrin and transferrin receptors does not appear to be altered in Alzheimer's disease (Morris et al 1987), it has recently been suggested by Farrar et al (1990) that the binding of ^{67}Ga, and by implication the binding of Al, to plasma transferrin is reduced in AD and Down's syndrome. It was suggested that defective binding of Al to transferrin in these disorders may result in the formation of low molecular weight complexes, such as aluminium citrate, which could cross the blood–brain barrier more readily than transferrin-bound aluminium. We have investigated this claim using both size-exclusion chromatography and rapid ultrafiltration to separate protein-bound from free ^{67}Ga. No difference between AD and control subjects was observed using either procedure. Ultrafiltration revealed that ^{67}Ga was $> 98\%$ protein bound in serum. With size-exclusion chromatography the serum-bound ^{67}Ga in Alzheimer's disease subjects (mean 8.3%, range 5.8–14.0, $n = 12$) was not significantly different from that observed in young (mean 8.5%, range 4.6–14.2, $n = 13$) or old (mean 10.7%, range 6.0–18.5, $n = 11$) control groups. The apparent low binding of ^{67}Ga to serum with this latter method is clearly an

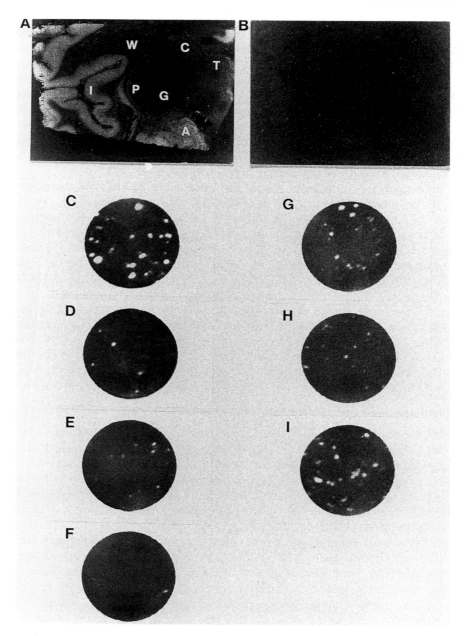

FIG. 1.   *(Caption opposite)*

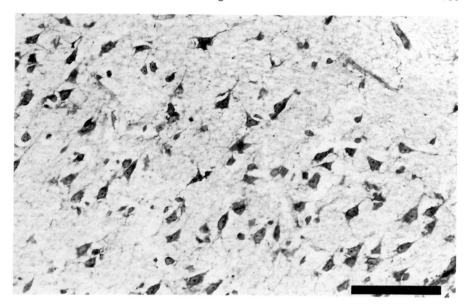

FIG. 2.   Transferrin receptor immunoreactivity associated predominantly with pyramidal neurons in the entorhinal cortex from a normal patient. The calibration bar represents 100 µm.

artifact of the separation procedure and is influenced by several factors, including the presence of bicarbonate, which is essential for metal-ion binding to transferrin, and the number of samples run on the column (Taylor et al 1991). These data suggest that there is no impairment in the transferrin-mediated transport of aluminium in AD.

Brain ferritin and aluminium

Ferritin has an essential role in intracellular iron homeostasis. It provides a storage pool of Fe for the biosynthesis of iron-containing proteins and detoxicant mechanism which limits free radical production through the sequestration of excess iron. Apoferritin consists of 24 subunits, designated either heavy (H) or low (L) molecular weight, which can assemble as a hollow sphere and store

FIG. 1.   *(previous page)* A and B show the total and non-specific [^{125}I]-Fe-transferrin binding in serial sections of the basal forebrain of a renal dialysis patient. I, insular cortex; G, globus pallidus; P, putamen; C, caudate nucleus; T, ventral thalamus; A, amygdala; W, white matter. Below are shown SIMS ion maps of the distribution of aluminium. C. Insular cortex. D. Globus pallidus. E. Putamen. F. White matter. G. Caudate nucleus. H. Ventral thalamus. I. Amygdala.

up to 4500 Fe atoms/molecule as crystalline ferrihydrite (Ford et al 1984, Theil 1987). We have found that ferritin isolated from cerebral cortex has a high H subunit content (65%) and relatively low iron content (1500 Fe atoms/molecule), similar to that of forms in the heart, for example, which are associated with active iron metabolism rather than long-term storage functions (D. J. Dedman et al, unpublished). Fleming & Joshi (1987) reported an increased content of both Fe and Al in ferritin in Alzheimer's disease. However, we have found that ferritin isolated from the cerebral cortex in normal subjects, Alzheimer's disease patients and renal dialysis patients has a similar Al content, of fewer than nine atoms per molecule, and these data suggest that neurons are unable to sequester Al within ferritin and that intracellular Al may be more toxic than Fe for this reason.

Brain content of aluminium in Alzheimer's disease

There has been a longstanding controversy over bulk analytical measurements of aluminium in the brain in AD, ever since the first reports by Crapper et al (1973, 1976), who found, using graphite furnace atomic absorption spectrometry (GFAAS), evidence for elevated concentrations from a number of brain regions in this disorder. In some cases the brain content of aluminium was reported to be elevated 2–3-fold, to levels which are encephalopathic in susceptible species. Subsequent studies by McDermott et al (1979) and Traub et al (1981) using GFAAS, and by Markesbery et al (1981) using neutron activation analysis, failed to find a significant elevation in the bulk aluminium content of the brain in AD. Potential flaws in the earlier reports (Crapper et al 1973, 1976) which may have led to this discrepancy were the lack of adequately age-matched control patients, the disparity between the numbers of samples measured in the control and AD groups, and lack of standardization of the brain regions sampled (see Table 1). A further deficiency present in all the studies to date is the lack of adequate neuropathological data; this is important, because

TABLE 1   Aluminium concentrations in frontal cortex in normal and Alzheimer's disease patients

Study	Aluminium concentration (μg/g dry weight) $\pm$ SD	
	Normal patients	Alzheimer patients
Crapper et al (1976)[a]	1.6 ± 0.6 (A, 47 yr; B, 6; C, 48)	3.5 ± 0.7 (A, 70 yr; B, 8; C, 175)
McDermott et al (1979)	2.4 ± 0.4 (A, 73 yr; B, 9; C, 56)	2.6 ± 0.3 (A, 81 yr; B, 10; C, 58)
Oslo study	2.6 ± 0.7 (A, 82 yr; B, 8; C, 28)	2.2 ± 0.5 (A, 81 yr; B, 8; C, 28)

[a]Data calculated from results.
A, mean age; B, number of cases; C, number of determinations.

almost two-thirds of normal elderly patients show a varying degree of AD-type neuropathological changes (Tomlinson et al 1968).

We have examined the brains from 92 demented and non-demented elderly individuals collected in Oslo. The left hemisphere was fixed for neuro-pathological investigation while the right hemisphere was snap-frozen for aluminium analysis using GFAAS and imaging SIMS. Bone and liver samples were also analysed for aluminium. The aluminium content in the frontal cortex for the entire group of patients was $2.4 \pm 0.6 \, \mu g/g$ dry weight. Within this group there were eight non-demented patients (mean age 81.6 yr, range 70–92 yr) who did not have senile plaques or neurofibrillary tangles in the frontal cortex, and eight patients with clinical AD (mean age 81.5 yr, range 71–91 yr) who had the maximum possible rating for senile plaques and neurofibrillary tangles, using the criteria formulated by the consortium to establish a registry for Alzheimer's disease (Mirra et al 1991). Table 1 shows that there was no significant difference in the aluminium concentrations between the two groups and that the values obtained are in agreement with those reported previously by McDermott et al (1979). These data indicate that the presence of extensive AD-type neuropathological features is not associated with an increase in the bulk cortical content of aluminium. However, as studies on renal dialysis patients using imaging SIMS have shown (Candy et al 1992), the absence of any changes in bulk content does not exclude the possibility that alterations in the focal accumulation of aluminium may occur.

Association of aluminium with plaques and tangles

It has been claimed that the intracellular content of aluminium is increased in neurofibrillary tangle-bearing neurons (Perl & Brody 1980), and it was proposed that aluminium plays an active role in the pathogenesis of this cytoskeletal lesion (see this volume: Perl & Good 1992). However, recent studies on chronic renal dialysis patients (Candy et al 1992) have failed to detect neurofibrillary tangles in the cerebral cortex when silver staining and immunocytochemical procedures were used that consistently revealed tangles in control sections from AD patients. The renal disease patients had received dialysis for periods ranging between eight months and 16 years and most had been uraemic for several years before dialysis commenced. Imaging SIMS revealed numerous, focal accumulations of aluminium in frontal and temporal cortex of 14 out of 15 dialysis patients, and the laminar distribution of these foci indicated a predominant association with pyramidal neurons. Measurement of the intracellular aluminium content of these foci using imaging SIMS showed concentrations up to 300 p.p.m., higher than those reported to occur in neurofibrillary tangles in Alzheimer's disease (this volume: Perl & Good 1992). Given the duration and extent of the accumulation of aluminium by cortical neurons in renal dialysis patients, these results suggest that aluminium does not play a direct role in neurofibrillary degeneration. It is

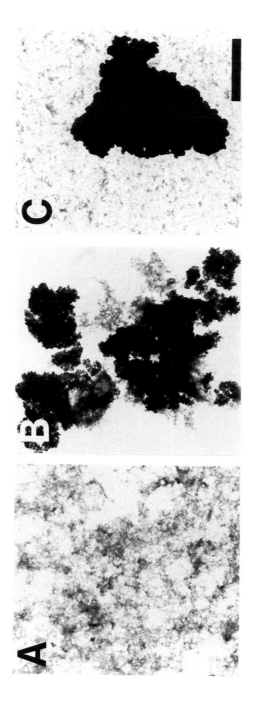

FIG. 3. Fibrillary aggregates of synthetic β-amyloid 1-42 in (A) distilled water and (B) and (C) after the addition of amorphous aluminosilicate and magnesium trisilicate to (A), respectively. Note the random appearance of the β-amyloid fibrils in (A) and (C), while in (B) the background fibrils have disappeared and β-amyloid fibrils are associated in an aggregated form with the aluminosilicate. The calibration bar represents 2 μm.

possible that the presence of intracellular tangles interferes with energy metabolism and that an increased need for iron in response to this burden may lead to secondary uptake of aluminium via the transferrin receptor system.

There have been several reports of aluminium associated with senile plaques (see Candy et al 1988 for review) and we have reported that a focal accumulation of aluminium in the form of aluminosilicate is present in the central core region of mature senile plaques in AD, in Down's syndrome and in normally elderly subjects (Candy et al 1986, Edwardson et al 1986). The core of these plaques consists of a compact, roughly spherical accumulation of bundles of radiating β-amyloid fibrils surrounded by reactive glial cells. The aluminosilicate may play a role in the nucleation of the β-amyloid protein, leading to its pathological crystallization in the characteristic form of the plaque core. A range of minerals have been shown to promote both heterogeneous and epitaxial nucleation of protein crystals (McPherson & Shlicta 1988). We have previously shown that an amorphous aluminosilicate can promote the formation of fibrillary aggregates of peptides which form fibrils in aqueous solutions, for example substance P (Edwardson et al 1991). In aqueous solution, synthetic β-amyloid peptide 1–42 forms linear fibrils which have a random appearance (Fig. 3A). We have recently found that amorphous aluminosilicates can cause marked aggregation of synthetic β-amyloid peptide 1–42 (Fig. 3B), while magnesium trisilicate does not promote aggregation (Fig. 3C). The loss of synapses and neurofibrillary degeneration are evident in the neuropil surrounding these compact deposits of β-amyloid and such changes are not seen in the more diffuse deposits of β-amyloid present in immature plaques (Masliah et al 1990). There is evidence from studies both *in vitro* and *in vivo* that β-amyloid is neurotoxic (Yankner

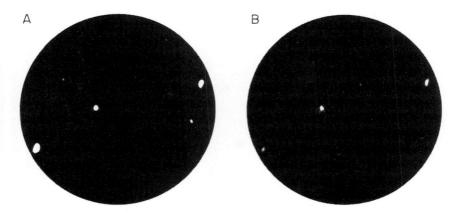

A   B

FIG. 4.   SIMS ion maps of (A) aluminium and (B) silicon in a 400 μm diameter field from an unfixed, unstained cryostat section from the deep layers of frontal cortex of a normal patient who had mature senile plaques with cores in an adjacent silver-stained section.

174

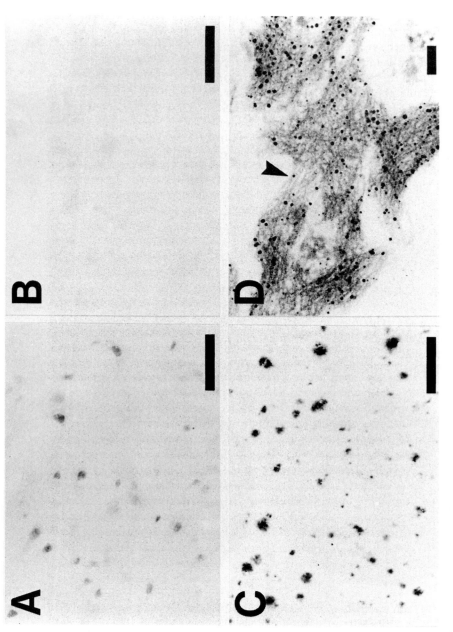

FIG. 5. A. β-Amyloid precursor protein immunoreactivity (GP3 antibody, G. Perry) associated with neuronal nuclei in the cerebral cortex of a renal dialysis patient. B. Absence of β-amyloid precursor immunoreactivity in the cerebral cortex of a control patient. C. Amorphous senile plaques showing β-amyloid immunoreactivity in the cerebral cortex of a renal dialysis patient. D. Electron micrograph of a silver-stained amorphous plaque in the cerebral cortex of a renal dialysis patient. Note the fibrillary appearance of

et al 1990, Kowall et al 1991) and it is possible that the association of aluminosilicate with β-amyloid in plaque cores may affect the neurotoxicity of such deposits, for example by altering the response of microglia.

The presence of focal deposits of aluminosilicate in plaque cores has proved a controversial issue. Some groups have reported finding aluminium and silicon associated with a lower proportion of plaque cores (Moretz et al 1990), while others have failed to find such an association (Jacobs et al 1989, Chafi et al 1991). However, there are many methodological problems associated with the microanalytical procedures used in these studies, which include the uncertainty in accurately localizing plaque cores in the absence of topographical or morphological imaging facilities, and the inadequacy of the analytical parameters used to determine the presence of aluminium and silicon in plaque cores *in situ*. It has been implied that the presence of focal aluminium in brain sections from AD patients is due to sample contamination. We have rigorously addressed this issue by using imaging SIMS to analyse unfixed, unstained sections prepared under the most stringent conditions in order to avoid contamination. Focal deposits of aluminium and silicon have been observed in sections known from histological examination of immediately adjacent sections to have a high density of core-containing plaques (Fig. 4), whereas such deposits were not observed in control sections where plaques were absent, eliminating contamination as an explanation for these findings. However, even the demonstration of a consistent association of aluminium with plaque cores and neurofibrillary tangles does not necessarily indicate a pathogenic role for aluminium in the formation of these neuropathological features.

β-Amyloid deposition in chronic renal failure

A prediction arising directly from the 'aluminium hypothesis' for AD is that the brains of patients exposed chronically to elevated blood levels of aluminium should show evidence of β-amyloid deposition and neurofibrillary tangle formation. In a post mortem study of 15 renal dialysis patients we have shown (Fig. 5A) immunostaining in cortical pyramidal neurons with an antibody directed against the N-terminal region of APP (Candy et al 1992). Comparable staining was absent in age-matched control cases (Fig. 5B). It is notable that the neuronal cell population showing evidence of increased synthesis or altered processing of APP in dialysis patients is the same population for which evidence of intracellular accumulation of aluminium has been obtained (Morris et al 1989, Candy et al 1992). Furthermore, one-third of the patients had β-amyloid-positive, amorphous senile plaques in the cortex (Fig. 5C and D), including three patients in the age range in which the prevalence of senile plaques is extremely low (Davies et al 1988). These plaques were identical to the immature plaques seen in Alzheimer's disease, in younger cases of Down's syndrome and, less frequently, in elderly, mentally normal subjects (Gibson 1983). There was no obvious relationship between the presence of β-amyloid plaques in the cerebral cortex in the dialysis patients and the level of aluminium measured by GFAAS, suggesting that if there is a causal relationship between aluminium and extracellular β-amyloid deposition, genetic or other constitutional factors must partly determine this response.

Discussion

Aluminium plays no known role as an essential trace element and is unquestionably neurotoxic in certain situations. This raises the question of whether the lifetime accumulation of aluminium within pyramidal and other neurons in the CNS could lead to, or accelerate, processes involved in the aetiopathogenesis of AD. Evidence suggests that the accumulation of aluminium by brain is largely, if not exclusively, determined by transferrin-mediated transport. Cell populations targeted through this mechanism are principally pyramidal and other large neurons which have high densities of transferrin receptors on their surfaces and are known to be selectively vulnerable in AD. Transferrin-mediated uptake of aluminium is not a complete explanation for selective neuronal vulnerability, because other populations such as spinal motor neurons also have high densities of transferrin receptors and are spared in AD. However, the intracellular actions of aluminium include effects on calcium homeostasis (this volume: Petersen et al 1992), and it is possible, for example, that differences in the major intracellular calcium-buffering systems may also determine vulnerability to aluminium.

We have been unable to find evidence that mechanisms in the transport of aluminium in blood or in brain are defective in Alzheimer's disease; nor do our studies support claims that the bulk concentration of aluminium in the brain is increased in this disorder or that there is increased uptake of aluminium into ferritin. The inability of neurons to sequester aluminium within ferritin is likely to contribute significantly to the neurotoxicity of aluminium.

The evidence of increased staining for APP in pyramidal neurons in renal dialysis patients, and the deposition of extracellular β-amyloid in a third of such patients, suggests that, if aluminium has a role in the aetiopathogenesis of AD, this is likely to involve increased or accelerated β-amyloidosis. While we have been unable to find evidence of mature senile plaques or neurofibrillary tangles in renal dialysis patients, it must be remembered that even in the presence of trisomy 21 or an autosomal dominant gene for AD it may require 50 years before such lesions develop. It is also possible that the association of aluminium, in the form of aluminosilicate, with 'dense' deposits of β-amyloid—which is still a disputed issue—could affect the neurotoxicity of β-amyloid and play some role in the neurodegenerative changes which occur in the vicinity of these deposits.

Little is known about the processes which underlie the age-related attrition of neuronal systems, especially those concerned with cognitive function and those which give rise to age-related memory impairment. It is possible that the lifetime accumulation of aluminium via transferrin-mediated uptake into cortical and hippocampal neurons, for example, may partly contribute to these age-related changes. Such a process would tend to reduce the adaptive capacity of cognitive systems to compensate for other disease-related changes and would thereby lower the threshold for the appearance of clinical symptoms. Thus, while not necessarily having a direct aetiopathogenic role in AD, aluminium may contribute significantly to age-related cognitive dysfunction and may influence the prevalence of AD.

Acknowledgements

The authors are indebted to the renal physicians and pathologists at the Royal Victoria Infirmary, Newcastle General Hospital, and Freeman Hospital, Newcastle upon Tyne, Sunderland General Hospital and Ashington Hospital for their help with these studies. Gifts of antibodies from K. Beyreuther, P. Davies, M. Landon and G. Perry are gratefully acknowledged. Mrs D. Hinds and Mrs M. Middlemist provided expert secretarial assistance.

References

Candy JM, Oakley AE, Klinowski J et al 1986 Aluminosilicates and senile plaque formation in Alzheimer's disease. Lancet 1:354–357

Candy JM, Oakley AE, Gauvreau D et al 1988 Association of aluminium and silicon with neuropathological changes in the ageing brain. Interdiscipl Topics Gerontol 25:140–155

Candy JM, McArthur FK, Oakley AE et al 1992 Aluminium accumulation in relation to senile plaque and neurofibrillary tangle formation in the brains of patients with renal failure. J Neurol Sci 107:210–218

Cannata JB, Fernandez Soto I, Fernandez-Menendez MJ et al 1991 Role of iron metabolism in absorption and cellular uptake of aluminum. Kidney Int 39:799–803

Chafi AH, Hauw J-J, Rancurel G et al 1991 Absence of aluminium in Alzheimer's disease brain tissue: electron microprobe and ion microprobe studies. Neurosci Lett 23:61–64

Chartier-Harlin MC, Crawford F, Houlden H et al 1991 Early-onset Alzheimer's disease caused by mutations at codon 717 of the β-amyloid precursor protein gene. Nature (Lond) 353:844–846

Crapper DR, Krishnan SS, Dalton AJ 1973 Brain aluminum distribution in Alzheimer's disease and experimental neurofibrillary degeneration. Science (Wash DC) 180:511–513

Crapper DS, Krishnan SS, Quittkat S 1976 Aluminum, neurofibrillary degeneration and Alzheimer's disease. Brain 99:67–80

Davies L, Wolska B, Hibich C et al 1988 A4 amyloid protein deposition and the diagnosis of Alzheimer's disease. Neurology 38:1688–1693

Edwardson JA, Klinowski J, Oakley AE, Perry RH, Candy JM 1986 Aluminosilicates and the ageing brain: implications for the pathogenesis of Alzheimer's disease. In: Silicon biochemistry. Wiley, Chichester (Ciba Found Symp 121) p 160–179

Edwardson JA, Ferrier IN, McArthur FK et al 1991 Alzheimer's disease and the aluminium hypothesis. In: Nicolini M, Zatta P, Corain B (eds) Aluminum in chemistry biology and medicine. Cortina international, Verona & Raven Press, New York, p 85–96

Farrar G, Altmann P, Welch S et al 1990 Defective gallium-transferrin binding in Alzheimer disease and Down syndrome: possible mechanism for the accumulation of aluminium in brain. Lancet 335:747–750

Fleming J, Joshi JG 1987 Ferritin: isolation of aluminum–ferritin complex from brain. Proc Natl Acad Sci USA 84:7866–7870

Ford GC, Harrison PH, Rice DW et al 1984 Ferritin: design and formation of an iron-storage molecule. Philos Trans R Soc Lond B Biol Sci 304:551–565

Gibson PH 1983 Form and distribution of senile plaques seen in silver-impregnated sections in the brains of intellectually normal elderly people and people with Alzheimer type dementia. Neuropathol Appl Neurobiol 9:379–389

Hardy JA, Allsop D 1991 Amyloid deposition as the central event in the aetiology of Alzheimer's disease. Trends Pharmacol Sci 12:383–388

Jacobs RW, Duong T, Jones RE, Trapp GA, Scheibel AB 1989 A re-examination of Al in AD: analysis by energy dispersive X-ray microprobe and flameless atomic absorption spectrophotometry. Can J Neurol Sci 16:498–503

Kowall NW, Beal MF, Busciglio J, Duffy LK, Yankner BA 1991 An *in vivo* model for the neurodegenerative effects of β-amyloid and protection by substance P. Proc Natl Acad Sci USA 88:7247–7251

Lee VMY, Blain BJ, Otvos L, Trojanoswki JQ 1991 A68: a major subunit of paired helical filament and derivatized forms of normal tau. Science (Wash DC) 251:675–678

Markesbery WR, Ehmann WD, Hossain TIM, Alauddin M, Goodin DT 1981 Instrumental neutron activation analysis of brain aluminum in Alzheimer's disease and aging. Ann Neurol 10:511–516

Martin RB 1986 The chemistry of aluminium as related to biology and medicine. Clin Chem 32:1797–1806

Masliah E, Terry RD, Mallory M, Alford M, Hansen LA 1990 Diffuse plaques do not accentuate synapse loss in Alzheimer's disease. Am J Pathol 137:1293–1297

McDermott JR, Smith AI, Iqbal K, Wisniewski HM 1979 Brain aluminum in aging and Alzheimer's disease. Neurology 29:809–814

McPherson A, Shlicta P 1988 Heterogeneous and epitaxial nucleation of protein crystals on mineral surfaces. Science (Wash DC) 239:385–387

Mirra SS, Heyman A, McKeel D et al 1991 The consortium to establish a registry for Alzheimer's disease (CERAD). Part II. Standardization of the neuropathologic assessment of Alzheimer's disease. Neurology 41:479–486

Moretz RC, Iqbal K, Wisniewski HM 1990 Microanalysis of Alzheimer disease NFT and plaques. Environ Geochem Health 12:15–16

Morris CM, Court JA, Moshtaghie AA et al 1987 Transferrin and transferrin receptors in the normal brain and in Alzheimer's disease. Biochem Soc Trans 15:891–892

Morris CM, Candy JM, Oakley AE et al 1989 Comparison of regional distribution of transferrin receptors and aluminium in the forebrain of chronic renal dialysis patients. J Neurol Sci 94:295–306

Morris CM, Candy JM, Bloxham CA, Edwardson JA 1990 Brain transferrin receptors and the distribution of cytochrome oxidase. Biochem Soc Trans 18:647–648

Muller-Hill B, Beyreuther K 1989 Molecular biology of Alzheimer's disease. Annu Rev Biochem 58:287–307

Perl DP, Brody AR 1980 Alzheimer's disease: x-ray spectrometric evidence of aluminum accumulation in neurofibrillary tangle-bearing neurons. Science (Wash DC) 208:297–299

Perl DP, Good PF 1992 Aluminium and the neurofibrillary tangle: results of tissue microprobe studies. In: Aluminium in biology and medicine. Wiley, Chichester (Ciba Found Symp 169) p 217–236

Petersen OH, Wakui M, Petersen CCH 1992 Intracellular effects of aluminium on receptor-activated cytoplasmic Ca^{2+} signals in pancreatic acinar cells. In: Aluminium in biology and medicine. Wiley, Chichester (Ciba Found Symp 169) p 237–253

Pullen RGL, Candy JM, Morris CM, Taylor GA, Keith AB, Edwardson JA 1990 [67]Gallium as a potential marker for aluminium transport in rat brain: implications for Alzheimer's disease. J Neurochem 55:251–259

St George-Hyslop PH, Haines JL, Farrer LA et al 1990 Genetic linkage studies suggest that Alzheimer's disease is not a single homogeneous disorder. Nature (Lond) 347:194–197

Taylor GA, Morris CM, Fairbairn AF, Candy JM, Edwardson JA 1991 Transferrin-gallium binding in Alzheimer's disease. Lancet 338:1394–1395

Theil EC 1987 Ferritin: structure, gene regulation, and cellular function in animals, plants and microorganisms. Annu Rev Biochem 56:289–315

Tomlinson BE, Blessed G, Roth M 1968 Observations on the brains of non-demented old people. J Neurol Sci 7:331–356

Traub RD, Rains TC, Garruto RM, Gajdusek DC, Gibbs CJ 1981 Brain destruction alone does not elevate brain aluminum. Neurology 31:986–990

Yankner BA, Duffy LK, Kirschner DA 1990 Neurotrophic and neurotoxic effects of amyloid β-protein: reversal by tachykinin neuropeptides. Science (Wash DC) 250:279–282

DISCUSSION

Garruto: Two studies (Brun & Dictor 1981, Scholtz et al 1987) have reported neurofibrillary degeneration in dialysis encephalopathy. What is your view of these studies?

Edwardson: There are also three studies which find no such degeneration and two that report spongiform changes, and one reporting the presence of both plaques and tangles. There are two problems. One is the age of the patients who were exposed to Al. Many of the studies involve much older patients who would have general age-related pathology in any case.

Secondly, we are not looking at encephalopathic patients. These are patients who have been on chronic renal dialysis and whose serum Al levels are higher than normal, but they are monitored regularly and steps are taken to control their intake of aluminium. Their exposure is much more chronic and less severe than that which occurred in the earlier encephalopathic patients.

Garruto: I raise this issue because of your argument that Al cannot be causally related to neurofibrillary tangle formation in humans. If there are two studies that *do* demonstrate neurofibrillary degeneration and three that don't, you should at least consider the possibility that in some dialysis encephalopathy patients aluminium may be causal in neurofibrillary tangle formation. The patients in both of the studies I cited were in their thirties, forties and fifties.

Candy: In the report of Brun & Dictor (1981) they don't discriminate whether the tangles in the encephalopathic and non-encephalopathic dialysis cases were in the hippocampus or in the cortex, and I suspect that they may have been present in the hippocampus, where they may simply reflect age-related changes.

Garruto: I gather that in one of your control cases the brain was loaded with Al. Can you tell us more about this patient?

Edwardson: This was a 74-year-old non-demented individual with a CERAD rating of zero, so there is no evidence neuropathologically of Alzheimer's disease. This patient received intravenous fluids a few days before death, which may have contributed to the high density of low intensity focal Al accumulations.

Garruto: What was the cause of death?

Candy: Pneumonia.

Edwardson: The controls were chosen from a sequential autopsy series on the basis of being non-demented and having little or no evidence of Alzheimer-type neuropathological changes. This particular case illustrates the problem of selecting appropriate controls for studies in relation to the brain content of aluminium.

Perl: Dr Edwardson, I gather that your SIMS images are obtained on unfixed, unstained sections. My problem is that at the resolution of these images, I can't be certain that I am looking at neurons in this pattern, rather than other cell types within the cortex. I wonder whether you have scanned hippocampus, where you could better see the pyramidal neurons in the absence of the other surrounding cells, and whether you get the same picture? Also, you show many images of Al; I wondered what you have in terms of Si distribution?

Candy: In dialysis patients we have looked in the hippocampus, using SIMS, and have observed an association of Al with pyramidal neurons. We have failed to find evidence, in the dialysis patients, of focal Si accumulation. We do find some evidence in elderly normal patients and patients with Alzheimer's disease of co-localized deposits of Al and Si, which are consistent with the presence of senile plaque cores.

Edwardson: But we see very few plaque cores in these dialysis patients.

Wisniewski: The finding that in patients after 15–16 years of dialysis, with high levels of Al in the brain, there is no accelerated rate of formation of neurofibrillary tangles, is extremely important. This is because, from the findings in parkinsonism-dementia complex of Guam, a high level of Al would, if anything, have something to do with the pathogenesis of neurofibrillary tangles. If this is not so, then this is an important finding, if we bear in mind that the neurofibrillary pathology is the lesion which tilts the scales and drives us towards dementia in Alzheimer's disease. This observation also undermines what Dr Strong brought up—the general mechanism of the role of Al in abnormal protein phosphorylation (p 101, 162).

Zatta: We have preliminary data concerning variation in the sialylation of transferrin in AD; however, I don't have enough evidence yet as to whether aluminium plays any role in the glycosylation of this protein. I do know that aluminium speciation has some effect on the inhibition of β-1,4-galactosyltransferase in human serum. We have recently shown how aluminium can modify the glycosylation of the brain microcapillaries in experimental animals (Zatta et al 1991). What I am trying to say is that aluminium can reach the brain in different ways in different pathological conditions, such as in dialysis dementia, Alzheimer's disease, or other diseases. What I strongly believe is that the biology of aluminium is just at the beginning.

Edwardson: I accept that completely, but although in theory there are many mechanisms by which Al can get into cells, after the study of very large numbers of brains using imaging SIMS, in cases of Al overload and in control cases such as the one we were discussing, and in normal subjects, the main distribution of Al is in these transferrin-rich areas of the brain and transferrin-rich populations of cells. I accept your hypothesis, but you have to produce evidence that Al is getting into the human brain by these other routes.

Zatta: In the dialysis patients, do you see differences in transferrin abundance, or other differences?

Edwardson: Absolutely, and the aluminium is going into transferrin receptor-rich cells.

Zatta: So what is the difference between transferrin in AD patients and transferrin in dialysis patients, with respect to Al internalization?

Edwardson: I don't think there is any difference; the more Al there is in the blood, the more is transported into brain via transferrin-mediated uptake.

Zatta: I think we need more data on metal uptake in the brain in pathological conditions.

Harrison: I should like to comment further on the uptake of aluminium by brain ferritin. There are low amounts of Al in brain ferritin (6–9 Al atoms per ferritin molecule, as compared to 1500 iron atoms), but this does not in itself show that ferritin is unable to take up Al. What it may show is that the pool of Fe that is available to ferritin may have a very low Al content (and lower than in the tissue as a whole, in which the Al:Fe ratio is much greater than that of the isolated ferritin). We have carried out experiments (on horse spleen ferritin rather than human brain ferritin, because we didn't have enough of the latter), to see whether we could get ferritin to take up Al, and, if so, under what conditions (D. J. Dedman, A. Treffry & P. M. Harrison, unpublished work). We found that on simply incubating ferritin with Al citrate in equilibrium dialysis experiments, very little Al was bound. Ferritin iron-cores are formed by adding Fe(II) to the protein, and the iron then becomes oxidized and deposited as ferrihydrite. If we added Al along with Fe(II), we found the molecules took up Al. When we added 3000 Fe atoms per apoferritin molecule along with 400 Al, virtually all the iron was taken up under the conditions of the experiments (pH 7.0) and a third of the Al was taken up. So ferritin incorporated 120 atoms under these conditions, along with 3000 Fe. In other words, in certain circumstances, ferritin can take up Al. If Al is not present in isolated ferritin, it suggests not an inability to take up Al, but the unavailability of Al to ferritin.

Edwardson: The ratio of Al to Fe atoms in ferritin molecules was unchanged in the renal dialysis cases, yet we know that their brains are loaded with Al. This strongly support your view that, because of differences in the intracellular handling of Fe and Al, the latter is not available for sequestration by ferritin.

Williams: Does Al ever come out of ferritin?

Harrison: I am sure it does, but what do you mean by 'ever'?

Williams: I mean in a reasonable length of time; once it's in there, is it gone for good? That's not the case for iron.

Harrison: I think Al which is associated with the iron-core must turn over with the iron, but I haven't done these experiments.

Day: In renal dialysis patients treated with desferrioxamine, one clears as much of the Al through faecal excretion as through release into the plasma and subsequent dialysis. This is very similar to the pattern of release of iron from patients in chronic iron overload, treated with desferrioxamine; most of the iron is removed from the liver by direct secretion into the bile, rather than by release into the blood plasma.

The comparison of Al and Fe behaviour is very interesting. When we treat renal dialysis patients who have both iron and Al overload, we observe similar kinetics for the release of Fe and Al through the faecal route. The simplest explanation is that Al is stored along with the iron in the ferritin of the liver, and moreover that the action of desferrioxamine is to release both elements by essentially the same mechanism. I think this observation is particularly interesting, because current wisdom has it that iron needs to be reduced in the first step of its release from ferritin. Since Al cannot be reduced, the implication of our observations is that iron may not require a reduction step either.

Edwardson: Analysis using imaging SIMS shows that Al is distributed widely within the cytoplasm of cells, where we presume it is attached largely between vicinal phosphate groups. In post mortem tissue there is evidence of nuclear accumulation of Al, but this may result from redistribution following cell death and acidification of the intracellular pH.

Blair: Over a period of years we have established that in Alzheimer's disease there is a defect in biochemical processes that are crucial to human cognition. This defect doesn't arise simply from the death of cells, but from the inhibition of a particular metabolic system within them. This leads me to the position that in searching for an agent that brings about these changes, we find a positive effect of Al. I believe that if Al is absent from a particular clinical situation, then you do not have Alzheimer's disease. This view does not commit me to the belief that Al *alone* is the instigator of Alzheimer's disease.

The feature that we have been looking at is the bioavailability of Al, and Al speciation within the plasma. These are crucial processes which must be examined in all situations of this kind. We (Dr Farrar and others) find that the speciation of Al (using gallium as a surrogate) is different in Alzheimer's disease and controls. In many cases of Alzheimer's disease (bearing in mind the difficulty of diagnosis), less gallium was bonded to a high molecular weight protein species (most likely transferrin) than in the control group. We found this also for Down's syndrome. Professor Edwardson has claimed that we are wrong, but the Institute of Psychiatry have repeated our work with similar results (Brammer et al 1990).

We subsequently re-examined this theme using modified systems in which physiological levels of bicarbonate are present. In some 68 control subjects the binding of Ga to high molecular weight transferrin is just short of 90%, a level which agrees with the findings of others. In Alzheimer's disease patients we find a significant depression in the mean binding, to 64.2%, although with a marked spread in the results (20–100%).

If our hypothesis is right, that Al speciation is the key to the accumulation process, we might expect to find similar effects in patients with other neurological diseases. The obvious one to investigate is Down's syndrome. In 29 patients, the level of gallium binding was down to 56%, a significant drop from the controls, not significantly different from senile dementia, with again some spread, but the younger patients have the higher values and the older ones in this group (which includes Down's syndrome subjects up to 40 and 50 years of age) have lower values for binding to transferrin; so that pattern fits very well.

Parkinson's disease is also a serious problem, with 125 000 sufferers in the UK; it too involves metal accumulation within the brain, namely iron, but this does not follow the pattern of transferrin receptors and it may lead, in certain circumstances, to dementia of the Alzheimer type. In Parkinson's patients without any medication we found a striking depression in gallium binding, down to 20%. Unlike the broad dispersion in the other groups, they show a very narrow dispersion indeed.

In neonates, only a minute proportion of the added Al was bound to high molecular weight species. This lack of binding is due to an almost 100% saturation of transferrin by iron in the plasma of these subjects, established by independent measurements.

So we have a series of distinct groups. We can begin to look at them another way. If you consider the control observations, you find an interesting pattern—a bimodal distribution. So the control subjects include two groups of people, a dominant group with binding of Al between 70% and 100%, and a small group where binding doesn't exceed 40%. If you analyse these people further, the high-binding normal subjects show a significant curvilinear effect with a maximum binding at about 40 years of age, and a decline with age. The analysis is significant at the 1% level. Whereas the low-binding group, including the neonates, rises steadily from 10% to about 20% binding and stops at about 65 years. I wonder whether this means that people in the low-binding group have now 'fallen off' the rising correlation to become sick, pathologically, because they are within the Parkinson's region.

These are our current observations. We suggest that they can be put together to make an intriguing pattern of Alzheimer's disease and related conditions. The low transferrin binding group can be identified in life as people vulnerable to metal intoxication. If they are exposed to an environment with a high level of Al in the plasma, they show Alzheimer's disease. If they are not in a high Al environment, they appear with Parkinson's disease, and there is no further complication. These are the 'pure' systems.

Let me remind you that Al is a potent agent, attacking all manner of biological systems. All the enzyme systems in brain which use Mg^{2+} are potentially capable of being affected by Al. In Parkinson's disease, you have a different pattern, because here there is a classical hydroxyl radical attack on lipid membranes, leading to their destruction, so there is plenty of opportunity for neuronal degeneration. In the other group, with the high binding of gallium to transferrin, you can approach Alzheimer's disease or Parkinson's disease with age in a smooth fashion. In a high Al environment, you pass into Alzheimer's disease first, because Al is squeezed out as transferrin levels in plasma fall. Then, as they fall, you pass into Parkinson's disease. The one detailed study relevant to this shows that patients with Alzheimer's disease have a greater risk of developing Parkinson's disease (if followed for a long enough time) than control populations—something like 50% of Parkinson's patients will develop Alzheimer's disease (van Duijn & Hofman 1991). In a situation in which they are no longer exposed to Al, they then pass into Parkinson's disease.

To conclude, when iron or aluminium is complexed to transferrin, there is a slow controlled process for the entry of iron or aluminium into the brain. Transferrin binding is a process which controls the uptake of iron, uptake depending on transferrin and the transferrin receptor. When these metals are displaced from transferrin they appear probably as the citrate. Both penetrate any membrane much faster than when bound to transferrin; it's no longer controlled and there is no longer any mechanism of turning transport on or off, as there is in this particular mechanism.

Day: Have you any evidence for a low molecular weight *protein* that binds Al, or was all the low molecular weight Al bound in essentially inorganic form?

Blair: So far as we can identify it, the low molecular weight species is probably Al citrate. It is possible that among the high M_r portion there are proteins that bind Al, but that's a very complex issue. So far as I understand it, Al binding by albumin is relatively weak, and what we observe in the high molecular weight species is, on the whole, transferrin.

Candy: In these studies you could also use ^{55}Fe to show that your methods are valid. I also think there is a problem of using gel-filtration chromatography for separation of gallium-loaded transferrin. The resin appears to affect gallium binding and retain gallium (Taylor et al 1991); you should therefore use another technique, such as ultrafiltration, for separation.

Farrar: We didn't have any problem in the recovery of ^{67}Ga when we did gel-filtration columns. We have done approximately 250 columns for speciation analysis in plasma or gut washings, and have not found problems with the recovery of radiolabelled [67]Ga.

Kerr: In your neonates the low Al binding was accounted for by the saturation of transferrin with Fe. What explanation have you for the low binding in the other group and the straight line rising with age?

Blair: We don't know. The neonatal process may reduce the level of transferrin in a plasma sample and increase the Fe a little. It's not unusual for genetic effects in the neonate to transfer into the adult phase, and produce pathology.

References

Brammer M, Richmond S, Burns A, Förstl H, Levy R 1990 Gallium–transferrin binding in Alzheimer's disease. Lancet 336:635

Brun A, Dictor M 1981 Senile plaques and tangles in dialysis dementia. Acta Pathol Microbiol Scand Sect A 89:193–198

Scholtz CL, Swash M, Gray A, Kogeorgos J, Marsh F 1987 Neurofibrillary neuronal degeneration in dialysis dementia: a feature of aluminum toxicity. Clin Neuropathol 6:93–97

Taylor GA, Morris CM, Fairbairn AF, Candy JM, Edwardson JA 1991 Transferrin-gallium binding in Alzheimer's disease. Lancet 338:1395–1396

van Duijn CM, Hofman A 1991 Relation between nicotine intake and Alzheimer's disease. Br Med J 302:1491–1494

Zatta PF, Nicolini M, Corain B 1991 Aluminum(III) toxicity and blood–brain barrier permeability. In: Nicolini M, Zatta PF, Corain B (eds) Aluminum in chemistry biology and medicine. Cortina international, Verona & Raven Press, New York, p 97–112

Aluminium(III) in experimental cell pathology

P. F. Zatta*, M. Nicolini** and B. Corain†

*Centro CNR per lo Studio della Biochimica e della Fisiologia delle Emocianine ed altre Metalloproteine and Dipartimento di Biologia, **Dipartimento di Scienze Farmaceutiche and †Dipartimento di Chimica Inorganica Metallorganica Analitica, Università di Padova, Padova, Italy

Abstract. Controversy over the relevance of aluminium to certain human encephalopathies has emphasized the importance of *in vivo* and *in vitro* models as tools for shedding light on the biological and molecular aspects of the aluminium toxicity. The search for an experimental model in animals or in cultured cells able to reproduce specific pathological human conditions may prove to be an unattainable aim; nevertheless, *in vivo* and *in vitro* models should be actively sought and the pathological changes induced in experimental animals should always be evaluated at the cellular level, just as for changes produced directly in cultured cells. These toxicological aspects are outlined with particular emphasis on the role played by the molecular form of aluminium (metal speciation) in determining the quality and intensity of the metal's biological effects.

1992 Aluminium in biology and medicine. Wiley, Chichester (Ciba Foundation Symposium 169) p 186–200

Aluminium is present in small amounts in mammalian tissues, yet it has no recognized physiological role. On the contrary, its neurotoxic effect on living organisms is becoming clear, aluminium being implicated as interfering with a variety of cellular metabolic processes in the nervous system and in other systems.

The problem of how aluminium causes brain encephalopathies both in experimental animals and in humans is far from being understood and the information is mostly based on a large body of phenomenological evidence. None of the proposed experimental models has so far satisfactorily explained the toxicity of the metal, especially at the molecular level. Although the molecular mechanisms by which aluminium exerts its neurotoxicity remain to be established, several pieces of evidence suggest that Al(III) can interfere with cellular metabolism in terms of biological stimulation, inhibition, or metal accumulation and compartmentation.

In vivo and *in vitro* models of Al(III) toxicity undoubtedly remain prominent tools for shedding light on the biological and, eventually, the molecular bases of metal toxicity. However, it should be emphasized that the induction in animals and/or in cultured cells of specific pathological patterns typical of humans might in principle be an unattainable goal. In spite of this, understanding the toxic action of Al(III) is a socially relevant aim, and more information and more basic research are needed.

Whereas for pathological conditions such as dialysis dementia, osteomalacia and microcytaemia in patients on long-term dialysis treatment the involvement of aluminium is now well established, the aetiological relevance of the metal's accumulation in selected brain areas is far from being proved for Alzheimer's disease (AD). However, it should be stressed that AD is likely to be a multistep event with multifactorial causation, and the accumulation of aluminium might be relevant to one facet of the unknown aetiopathogenesis of this most elusive disease.

The various aetiological hypotheses proposed for AD include a multiple trace element metabolic imbalance. In fact, alterations in trace element concentrations for aluminium, silicon, lead, chlorine, phosphorus, sodium, bromine, rubidium, and recently for mercury have been recognized in AD patients (Wenstrup et al 1990, Basun et al 1991). Among these, aluminium is receiving more attention than other trace elements on the grounds of its accumulation in features characteristic of AD, such as neurofibrillary tangles (NFT) (Perl & Brody 1980) and senile plaques (Candy et al 1986).

However, arguments against the so called 'aluminium hypothesis' stem, among other things, from the observation that neurofilaments accumulated in neurons in Alzheimer's disease patients occur in a double helical array, never observed in Al(III)-induced neurofibrillary degeneration, either in experimental animals or in human cell cultures.

More recently, Rapoport (1988) suggested that AD is a disease specific to the human, as the result of molecular genetic events which promoted the rapid evolution of the hominid brain, especially of those areas which are affected by AD, and that, consequently, the pathology has to be considered as 'phylogenetically' caused. This hypothesis could explain the difficulty, if not the impossibility, of reproducing a complete AD scenario in cellular and/or animal models.

Whether aluminium is only a trivial marker or an aetiological factor or a secondary cause in AD is still a matter of intense investigation, and a challenging part of our task.

Speciation in Al(III) aqueous solutions: relevance to aluminium bioavailability and experimental pathology

In aqueous solutions, Al(III) possesses a seemingly simple chemistry dominated by a few major distinct species, namely: $[Al(H_2O)_6]^{3+}$, $[Al(H_2O)_5OH]^{2+}$, $[Al(H_2O)_4(OH)_2]^{2+}$, $Al(OH)_3$, and $[Al(OH)_4]^-$.

Oligomeric hydroxo-aquo Al(III) species are also known to occur but they represent a significant fraction of total soluble Al(III) only at higher metal concentrations ($>10^{-2}$ M) and are, therefore, not of great interest in toxicological experimentation.

The 'dominance' diagrams which provide information on the prevailing species at a given pH value have been reported and illustrated by us elsewhere (Corain et al 1991). In this chapter we shall emphasize that at physiological pH values, $[Al(OH)_4]^-$ and the barely soluble $Al(OH)_3$ are the expected major species under equilibrium conditions, with a total metal solubility of around 10^{-7} M.

It should also be stressed that a suspension of $Al(OH)_3$ prepared by neutralizing a solution of a typical Al(III) salt, such as $AlCl_3$, $Al_2(SO_4)_3$ or $Al(NO_3)_3$, is an ill-defined system from the chemical and the toxicological point of view. In fact, thermodynamic considerations, based on data collected in acidic solutions (pH < 4) (Corain et al 1991), where equilibria involving possible Al(III) hydroxo-aquo monomeric and oligomeric species are *fast*, make it possible to carefully predict the theoretical speciation of the aqueous solution in *equilibrium* with amorphous $Al(OH)_3$. However, the equilibration rates at a pH of around 7.0 are in fact very slow, so the actual detailed molecular composition (speciation) of the supernatant solution is largely unpredictable, in real-life conditions.

Until a few years ago, this circumstance was not considered by toxicologists as a basic drawback in studying the toxicology of aluminium, but this view has no longer been tenable since the discovery of dramatic speciation effects in the experimental pathology of Al(III) in these and other laboratories. These unavoidable speciation problems were surmounted by a new experimental attitude which developed in 1986 after the publication of two seminal papers in the field (Finnegan et al 1986; Bombi et al 1986). On the basis of this novel toxicological viewpoint (also relevant to the toxicology of all metal centres giving insoluble hydroxides at physiological pH values), hydrolytically stable, structurally defined and sufficiently water soluble neutral aluminium complexes have been chosen as the new generation of Al(III) toxins. $Al(acac)_3$ (acac = 2,4-pentanedionate; acetylacetonate) and hydrophilic $Al(malt)_3$ (malt = 3-hydroxy-2-methyl-4H-pyran-4-onate; maltolate) appeared to be convenient in that they share a similar molecular structure but possess opposite lipophilic properties, $Al(acac)_3$ being lipophilic and $Al(malt)_3$, hydrophilic in nature. Aqueous solutions, obtained by dissolving $Al(lac)_3$ (lac = lactate; 2-hydroxypropionate) in water followed by neutralization at pH 7.5, have also been used extensively. After a thorough investigation (Corain et al 1992), these solutions were shown to contain mainly $Al(OH)_3(H_2O)_3$—that is, the simplest (highly metastable) precursor of $Al(OH)_3$.

These three toxins have been used to reveal impressive examples of speciation effects in our (Zatta et al 1989) and other laboratories. Generally speaking,

$Al(acac)_3$ appears to be more biologically active than $Al(malt)_3$ and $Al(lac)_3$, probably as a consequence of its lipophilicity. However, the scope of the biological systems investigated is still too limited for conclusive statements to be made. Interestingly enough, $Al(OH)_3(H_2O)_3$, when enclosed in liposomes, turned out to be almost equally cardiotoxic to rabbits as aqueous $Al(acac)_3$ (Bombi et al 1990), though the clinico-chemical and histopathological details of the cardiac lesions are different in the two cases. The observed amplification of the general toxic effect of $Al(OH)_3(H_2O)_3$ resulting from the physical trapping of this hydrophilic toxin inside a lipophilic vesicle emphasizes again the role of lipophilicity in directing the effects of Al(III) on cellular targets.

Controversy over the presence of aluminium in human encephalopathies: intraneuronal compartmentation

A number of observations indicate that higher than normal concentrations of aluminium can reach the central nervous system, producing a heterogeneous and not yet fully understood pathological scenario.

Relatively elevated concentrations of aluminium in neurons bearing NFT from Alzheimer's disease patients were first reported by Perl & Brody (1980). The presence of aluminosilicates in the core of the plaques prepared from brains of AD patients was later reported by Edwardson and his colleagues in 1986. NFT in amyotrophic lateral sclerosis (ALS) and in parkinsonism-dementia complex of Guam (PD) were found to be associated with abnormally high concentrations of aluminium and silicon (Perl et al 1982, Garruto et al 1984).

AD appears to be more controversial than these other conditions as far as the question of aluminium accumulation is concerned. Bulk elevated concentrations of aluminium were claimed in the brain of patients affected by AD by some authors (Krishnan et al 1987), whereas other researchers have reported abnormal levels of aluminium in some but not all features of AD (namely plaques) (Moretz et al 1990). Others, more recently, have observed no accumulation of the metal in the characteristic AD features (plaques and tangles) (Chafi et al 1991).

These discrepancies have been partially explained by a lack of strict adherence to a proper sampling procedure (Krishnan et al 1987) or by an inappropriate analytical approach with respect to interference by the matrix (Edwardson et al 1989). In contrast, Chafi et al (1991) argue that these discrepancies could also be explained by contamination effects in the preanalytical and analytical stages of the aluminium measurements.

In our view, these discrepancies between the observed phenomena are producing enormous confusion and a final answer has become an imperative task requiring an unbiased interlaboratory exchange project (Zatta et al 1991).

Effects of aluminium on cultured cells

Neuroblastoma cells as a model for studying aluminium toxicity

Studies on the toxicity of Al(III) in cultured cells are scanty in comparison with the large body of data derived from experiments *in vivo*. The different effects produced by aluminium in cultured cells are summarized in Fig. 1.

Cellular experimental models have provided evidence that aluminium affects the neurofilament distribution in the perikaryon of neuronal cells, with the production and accumulation of 10 nm NFT (Terry & Peña 1965). Human cerebral cortical neurons growing in culture medium in the presence of added aluminium lactate for 2–3 weeks showed an accumulation of neurofilaments similar to that induced in experimental animals (De Boni et al 1980). NFT have been induced by aqueous $AlCl_3$ and $Al(malt)_3$ in cultivated rat brain neurons (Langui et al 1988). Aluminium-induced tangles appear to be not only morphologically but also immunologically distinct from the paired helical filaments of Alzheimer's disease.

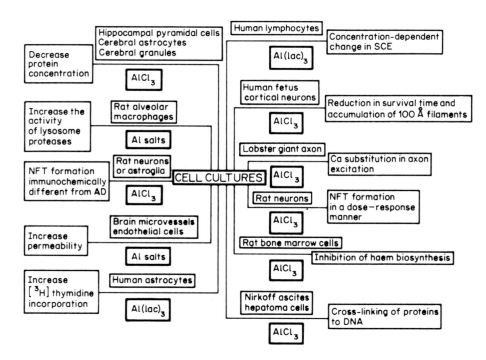

FIG. 1. Differential effects produced by aluminium compounds on cell cultures. Al(lac)$_3$, aluminium lactate.

Neuroblastoma cells are malignant tumour cells that strongly resemble embryonic neuroblasts. Several neuroblastoma cell lines able to retain the capacity to differentiate *in vitro* have now been established. During differentiation a variety of neuronal properties are expressed, including the formation of neurites and the presence of excitable membrane, as well as the production of neuron-specific enzymes and neurotransmitters. For these reasons, neuroblastoma cells are a convenient system for studying the neurotoxicity of aluminium *in vitro* in a stabilized neuron-like cell culture system. In fact, different lines of neuroblastoma cells have been used to study aluminium toxicity, and the relevant effects are summarized in Fig. 2.

Mouse neuroblastoma N1E-115, treated with aqueous $AlCl_3$, showed a premature onset of deterioration in fully differentiated cells (Roll et al 1989). These experiments demonstrated that neuroblastoma cells are less susceptible to aluminium toxicity during the process of development than after differentiation. Furthermore, in the differentiated human neuroblastoma cell line IMR32, aqueous $AlCl_3$ increased the density of muscarinic receptors as measured by [^3H]scopolamine binding (Gotti et al 1987). In neuroblastoma C1300 exposed to $AlCl_3$, 70% of the metal was found to be associated with post-mitochondrial components (endoplasmic reticulum, lysosomes and cytosol), while 20% was found in the nuclear pellet, and the remaining 10% was measured in the crude

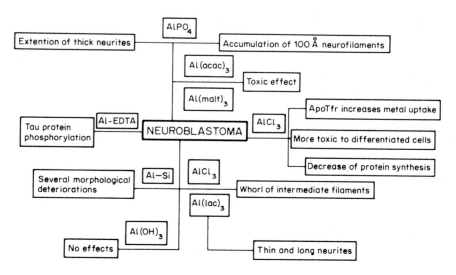

FIG. 2. Effects produced on neuroblastoma N2A cells by: $AlPO_4$; $Al(acac)_3$, aluminium acetylacetonate; $Al(malt)_3$, aluminium maltolate; $Al(lac)_3$, aluminium lactate; $Al(OH)_3$; $AlCl_3$; ApoTfr, apotransferrin; Al-Si, aluminosilicates; and Al-EDTA, Al-ethylenediaminotetraacetic acid.

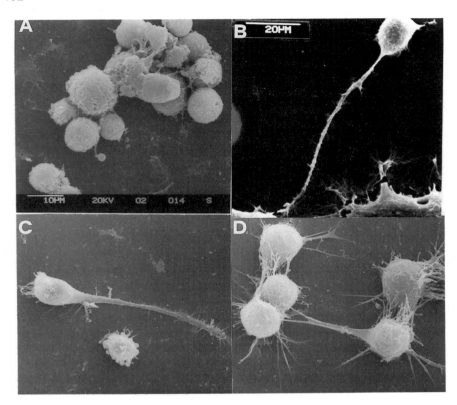

FIG. 3.   Scanning electron microscopy of neuroblastoma N2A cells. Differentiating effects produced by: A. Control; B. Al(lac)$_3$; C. AlPO$_4$; D. AlCl$_3$.

mitochondrial fraction (Shi & Haug 1989). No biological effects were observed in these conditions of very low dose exposure.

Neuroblastoma N2A in the presence of insoluble AlPO$_4$ did not show an abnormal incorporation of [^{14}C]thymidine. In contrast with this, incorporation of [^{3}H]leucine increased from that in controls, suggesting some stimulatory effect of aluminium(III) on protein synthesis (Miller & Levine 1974). Recently, Guy et al (1991) provided evidence that human neuroblastoma IMR 32, after treatment with Al–EDTA (Al-ethylenediaminetetraacetic acid) complex at a concentration of 250 μM, reacts positively with an antibody to phosphorylated tau protein which reacts specifically with AD NFT. This result is of great interest in that it strongly suggests neuroblastoma cells as candidates for further investigations on the 'aluminium hypothesis' for AD. With only two exceptions (Langui et al 1988, Guy et al 1991), the *in vitro* experiments described so far were based on Al(OH)$_3$ or AlPO$_4$, and were independent of the type of salts used.

Within the framework of our heuristic model (Zatta et al 1991) in which we used Al(acac)$_3$ and Al(malt)$_3$, we found no specific toxic metal effect in CHO cells and *Salmonella typhimurium* cells, where in both cases the ligands themselves were highly genotoxic. By contrast, N2A murine neuroblastoma cells provided a good model for demonstrating the importance of Al(III) speciation in the biological effects of the metal centre (Fig. 3).

Aqueous Al(lac)$_3$ in the 0.1–10 nM concentration range showed no cytotoxic effects on N2A neuroblastoma cells, by 48 hours, but, most remarkably, a clear cytostatic effect on neurite proliferation was observed. Experiments using suspensions of aluminium hydroxide in the 1.0–10 mM 'analytical' concentration range revealed that, even in this form, Al(III) may have a cytostatic action as well as a differentiating effect, although less pronounced than the effects expressed by Al(lac)$_3$. Neuroblastoma N2A cells treated with AlPO$_4$ at the analytical concentration (between 0.1 and 10 mM) produced a moderate cytostatic effect and the formation of neurites shorter and thicker than those produced by Al(lac)$_3$ (Fig. 3).

In contrast to these observations, Al(acac)$_3$ and Al(malt)$_3$ proved to be markedly cytotoxic in the 0.1–0.5 mM range, with a negligible differentiating effect.

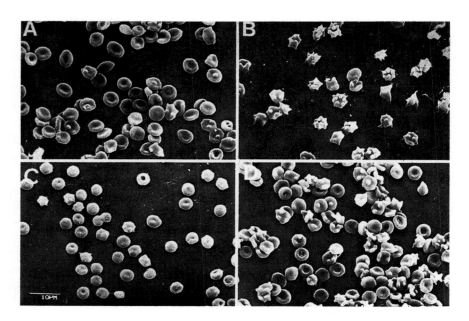

FIG. 4.   Scanning electron microscopy of rabbit erythrocytes. Differential effects produced by: A. Control; B. Al(acac)$_3$; C. Al(malt)$_3$; D. Al(lac)$_3$.

Erythrocytes as a model for studying aluminium toxicity

In 1987 we observed a peculiar effect of Al(III) on the stability of the red cell membrane (Zatta et al 1987). Rabbit erythrocytes suspended in 0.34 mM aqueous solutions of the lipophilic species Al(acac)$_3$ were found to undergo a dramatic morphological change—echinoacanthocytosis—coupled with increased osmotic fragility; Al(lac)$_3$ showed negligible effects (Fig. 4). Fe(acac)$_3$, which is structurally and thermodynamically close to Al(acac)$_3$, was inactive. On the other hand, Al(malt)$_3$ had a distinct morphological effect on erythrocytes treated at 0.34 mM (Fig. 4), producing doughnut-shaped cells.

The speciation-dependent action of Al(III) on membrane stability was found to be associated with a large increase in aluminium concentration in the relevant ghosts (from 11 to 5700 p.p.m., dry weight, for Al(acac)$_3$). The metal's biophysical effect was later investigated using electron spin resonance (ESR) spectroscopy after labelling with selected spin labels (Fontana et al 1992).

Rabbit and human erythrocytes were exposed to 8 mM solutions of Al(acac)$_3$, Al(malt)$_3$ and Al(lac)$_3$ and labelled with stable radicals able to inform on specific aspects of membrane biophysical status (Fontana et al 1992). ESR measurements revealed that: (i) Al(acac)$_3$ strongly reduces membrane fluidity in rabbit (Fig. 5) but not in human erythrocytes—Al(malt)$_3$ and Al(lac)$_3$ exhibit minor effects; and (ii) Al(acac)$_3$ causes a marked structural compacting effect on cytoskeleton and transmembrane proteins as well as a concomitant effect on the configuration of cell surface carbohydrates in human erythrocytes; no significant effect was observed in rabbit erythrocytes. No precise explanation has been put forward for this species-related biophysical effect but, again, a strong speciation effect of the toxin was clearly observed.

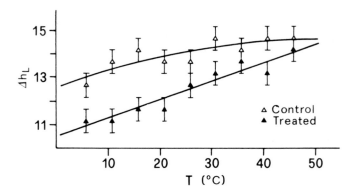

FIG. 5. Effect of Al(acac)$_3$ on rabbit erythrocyte membranes labelled with 5-deoxyl stearic acid. Δh_L values (half width at half height of $M_i = +1$ low field ESR line) (Butterfield 1986) are plotted versus T (°C). Δh_L values refer to a specific sample of ghosts obtained from rabbit erythrocytes exposed to 8 mM Al(acac)$_3$. A reduction in membrane fluidity is revealed by slopes higher than in the controls.

These results probably raise more questions than they offer unambiguous answers on the effect of Al(III) on membrane stability. What appears to be certain is that the aluminium toxicity is a direct function of the molecular nature of the synthetic toxin.

Moreover, virtually nothing is known about the molecular mechanisms by which the Al(III) enters the cell. Potentially, it may cross the plasma membrane lipid bilayer by diffusion, or through mechanical and biophysical lesions or channels, or via receptors. Al(III) speciation for instance, plays an important role in the physiological properties of the voltage-dependent anion-selective channels (VDAC) (Colombini 1991; P. F. Zatta, unpublished results). At neutral pH values, non-esterified fatty acids facilitate aluminium uptake into cultured neuroblastoma cells (Shi & Haug 1990). It has also been reported that an Al–phosphatidylcholine complex is very active in causing an increase in the membrane surface potential and so reducing the absorption and uptake of other cations (Akeson & Munns 1989). A reaction between aluminium and phosphatidylcholine could contribute to the metal's uptake by endocytosis (Akeson & Munns 1989).

It is worth mentioning that alterations in phospholipid metabolism and in the hippocampal membrane have been reported in AD (Zubenko 1986), and that Bosman et al (1991) showed an increased breakdown of the anion transport protein, band 3, in the majority of red blood cells from AD patients.

In view of the well-known relevance of membrane integrity to human health, our results all indicate further aspects of aluminium toxicity.

Conclusions

The molecular bases of aluminium toxicity are open questions. However, cellular models are useful probes for defining speciation-controlled dose–response relationships. Among many candidates for *in vitro* research, neuroblastoma cells and erythrocytes appear to be promising biological systems for directing toxicological investigation towards animal models tailored for particular neurodegenerative diseases.

Acknowledgements

Fidia Research Laboratories, Abano Terme, Italy and Parke-Davis, Italy are gratefully acknowledged for partial financial support of this research.

References

Akeson MA, Munns DN 1989 Lipid bilayer permeation by neutral aluminum citrate and by three α-hydroxyl carboxylic acids. Biochim Biophys Acta 984:200–206
Basun H, Forssell LG, Wetterberg L, Winblad B 1991 Metals and trace elements in plasma and cerebrospinal fluid in normal ageing and Alzheimer's disease. J Neural Transm Parkinson's Dis Dementia Sect 3:231–258

Bombi GG, Corain B, Giordano G, Pagani D, Sesti AG, Zatta P 1986 Chemical and biological investigations on aluminum neurotoxicity. In: Vezzadini P, Facchini A, Labo' G (eds) Neuroendocrine system and aging. EURAGE (Proc Int Symp on New Trends in Aging Research) p 253–258

Bombi GG, Corain B, Favarato M et al 1990 Experimental aluminum pathology in rabbits: effects of hydrophilic and lipophilic compounds. Environ Health Perspect 89:217–223

Bosman GJCGM, Bartholomeus IGP, De Man AJM, Van Kalmthout PJC, De Grip WJ 1991 Erythrocyte membrane characteristics indicate abnormal cellular aging in patients with Alzheimer's disease. Neurobiol Aging 12:13–18

Butterfield DA 1986 Spectroscopic methods in degenerative neurological diseases. Crit Rev Clin Neurol 2:169–240

Candy JM, Oakley AE, Klinowski J et al 1986 Aluminosilicates and senile plaque formation in Alzheimer's disease. Lancet 1:354–357

Chafi AH, Hauw J-J, Rancurel G, Berry J-P, Galle C 1991 Absence of aluminum in Alzheimer's disease brain tissue: electron microprobe and ion microprobe studies. Neurosci Lett 123:61–64

Colombini M 1991 Aluminum and membrane channels. In: Nicolini M, Zatta PF, Corain B (eds) Aluminum in chemistry biology and medicine. Cortina international, Verona & Raven Press, New York, p 33–42

Corain B, Tapparo A, Sheik-Osman A, Bombi GG, Zatta P, Favarato M 1991 The solution state of aluminum (III) as relevant to experimental toxicology: recent data and new perspectives. Coord Chem Rev 112:19–32

Corain B, Longato B, Sheik-Osman A, Bombi GG, Maccà C 1992 The speciation of aluminum in aqueous solution aluminum carboxylate. Part II. Solution state of the metal center in the AlIII/lactate/OH⁻/H₂O system. J Chem Soc Dalton Trans, p 169–172

De Boni U, Seger M, Crapper McLachlan DR 1980 Functional consequences of chromatin bound aluminum in cultured human cells. Neurotoxicology 1:65–81

Edwardson JA, Klinowski J, Oakley AE, Perry RH, Candy JM 1986 Aluminosilicates and the ageing brain: implications for the pathogenesis of Alzheimer's disease. In: Silicon biochemistry. Wiley, Chichester (Ciba Found Symp 121) p 160–179

Edwardson JA, Oakley AE, Pullen RGL et al 1989 Aluminum and pathogenesis of neurodegenerative disorders. In: Massey R, Taylor D (eds) Aluminium in food and the environment. Royal Society of Chemistry (Distribution Centre, Blackhorse Road, Letchworth, Herts, UK) p 20–36

Finnegan MM, Rettig SJ, Orwig C 1986 A neutral water soluble aluminum complex of neurological interest. J Am Chem Soc 108:5033–5035

Fontana L, Perazzolo M, Zatta P, Corvaja C, Corain B 1992 Aluminum(III) induces speciation-dependent alterations of the physical state of human and rabbit membrane: an ESR evaluation. Biochim Biophys Acta (submitted)

Garruto RM, Fakatsu R, Yanagihara R, Gajdusek DC, Hook G, Fiori CE 1984 Imaging of calcium and aluminum in neurofibrillary tangle-bearing neurons in Parkinsonism-dementia of Guam. Proc Natl Acad Sci USA 81:1875–1879

Gotti C, Cabrini D, Sher E, Clementi F 1987 Effects of long-term *in vitro* exposure to aluminum, cadmium or lead on differentiation and cholinergic receptor expression in a human neuroblastoma cell line. Cell Biol Toxicol 3:431–440

Guy SP, Jones D, Man DMA, Itzhaki RF 1991 Human neuroblastoma cells treated with aluminum express an epitope associated with Alzheimer's disease neurofibrillary tangles. Neurosci Lett 121:166–168

Krishnan SS, Harrison JE, Crapper McLachlan DR 1987 Origin and resolution of the aluminum controversy concerning Alzheimer's neurofibrillary degeneration. Biol Trace Elem Res 13:35–42

Langui D, Anderton BH, Brion J-P, Ulrich J 1988 Effects of aluminum chloride on cultured cells from rat brain hemispheres. Brain Res 438:67–76

Miller CA, Levine EM 1974 Effects of aluminum salts on cultured neuroblastoma cells. J Neurochem 22:751–758

Moretz RC, Iqbal K, Wisniewski HM 1990 Microanalysis of Alzheimer's disease: NFT and plaques. Environ Geochem Health 12:15–16

Perl DP, Brody AR 1980 Alzheimer's disease: x-ray spectrophotometric evidence of aluminum accumulation in neurofibrillary tangle-bearing neurons. Science (Washington DC) 208:297–299

Perl DP, Gajdusek DC, Garruto RM, Yanagihara RT, Gibbs CJ Jr 1982 Intraneuronal aluminum accumulation in amyotrophic lateral sclerosis and parkinsonism-dementia of Guam. Science (Washington DC) 217:1053–1055

Rapoport SI 1988 Brain evolution and Alzheimer's disease. Rev Neurol (Paris) 144:79–90

Roll M, Banin E, Meiri H 1989 Differentiated neuroblastoma cells are more susceptible to aluminum toxicity than developing cells. Arch Toxicol 63:231–237

Shi B, Haug A 1989 Aluminium uptake by neuroblastoma cells. J Neurochem 28:3911–3915

Terry RD, Peña C 1965 Experimental production of neurofibrillary degeneration. (2). Electron microscopy, phosphatase histochemistry and electron probe analysis. J Neuropathol & Exp Neurol 24:200–210

Wenstrup D, Ehmann D, Markesbery WR 1990 Trace element imbalances in isolated subcellular fractions of Alzheimer's disease brain. Brain Res 533:125–131

Zatta P, Giordano R, Corain B, Favarato M, Bombi GG 1987 A neutral lipophilic compound of aluminum(III) as a cause of myocardial infarct in the rabbit. Toxicol Lett 39:185–188

Zatta P, Perazzolo M, Corain B 1989 Tris acetylacetonate aluminum(III) induces osmotic fragility and acanthocyte formation in suspended erythrocytes. Toxicol Lett 45:15–21

Zatta PF, Nicolini M, Corain B 1991 Aluminum(III) toxicity and blood–brain barrier permeability. In: Nicolini M, Zatta PF, Corain B (eds) Aluminum in chemistry biology and medicine. Cortina international, Verona & Raven Press, New York, p 97–112

Zubenko GS 1986 Hippocampal membrane alteration in Alzheimer's disease. Brain Res 385:115–121

DISCUSSION

Perl: I totally agree with you that in looking at the toxicity of Al the issue of Al speciation is very important. So much of the experimental work has been done using Al chloride, which seems to be the wrong choice in terms of many of the phenomena we want to look at. We have followed your line of reasoning on this and have introduced a model compound similar to Al maltol, namely Al flavonol (tris(flavonoto)aluminium) which is even more lipophilic. It shows very different patterns of pathological response, forming tangles, but in very selected regions of the brain. We hope to make additional compounds of this type and to develop an even better model of developmental pathology.

I also agree that we need cross-laboratory studies. Already some laboratories have begun to collaborate. In my paper, I shall show data on Al concentration standards that could be used among laboratories. More of this should be done, particularly in the preparation of standards. The variability in the instrumentation for detecting Al by various microprobe techniques is great, and one cannot rely on the instrumental detection limits provided by the manufacturers. In fact, particularly with biological tissues, detection limits are very variable. The development of reliable elemental standards to be shared among laboratories is critical if we are to settle these issues.

Zatta: Now we are able, by exploiting the tool of Al speciation, to introduce the metal wherever we want to, it is a matter of fact that we can direct Al to the cytoplasm of the neuron, inside the nucleus, or to the nucleolus. It is clear to this end that the appropriate choice of the coordination sphere is very important, also in terms of size. Most authors have used (and are still using) Al chloride at autogenous pH, in toxicological experimentations, so one is in fact injecting hydrochloric acid!

Perl: We have looked at that, studying rabbits injected intraventricularly with Al chloride. Using our microprobe techniques, we detect most of the Al within macrophages. We presume that this is Al phosphate that is precipitated out of solution and is picked up by the macrophages. So the rabbit must be extremely sensitive to Al, and the tangles form after exposure to very low concentrations. We have also looked at some of the heart samples that you sent us. The extent of cardiac muscle damage is remarkable after exposure to these compounds. So far we have not found Al in these damaged cardiac muscles; however, these lesions are large and very complex, with many different cell types involved. One problem with LAMMA is that although it's a very sensitive and very specific analytical instrument, it is not particularly practical for surveying large areas of tissue. This is probably a better problem for SIMS!

Candy: Yes. The SIMS (secondary ion mass spectrometry) technique would be better suited for this type of study, because it is possible to analyse, at high sensitivity, areas up to 400 μm in diameter.

Williams: Dr Zatta, if you inject Al(acac)$_3$ into water at pH 7.0 containing ATP, what do you have at the end of the experiment?

Zatta: You would have to take into account two parameters of the reaction between Al(acac)$_3$ and ATP: the kinetics and the thermodynamics of the aluminium species.

Williams: If you let the solutions stand, and just wait, it doesn't usually take very long for Al to equilibrate in such a system. I am afraid that introducing Al in a new form, such as Al(acac)$_3$, may lead you into complicated problems. When Al is introduced as a lipophilic compound, such as Al(acac)$_3$, in a homogeneous model system we might know where we were, but once it goes into a heterogeneous biological system, Al could go into a membrane from one side and come out on the other side. I don't think it will be Al(acac)$_3$ after it emerges. We have shown that the pH relative to the pK of the various ligands, and the presence of Ca or Mg as other metals, could affect the Al-binding partner (Tam & Williams 1986). Some of these elements are able to take acac away from

Al, in the presence of a phosphate (or something similar), because Al goes to the phosphate while the other metal takes the acac.

Zatta: We believe in fact that administered aluminium should be 'capped' by convenient coordination spheres, which will control both the hydrolytic stability and the lipophilic or hydrophilic character of the toxin employed. As to the specific question, at lower analytical concentrations, such as in the millimolar range, Al(acac)$_3$ undergoes a very slow spontaneous hydrolysis and it will react more or less rapidly with ATP, if this ligand is present at an actively high concentration (log K = 9.8 at pH 7.4). However, the kinetics of these reactions cannot be neglected, as shown by observations in our laboratories which reveal that the species Al(acac)$_3$ and Al(malt)$_3$ do survive hydrolysis at pH 7 for many hours at analytical concentrations as low as 100 mM. The major role of kinetics in aluminium toxicology is illustrated by the case of Al(lac)$_3$. The carboxylate is a coordination compound in the solid state, which undergoes, at physiological pH, a complete conversion into a very metastable hydrolytic toxin, namely Al(OH)$_3$(H$_2$O)$_3$. In conclusion, the model in which I believe starts from employing different aluminium complexes, which should be adequately stable towards chemical modifications in biological environments, but not stable to such an extent as to make the metal centre in fact not bioavailable.

Martin: Acac and maltol bind Al^{3+} with about the same strength in neutral solutions. The two ligands display similar conditional stability constants (Martin 1991). As indicated by comparing Figs. 2 and 3 in my paper in this volume (p 10 and 11), the pAl value for ATP is significantly higher than that of maltol. So Al^{3+} should pass easily from either maltol or acac to ATP.

Williams: That's what I feared could happen. The point is that silicate and phosphate are very good binding agents for Al; phosphate compounds, especially pyrophosphates or carboxyphosphates, will pick Al off most of the things we try to inject; that's the trouble.

Candy: You can poison cultured cells in many ways; why are you using such high Al concentrations, Dr Zatta?

Zatta: Dealing with experimental pathology, one is expected to investigate biological phenomena (effects) produced by an appropriate external cause (dose). In our case the effects of aluminium appear in the millimolar range using Al(acac)$_3$. Therefore, the choice of higher aluminium concentrations for Al(lac)$_3$ was dictated by its lower toxicity. In addition, we used Al(acac)$_3$ and Al(malt)$_3$ because they are thermodynamically comparable, so they have practically the same structure. However, one is more lipophilic and the other is more hydrophilic—a useful combination for producing a model that concerns the lipophilicity and the hydrophilicity of aluminium compounds. We know that these molecules are not 'natural' compounds, but we are dealing with a heuristic model. We think that this is a correct approach for a better understanding of the biology of aluminium.

Williams: I am surprised that a diester phosphate could displace acetylacetonate (acac) from Al; a diester phosphate has almost no affinity for Al.

Edwardson: This kind of approach is important from a mechanistic and neurotoxicological point of view, but an even more important issue is the possibility of exposure to such compounds in the human environment. What is the evidence that such complexes may occur in Nature, and under what theoretical conditions could they be formed and perhaps be absorbed?

Williams: The nearest thing to Al(acac)$_3$ in the environment is the sort of chelating agent that occurs in tea, such as a dihydroxyphenol. If it became free, it would carry Al through membranes just as fast as does acac. So Dr Zatta could say that using acac is a way to mobilize Al so that it goes through membranes very quickly. What happens after that is a mystery, because the experiment has just introduced a very high dose past a membrane, very quickly. So if you absorb a dihydroxyphenol, it will work exactly like acac, if not better, going through a cell membrane. When you absorb some plant phenolates in the presence of, or combined with Al, you could be in a great deal of trouble. Another compound that would facilitate Al transport is salicylic acid, from aspirin!

Birchall: Maltol is a very common food additive, too, and it forms complexes with aluminium.

Petersen: I wanted to ask about the membrane-stabilizing effect of Al. A number of multivalent ions are known to be membrane-stabilizing agents. I wonder therefore whether this effect is one that you regard as specific for Al.

Zatta: No, it is not specific at all.

Blair: Have you looked at any system in which there is a Mg^{2+} 'gate' in your cell membrane—that is, where water flow is controlled by Mg^{2+} gates, as in some neurons?

Zatta: Not yet.

Wischik: You mentioned phosphorylated tau protein immunoreactivity in neuroblastoma cells: which monoclonal antibody was used?

Zatta: These results have been published by Guy et al (1991).

References

Guy SP, Jones D, Mann DMA, Itzhaki RF 1991 Human neuroblastoma cells treated with aluminum express an epitope associated with Alzheimer's disease neurofibrillary tangles. Neurosci Lett 121:166–168

Martin RB 1991 Aluminum in biological systems. In: Nicolini M, Zatta PF, Corain B (eds) Aluminum in chemistry biology and medicine. Cortina international, Verona & Raven Press, New York, p 3–20

Martin RB 1992 Aluminium speciation in biology. In: Aluminium in biology and medicine. Wiley, Chichester (Ciba Found Symp 169) p 5–25

Tam SC, Williams RJP 1986 One problem of acid rain: aluminium. J Inorg Biochem 26:35–44

Elemental analysis of neurofibrillary tangles in Alzheimer's disease using proton-induced X-ray analysis

F. E. S. Murray*, J. P. Landsberg**, R. J. P. Williams*, M. M. Esiri† and F. Watt**

*Inorganic Chemistry Laboratory, South Parks Road, Oxford, OX1 3QR, **Nuclear Physics Laboratory, Keble Road, Oxford, OX1 3RH, and †Department of Neuropathology, Radcliffe Infirmary, Woodstock Road, Oxford, OX2 6HE, UK

Abstract. We have investigated the elemental content of hippocampal slices from normal human brain and from brains of Alzheimer's disease patients by X-ray fluorescence using both electron and proton beam microprobes. The sections have been stained with a dye—toluidine blue—which contains sulphur so that the X-ray fluorescence map can be correlated with known intracellular sites as seen under the light microscope. The results show that associated with neurofibrillary tangles and Hirano bodies (the distinctive internal visual features of cells from Alzheimer's disease patients) there is increased calcium. We cannot confirm that there are peculiarities in the distribution of aluminium in cells.

1992 Aluminium in biology and medicine. Wiley, Chichester (Ciba Foundation Symposium 169) p 201–216

The hippocampus is a key site of damage in Alzheimer's disease and many of the significant changes in the cognitive and emotional patterns of the sufferers can be related to the neuropathological changes that take place there. The clinical features of the disease are complex and can be difficult to distinguish from other causes of dementia. However, the pathological signs are easier to observe. In particular, neurofibrillary tangles have been identified in large numbers throughout the brains of those with Alzheimer's disease. They are found to lie within the pyramidal cell bodies of the neurons and are composed of paired helical filaments (PHF) whose neurofilamentous and microtubule-associated protein (tau) components are abnormally phosphorylated (Ball 1976). Hirano bodies are also found in the hippocampus in Alzheimer's disease and are, as we have confirmed using immunohistochemical staining procedures, composed of actin. Plaques, which occur external to cells, and are often said to be associated with aluminium, are not studied here.

In some experiments relating to the inorganic components of the neuro-fibrillary tangles, predominantly using electron dispersed X-ray analysis (EDXA), aluminium has been identified in those neurons that contain neurofibrillary tangles (see Perl et al 1982). Other recent techniques used to study neurofibrillary tangles have included laser microprobe mass analysis (LAMMA), and this work also suggests that aluminium is localized in the nuclei of tangle-bearing neurons (Perl & Brody 1980, Perl et al 1982, Perl & Good 1987). However, the most recent EDXA experiments have found these results to be irreproducible (Jacobs et al 1989).

One of the major problems with elemental analyses by any method is that the tissue preparation, and particularly the staining, can introduce a wide range of chemical components into the tissue, thereby changing the nature of the intrinsic inorganic content. The fixing, embedding and staining processes will introduce elements into the samples that are extraneous and may distort the elemental analysis. Furthermore, any impurities in the stains may be magnified in regions where a high affinity for the stain exists—such as the neurofibrillary tangles—and thereby provide increased distortion of the data. The use of inorganic salt precipitation as a means of cell observation and identification always carries the risk of co-precipitation of other metal elements which have insoluble salts.

Despite there being some evidence suggesting the presence of aluminium in Alzheimer's disease and its role in the disease aetiology, there has been limited confirmation of the presence of aluminium and other inorganic elements in the nucleus and in the neurofibrillary tangles, by means of electron or proton microprobe methods that aim to minimize the presence of contaminants in the samples. This study attempts to minimize metal contamination through the use of organically stained samples combined with the use of the Scanning Proton Microprobe (SPM), which generates two-dimensional elemental maps. In order to orient the brain sections and relate elemental data to cell pathology, the SPM utilizes a light microscope viewing system. The sections must therefore be stained. In place of the silver stains commonly used in neuropathology, toluidine blue, an organic dye, has been used here. This general pathological stain gives a positive stain for nuclei, neurofibrillary tangles and Hirano bodies. It is a large organic molecule composed of several aromatic rings, one of which contains a sulphur atom. It is through the mapping of the sulphur concentrations by the microprobe across the tissue that the elemental composition and cell structure may be related, while minimizing the introduction of potentially contaminating elements. The sulphur content can then be used to give a series of element ratios in any part of the tissue. The finding of particular ratios associated with any particular intracellular body not only should help to identify that body, but should reveal abnormal accumulations of elements in diseased as compared with normal cells.

Materials and methods

Sample preparation

Initial work was undertaken as an exploratory investigation of the value of the physical methods involved. For this reason, and to provide analytical consistency, all the tissue which was analysed in detail came from blocks of hippocampal tissue from a single diagnosed case of Alzheimer's disease and from normal control subjects. These blocks were fixed in 4% formalin for up to three weeks. They were then transferred into a 2% glutaraldehyde solution and finally embedded in E-Mix, an epoxy resin matrix.

Sections of 10 µm were cut from these blocks onto glass slides using a microtome, stained with methylene blue (a routine neuropathological stain similar to toluidine blue), and studied under the light microscope while searching for regions of well-preserved tissue containing typical Alzheimer's disease pathology (neurofibrillary tangles, Hirano bodies and senile plaques). A series of sections cut from a well-characterized Alzheimer's disease case known to contain large numbers of neurofibrillary tangles were used. The best areas were chosen for study by proton-induced X-ray emission (PIXE). Note that no attention was paid to plaques or their location. Control samples consisting of hippocampal tissue blocks from subjects with no neurological degeneration and no neurofibrillary tangles were also examined.

Preparation of samples for PIXE

Five-µm sections of the hippocampus, identified through the light microscope investigations, were cut on a microtome and picked up onto Pioloform-coated glass slides. They were stained with toluidine blue on a hotplate for three minutes and washed with distilled water to remove excess reagent. In order to mount the sections onto the PIXE holders, which were backed with another layer of Pioloform, the sections were cut from the slides and floated off by immersion in a tank of Milli-Q water, used because it contains virtually no trace elements, including aluminium.

Light microscopy for PIXE

The cell components of the prepared sections were observed using a light microscope and photographed at 20× and 40× magnification, to produce a detailed map of the whole section and detailed high magnification photographs of areas of special interest containing well-defined neurofibrillary tangles, Hirano bodies and nuclei. These maps and photographs provide locations of the regions of the section to analyse for elements by EDXA and PIXE and were used for correlation with the elemental maps.

Proton-induced X-ray emission (PIXE)

PIXE is a method of elemental X-ray fluorescence analysis linked to a scanning microprobe. The Scanning Proton Microprobe Unit equipment in Oxford is routinely used for the nuclear microscopy of medical and biological specimens. The technique allows us to detect very low concentrations of elements to a sensitivity in the parts per million range. The high energy proton beam (3 MeV) is scanned across the sample. The interaction of the beam with the specimen leads to the excitation of core electrons, and an X-ray of characteristic energy for each element in a given region (pixel) is emitted.

The viewing system in the Oxford proton microprobe is a light microscope and this can be used to select the areas for scanning and for detailed elemental point analyses, to obtain an accurate correlation between the cell pathology and the analysis. When spatial and analytical information is sought, the SPM beam is scanned across the specimen area and data are collected and recorded for each point. Data are obtained as an elemental map, but may also be obtained in the form of a point elemental analysis taken at a specific region of interest.

Electron dispersed X-ray analysis (EDXA)

The EDXA method is the standard procedure for elemental analyses in biological samples. We have therefore confirmed our PIXE observations using the EDXA facilities in Oxford. The detector system uses a TN-5500 Tracor Northern computer, and collects information over 100–1000 seconds, over the X-ray energy range 2–20 keV.

For EDXA analysis, sections from the hippocampus from the same subject as was used in the PIXE investigations were prepared, using standard procedures for electron microscopy. Sections of 0.5 μm thickness were cut and mounted onto plastic grids. They were not stained with the heavy metal and silver stains that are commonly used in this type of work. This lack of staining has the disadvantage of decreasing the quality of the microscope observations and increasing the sample sensitivity in the electron beam. With this in mind, a cold stage was used at $-175°C$, to try to minimize electron beam damage.

Point analyses were made in regions identified as containing tangle-bearing neurons, including the nuclei, tangles and Hirano bodies. The sensitivity of the unstained sections precluded the collection of data in the form of elemental maps of the tissue sections.

Results

General

Using a large-size scan (250 μm × 250 μm) we could distinguish the outline of the resin of the Pioloform and of the tissue in the resin (Fig. 1) (see colour plate).

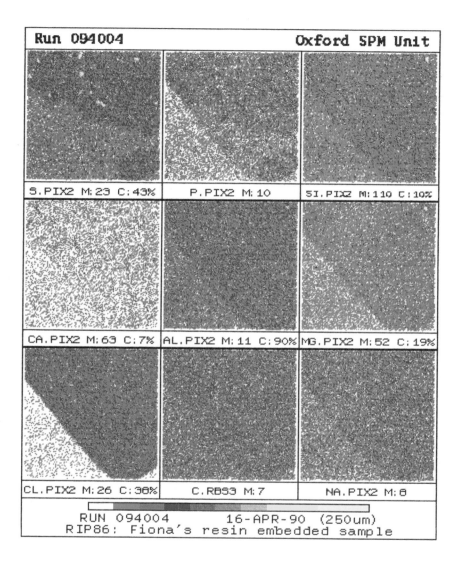

FIG. 1 *(Murray et al)* Scanning proton microprobe (SPM) maps of a corner of a typical hippocampal section scanned at 250 μm. The scan for chlorine *(bottom row, left)* clearly shows the Pioloform backing and the resin. The phosphorus and sulphur scans *(top row, centre and left)* can be used to distinguish the tissue and cellular features from the resin. Note the differential staining. The apparently high Al level is from the Pioloform.

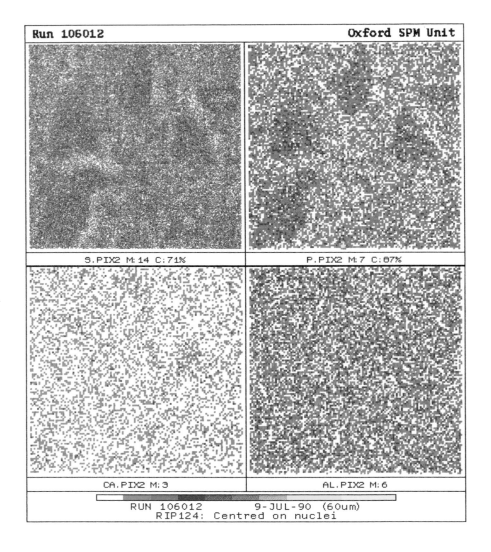

FIG. 3 *(Murray et al)* SPM maps of a section from the hippocampus of a normal subject. A series of neuronal nuclei stained with toluidine blue can be identified in the sulphur scan *(upper left)*. The high phosphorus concentrations *(upper right)* of the nuclei are also visible.

TABLE 1   Elemental ratios for the different features of Alzheimer's disease, from PIXE analysis

Feature	Al/Al	Ca/Ca	P/P	S/S
Background	1.00 (0.00)	1.00 (0.00)	1.00 (0.00)	1.00 (0.00)
Blood vessels	0.96 (0.14)	1.80 (0.94)	2.52 (0.47)	5.83 (2.62)
Normal nuclei	0.68 (0.27)	1.32 (0.72)	5.37 (3.07)	2.67 (1.52)
Tangle nuclei	0.59 (0.37)	2.30 (1.28)	4.26 (2.30)	2.70 (1.33)
Tangles	0.63 (0.28)	3.24 (1.41)	3.20 (0.99)	2.17 (0.43)
Tangle cell and normal cell cytoplasm	0.59 (0.08)	2.08 (0.47)	1.97 (0.31)	1.51 (0.15)
Hirano bodies	1.06 (0.80)	2.16 (0.76)	0.59 (0.52)	2.20 (0.80)

The table shows the elemental ratios of four elements in the hippocampal samples from Alzheimer's disease patients. Each cellular structure has its own characteristic ratio for each element (calculated as a ratio of the average concentration of the element in the structure to the average concentration of that element in the background). Standard deviations are given in brackets. The graph in Fig. 2 also presents this information.

The elemental distribution shows that the resin contained Na, Al, Mg, Cl and C, whereas the backing Pioloform was high in Na, Al and C. When the resin is compared to the tissue, a clear distinction can be seen of the sulphur from the stain and phosphorus from the cellular components (Fig. 1) (see colour plate). Tissues were examined from five Alzheimer's disease patients and two non-Alzheimer's subjects as controls, but sections from only one Alzheimer's case have been examined in detail as yet.

The results from the PIXE analyses have been combined statistically to give a series of elemental ratios which are used to compare the elemental distributions of the stain and the intrinsic elements, through the different cellular features. For each section, a background count for the different elements was taken as an average of the elemental levels for the large scans, and point analyses were taken in background areas. A series of ratios was then calculated using the point analysis data for the different identifiable features. These included normal neuronal nuclei, nuclei of tangle-bearing neurons, tangles and tangle cell cytoplasm, Hirano bodies

TABLE 2   Comparison of microprobe sensitivities with background concentrations

	Al	Ca	P	S
Microprobe sensitivity (p.p.m.)	50	10	20	20
Background levels (p.p.m.)	225.28 (50.09)	60.24 (34.67)	182.28 (56.82)	1069.66 (543.9)

The microprobe sensitivity is given in parts per million. This is compared with the average background concentration of each element in the samples measured (standard deviation in brackets).

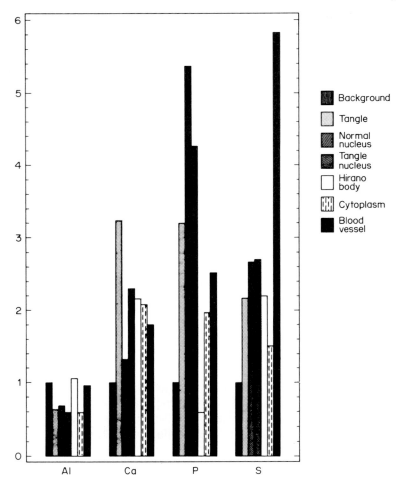

FIG. 2.   Elemental analyses, expressed as ratios with background levels, obtained from PIXE studies of Alzheimer's disease (AD) tissue (see text for the manner of calculation). For each element the order of analyses was background (= 1.0), tangle (AD), normal nucleus (AD), nucleus associated with tangles (AD), Hirano body (AD), cytoplasm (AD) and blood vessel (AD).

and blood vessels. For each feature the range of ratios was averaged and the standard deviation calculated. The results are presented in Table 1 and shown graphically in Fig. 2. The average background concentrations of aluminium, calcium, phosphorus and sulphur found in the scans are shown in Table 2, with the levels of sensitivity of the PIXE analysis for comparison.

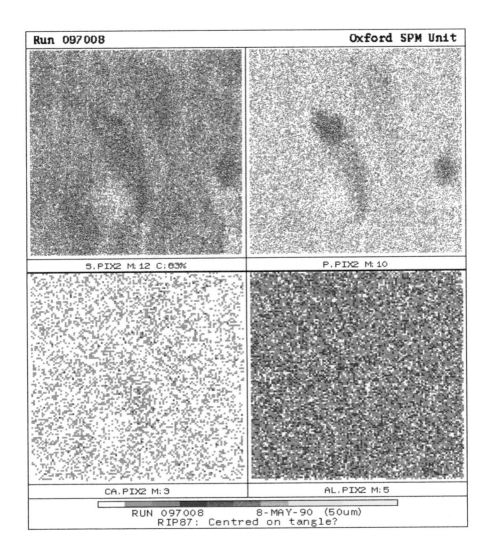

FIG. 4 *(Murray et al)* SPM maps of a tangle-bearing neuron from the hippocampus of an Alzheimer's disease subject. The region highest in sulphur *(upper left)* corresponds to the nucleus of the cell. Both the nucleus and the tangle can be observed in the phosphorus map *(upper right)* and the tangle is just observable in the calcium map *(lower left)*.

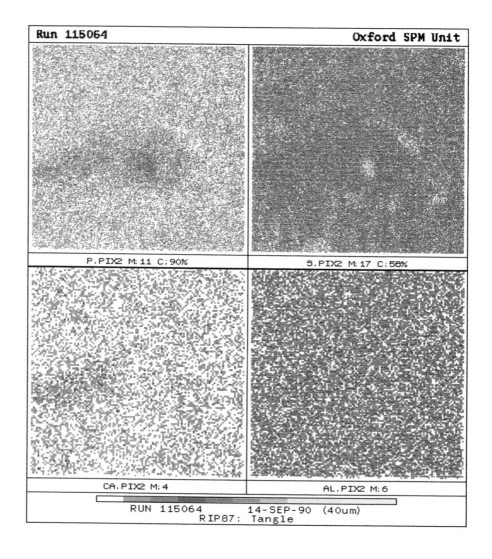

FIG. 5 *(Murray et al)* SPM maps of another tangle-bearing neuron from a hippocampal cell of an Alzheimer's disease patient, visualized in the sulphur and phosphorus maps. The tangle can also be identified in the calcium map. There is no evidence for high Al levels *(lower right)*.

Nuclei

Information on the elemental composition of the neuronal pyramidal nuclei was gathered from small-scale scans (50 μm × 50 μm) used in conjuction with light micrographs. Every nucleus studied took up the toluidine blue and therefore was seen as a region with a high local concentration of sulphur. Data from the point spectra and ratios of background to point elemental composition show that all hippocampal nuclei have characteristic elemental ratios as follows (ratios in brackets): high sulphur (2.67), high phosphorus (5.37), low aluminium (0.68) and medium calcium (1.32) (Fig. 3) (see colour plate).

Neurofibrillary tangles

Using similar small-scale scans and light micrographs as described above, we were able to visualize the tangles in AD cells as nuclei with a 'tail' region which contained the perinuclear neurofibrillary tangles. The whole structure stains with toluidine blue and is well defined in the sulphur map. In the phosphorus map the nucleus region has a very high concentration (ratio 4.26), while the tangle itself shows only a somewhat higher than background phosphorus concentration (ratio 3.20). The nucleus and tangle are not visible in the aluminium map. The calcium map shows a relatively high concentration in the tangle region and in the nucleus (ratios 2.30 and 3.24) (Figs. 4 and 5) (see colour plate). The similar sulphur ratios for normal and tangle-bearing nuclei suggest that they have an equal affinity for the stain (sulphur ratios of 2.67 and 2.70 respectively) and that therefore the other features are related to the different chemical compositions of the cells. The observations made from the 2D elemental maps are confirmed by the point spectra (Fig. 6).

Specifically, the detection of calcium in the tangles was confirmed by EDXA, and has been shown elsewhere (Garruto et al 1984).

Hirano bodies

Staining heavily with toluidine blue, the Hirano bodies were easily observed in the sulphur map. They show no increase in the aluminium concentration relative to background, but can be observed in the calcium, phosphorus and sulphur maps (Fig. 7) (see colour plate).

Impurities

Impurities on the section surface can be distinguished from the stained cellular components by an analysis of their elemental composition. Dust has a characteristic composition, including high concentrations of aluminium, silicon, magnesium and sodium. Regions of toluidine blue precipitation unrelated to

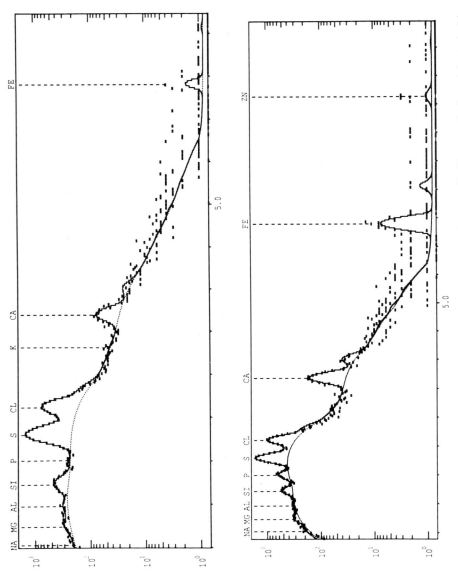

FIG. 6. SPM spectra showing the point analyses of a Hirano body (*upper*) and a neurofibrillary tangle (*lower*) from the hippocampus of an Alzheimer's disease patient. The high concentrations of calcium can be seen in the Hirano body, whereas the tangle shows high calcium and phosphorus concentrations.

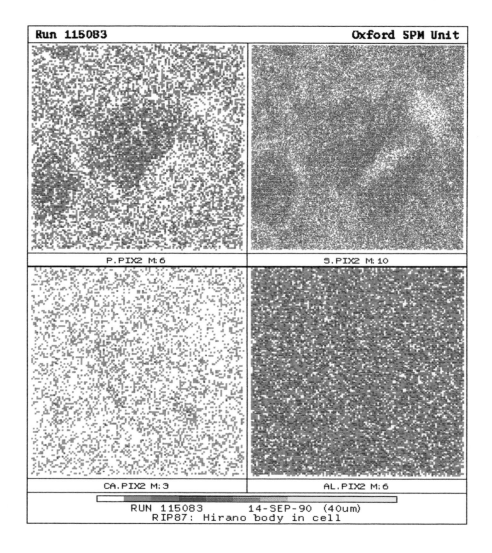

FIG. 7 *(Murray et al)* SPM maps of a hippocampal cell from an Alzheimer's disease patient. This cell contains a Hirano body. It is visualized in the phosphorus and sulphur scans *(upper left and right)* but not in the aluminium scan *(lower right)*.

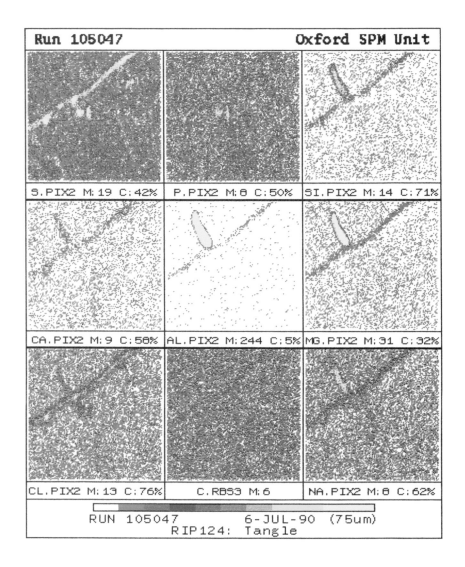

FIG. 8 *(Murray et al)* SPM maps of a hippocampal cell from a patient with Alzheimer's disease. This section has a crease through it which can be seen in all the elemental maps, thus distinguishing it from cellular features. There is also a large silicon-aluminium-magnesium-sodium impurity on the section.

cell components have high sulphur concentrations but no related phosphorus or calcium. Creases in the resin, which provide useful landmarks, have high concentrations of sulphur, phosphorus, chlorine, carbon and sodium, consistent with the resin composition (Fig. 8) (see colour plate).

Discussion

The proton microprobe method provides extremely sensitive and well-localized elemental analyses. However, we consider that while it is interesting to comment upon the local content of individual elements, it is more revealing to examine the ratios of elements at points in a particular region of the cell.

In the normal hippocampal tissue the neuronal nuclei were identified by the co-localization of P from the nuclear DNA and S from the toluidine blue stain. Other elements had a uniform distribution and originate mainly from the resin. For example, while calcium is quite high in the nucleus, we observe none above background elsewhere. In the hippocampal tissue from Alzheimer's disease patients the nuclei are similarly detectable. In addition, the tangles could be located in the sulphur, phosphorus and calcium scans, and Hirano bodies in the sulphur and calcium maps. There were no unusual variations found in the concentration of aluminium in either the normal or the diseased hippocampal tissue, except in relation to easily distinguishable impurities. These impurities had a consistent and most unusual set of element ratios.

These data provide the basis for an understanding of the stain distribution across the cellular structures, the chemistry of the stain, and the chemical composition of the abnormal pathology internal to cells associated with Alzheimer's disease.

The co-location of high concentrations of phosphorus and calcium with the tangles and of calcium with the Hirano bodies may provide an insight into the mechanism of neuronal degeneration in Alzheimer's disease and its associated pathology. Because variation in the calcium levels in cells can lead to irreversible protein changes, including abnormal phosphorylation and polymerization (found to have occurred in the neurofilamentous proteins that make up tangles: Ball 1976), which in turn may help to accumulate calcium (see Williams 1992: introduction to this volume), one possible initiating mechanism of Alzheimer's disease-related changes could be a mismanagement of the calcium metabolism. A possibility is that the known mutation of the β-amyloid protein in a membrane helix could cause leakage of calcium into the cell (see Wischik et al 1992: this volume). However, it should not be forgotten that changes in the levels and distribution of calcium in cells can arise during the fixing and staining processes (Van de Putte et al 1990).

Turning to the question of aluminium in Alzheimer's disease, a leaky damaged nerve cell with an abnormally high phosphorous content could readily pick up aluminium adventitiously, though this has not been observed in these experiments.

We were not able to detect abnormal aluminium levels inside cells, but our methods are not very sensitive procedures for the detection of very low levels of aluminium in the tissue, given the high concentrations of aluminium in the resin. On the other hand, we would expect exposed, heavily phosphorylated cellular zones to adsorb aluminium, and we have not found this. We have not concerned ourselves with plaques, which may well result from earlier cell damage.

Acknowledgements

Thanks must go to the technicians in the Neuropathology Departments of the Radcliffe Infirmary and John Radcliffe Hospitals for sample preparation, and to those in Nuclear Physics and Chemical Crystallography for running the microprobe facilities. We also acknowledge the support of the Wellcome Trust in funding the SPM.

References

Ball MJ 1976 Neurofibrillary tangles and the pathogenesis of dementia. J Neuropathol & Appl Neurobiol 2:395–410
Garruto R, Fukatsu R, Yanagihara R, Gajdusek D, Hook G, Fiori C 1984 Imaging of calcium and aluminium in neurofibrillary tangle-bearing neurons in parkinsonism-dementia of Guam. Proc Natl Acad Sci USA 81:1875–1879
Jacobs RW, Duong T, Jones RE, Trapp GA, Scheibel AB 1989 A reexamination of aluminum in Alzheimer's disease: analysis by energy dispersive X-ray microprobe and flameless atomic absorption spectrophotometry. Can J Neurol Sci 16 (4 suppl):498–503
Perl DP, Brody AR 1980 Alzheimer's disease: x-ray spectrometric evidence of aluminum accumulation in neurofibrillary tangle-bearing neurons. Science (Wash DC) 208: 297–299
Perl D, Good P 1988 Laser microprobe mass analysis (LAMMA) evidence that aluminium selectively accumulates in the neurofibrillary tangles. J Neuropathol & Exp Neurol 47:318 (abstr)
Perl DP, Garruto RM, Gajdusek DC, Yanagihara RT, Gibbs CJ Jr 1982 Intraneuronal aluminium accumulation in amyotrophic lateral sclerosis and parkinsonism-dementia of Guam. Science (Wash DC) 217:1053–1055
Van de Putte DE, Jacob WA, Van Grieken RE 1990 Influence of fixation procedures on the microanalysis of lead induced intranuclear inclusions in rat kidneys. J Histochem Cytochem 38:331–337
Williams RJP 1992 Aluminium in biology: an introduction. In: Aluminium in biology and medicine. Wiley, Chichester (Ciba Found Symp 169) p 1–4
Wischik CM, Harrington CR, Mukaetova-Ladinska EB, Novak M, Edwards PC, McArthur FK 1992 Molecular characterization and measurement of Alzheimer's disease pathology: implications for genetic and environmental aetiology. In: Aluminium in biology and medicine. Wiley, Chichester (Ciba Found Symp 169) p 268–302

DISCUSSION

Candy: You are using toluidine blue to stain expoxy resin embedded sections of formalin-fixed tissue. Wouldn't the staining procedure, which presumably involved heating, lead to extraction/redistribution of calcium?

Williams: Yes, the procedure involves heating on a hotplate for 3 min.

Candy: People have tried for a long time, using X-ray microanalysis, to study the intracellular distribution of calcium, and it has proved extremely difficult to immobilize Ca. The intracellular concentration of free Ca is approximately 10^{-7} M; it is 10^{-3} M outside the neuron and the same concentration in the intracellular storage pools. Since you are not using a freeze-substitution technique with chelation—you fix and embed, and hope that calcium stays where it was—I am not sure you can interpret your calcium results.

Williams: If you release calcium, one of the places where you expect it to go, in those preparative circumstances, is into the nucleus; we did *not* find it in the nucleus, and we didn't find calcium above blanks anywhere else, in the normal cells; we found it only in the cells from Alzheimer's disease cases. So I just wonder if anybody else has noticed that there is something funny about calcium!

Garruto: Yes! I shall present some data on calcium, both in Guamanian ALS and PD and in experimental aluminium intoxication (p 212, 230).

McLachlan: When lethal doses of aluminium are injected as soluble salts directly into the brains of rabbits, calcium concentrations in the brain slowly rise over 10–25 days to remarkably high levels (Farnell et al 1985). Therefore, aluminum disturbs calcium homeostasis.

Williams: There is nothing contrary to that in my results, but you do have to say, however, that this Al level caused the damage and allowed the calcium to play its part. However, if our results are right, the Al level is low. It couldn't be the sort of *local* level of Al which we have discussed, approaching millimolar. It has to be at least a factor of 10 below that (10^{-4} M).

McLachlan: This was 5 μg of Al per g dry weight of neocortex, so it is in the lower 1–10 μM range in wet tissue.

Williams: I have calculated the *local* concentration from what was mentioned earlier (p 102). The Al concentration is very close to millimolar in the tau region which Dr McLachlan mentioned, and in the DNA it's more like 10^{-4} M.

Petersen: You mentioned calcium-dependent phosphorylation in relation to neurodegeneration. We should not forget that there are also calcium-dependent phosphatases, so in the same way that you can get abnormal phosphorylation, you can also get abnormal dephosphorylation.

Williams: Yes. What normally happens in the cell is that calcium stimulates phosphorylation and then dephosphorylation takes over, to bring the cell back to its normal state, because that's the normal triggering pattern of cells.

Petersen: But there are also specific calcium-activated phosphatases.

Williams: What we are told is an increase in phosphorylation, and we see phosphate increased in the tangle region. We therefore have to explain why that happens. The immediate thing is to look for something that does get phosphorylated. We conjecture that calcium can stimulate that. We don't know whether Al can.

Garruto: We performed an acute Al intoxication experiment in rabbits using 1% Al chloride which we inoculated intracisternally once weekly. All rabbits received a total Al dose of 0.5 mg or greater. We generated elemental images of the spinal cord using computer-controlled X-ray microanalysis and wavelength dispersive spectrometry to chemically map the elemental deposition. Using unfixed fresh frozen tissue we observed what appeared to be the active or passive movement of Al from the central canal into the surrounding spinal cord tissue and ultimately into ventral horn region where large motor neurons were undergoing degeneration (Strong et al 1990). What was surprising to us was not the preferential accumulation of aluminium, but the co-deposition of Ca in the same region, a situation reminiscent of our findings in Guamanian ALS and PD (see also p 230–231).

Perl: Basically, I have no problem with what you have shown, Professor Williams! I think that for the detection of Al the Proton Microprobe Unit at Oxford is somewhat more sensitive than standard X-ray energy spectrometry, but not much more. That's been my experience, having taken samples that we have already probed by laser microprobe mass analysis (LAMMA) and subjecting them to Proton Beam microanalysis. We find considerable variation in the concentration of Al in individual tangled neurons. I am not surprised that in examining the numbers of cells that you did, you didn't find Al. However, it is nice that you have now confirmed our findings by using perhaps a better detector system. This is where I think the problem lies. Recently, I provided the Oxford group with AD cases that we had probed using LAMMA and where we had identified high concentrations of Al. We picked the cases for their higher concentrations, to see what the proton beam instrument would do. Then, indeed, on the Oxford instrument, we began to identify Al peaks in tangles.

Williams: We should remember that the EDXA method with light elements is very dependent on the 'windows' that you managed to get into the machine. Most people have not got the right sort of window to be able to measure Al. We also have a parallel standard and measurement procedure for magnesium. We see magnesium, expected to be above mM. It does drift about, but as a rule it is uniformly above millimolar. That concentration is not a real problem. You would see Al in that range (where the Mg is). Thus the Al level has to be a good bit lower than the Mg, if Al is there at all. You, Dr Perl, can easily pick this up.

Perl: We pick Mg up very nicely, using LAMMA. The other point is that in the mapping mode, as you said, it's virtually impossible to identify these low concentrations of elements.

Williams: It's a mistake to use the mapping mode, except to localize the probe; then you must do point analysis.

Perl: That has been our finding, too.

Jope: There's no question that Al increases the phosphorylation of proteins (Johnson & Jope 1988), just as there's an abnormal increase in the phosphorylation of tau protein in Alzheimer's disease. They key question is: what causes the increased phosphorylation? We have shown that Al increases cyclic AMP (Johnson & Jope 1987), which will increase cyclic AMP-dependent kinase. We also suggest that Al blocks Ca efflux from synaptosomes (Koenig & Jope 1987); this will cause an accumulation of Ca in neurons, so aluminium may cause increased Ca-dependent phosphorylation. I think the key question isn't whether proteins are abnormally phosphorylated, but why specific proteins are abnormally phosphorylated, *which* kinases are activated, and how Al activates those kinases.

Williams: What level of Al do you need to get into the cell to do that?

Jope: Aluminium causes these effects after *in vivo* administration or at a concentration of about 10^{-4} M *in vitro*.

Williams: It would be very important if, at 10^{-4} M Al, you could get increased phosphorylation inside the cell, because then you would start to have a mechanism and, as you say, the effect must be specific. What we have to do now is to extend our experimental attack to find what is it that disturbs the incoming Ca, or the incoming Al, so that either could get to the kinases. But note that we cannot really detect 10^{-4} M Al.

Martin: What is the concentration of Ca that you think you are observing, when you see the Ca? And since you don't see Al, what's the limit on that?

Williams: Without any doubt, we do not see 10^{-4} M Al, but we would see Al in the range from this to 10^{-3} M with relative ease. Anything below that, we would not see and nobody else would either, with EDXA or the proton microprobe. I don't believe you can see it; if you say you can, we want to test your sample in our lab.! If you want a better resolution than ours, you must use a different type of analytical device. It's no good suggesting that everybody looks at similar samples, if they are using apparatus which cannot measure down to the supposedly significant levels. It has to be that only those people who have apparatus able to measure with sensitivity should exchange data, e.g. at one level, ourselves; at a lower level, those with SIMS or LAMMA.

For calcium, the proton microprobe is a lot more sensitive, because Ca is heavier than Al. So, when we are using the microprobe, we are going down to 10^{-4} M.

Perl: We are mostly examining post mortem material which has been formalin fixed, so we are detecting whatever calcium is left after these procedures. Calcium values will depend on the pH of the formalin, the post mortem interval, the interval in fixative, how you process it, and how rapidly you dehydrate— other things that are extraneous to the biology of the disease process. To try to extrapolate the results to the biology of the system as it was in life is very difficult.

Candy: The use of post mortem tissue excludes microanalytical studies on the distribution of diffusible elements.

Edwardson: Some work has been done on biopsy tissue from Alzheimer's disease patients; such samples could be snap frozen and sectioned for microanalysis.

Williams: I know. If we had been using plant tissue rather than brain, we could do that. I am relying on people like my colleague Dr Esiri to inform me and to tell me how to handle the sample, of course. The 'abnormal' calcium is only seen in the Alzheimer brain cells, but all cells are treated the same way.

Edwardson: We heard from Dr Martin earlier about an effect of Al on the aggregation of tubulin, with responses down to 10^{-10} M Al. This low concentration causes a stable aggregation of microtubules. If what Don McLachlan was saying is right, there may be a coupling of histone linker proteins to DNA, via aluminium, in which case you would be dealing with a single atom of Al controlling the expression of a neurofilament gene. When we are considering biological reactions such as these, we are dealing with concentrations of Al far below 10^{-4} M. The gap between what we can measure and what may be happening biologically is so immense that we may be out by five or more orders of magnitude.

Williams: Then you should give up the subject, because you are not going to be able to deal with water solution at those concentration levels by our methods.

McLachlan: You could use crystallography?

Williams: You can't do crystallography, unless you can have a diffraction pattern, which means you must have a lattice; and you can't get a lattice out of a system like this.

Petersen: I accept that there is something 'wrong' with Ca, but you mustn't jump to the conclusion that there is an increase in cytosolic free Ca, from your data. You can't draw that conclusion. There is apparently more Ca in the cell as a whole. This could be because there's more Ca in certain stores. So in fact you could argue the other way round, that cytosolic Ca was particularly low in that case, because for some reason it was taken up more readily in certain compartments. This is perfectly possible, and what you say would not necessarily be incompatible, for example, with Derek Birchall's hypothesis that there is an inhibition of Ca signal generation. In fact, I shall show evidence on that in my paper (p 237). This is why you ought to think about Ca-dependent phosphatases, because you could argue that increased phosphorylation was due to a *decreased* Ca signal.

Williams: I accept that. I should have said that what I see associated with particular sites is Ca, above 10^{-4} M on the site. The free Ca opposite that I don't know at all, because there's no measurement of Ca in the cytoplasm from these sorts of approaches. We suspect that when cells get damaged, one of the early changes is that they start to leak Ca. So we suspect that the calcium we see has arisen through cell damage in the brain, since we do not see the calcium in normal cells. It is interesting that you don't need to increase the free Ca very much, only to 10^{-5} M, before the cell switches on proteases, so that the cell commits suicide. The damage we see is not that.

Petersen: If you maintain a level of Ca of 1 µM for 10 min, the cell will be dead. This could be a late effect.

Williams: With all these analytical procedures, you can always drive yourself into your grave by making problems more difficult! If we think that 10^{-10} M Al can damage a cell, then this is not a problem that we are likely to resolve.

Martin: I disagree with that! You can reliably set metal ion concentrations with metal buffers. Using nitrilotriacetate we set the Al^{3+} concentration in the region of 10^{-12} M and observed promotion of tubulin assembly (Macdonald et al 1987).

Williams: I am talking about measuring free Al in a cell.

Martin: And I am talking about our system, and you said earlier you couldn't deal with low concentrations in water, but we went down to 10^{-12} M in water!

Williams: Then I have made a mistake! I intended to discuss the cell or circulating fluid water in the body. I agree that you can set the concentration *in vitro* using buffers.

Martin: It is an important point. I wish more people would use metal ion buffers, to set reliably low, known concentrations of metal ions. Clean interpretations become possible when metal ion concentrations are accurately known.

Williams: It's readily possible to measure calcium down to the level of about 10^{-7} M. That is done regularly in cells today. It may be that one could find an intracellular dyestuff which would work in your manner, but 10^{-10} M is a tall order, with Mg^{2+} at 10^{-3} M.

Note added after the symposium: During the visit of Dr Perl to Britain, the research team who developed the Oxford proton microprobe (see my paper) have had an opportunity to examine samples supplied by him from Guam ALS/PD patients who have some brain cells some five- to ten-fold more contaminated with Al than Alzheimer's disease cases. In the Guam neurons, levels of aluminium were not 'just about detectable' as in the samples described in our paper, and using the same equipment, but were measurable at about mM. Another element detected in considerable amounts associated with the tangles was calcium (about mM). These results confirmed the impressions gained through the discussion. It is now essential to compare data with others concerning several elements and Alzheimer brain samples.

References

Farnell BJ, Crapper McLachlan DR, Baimbridge K, De Boni U, Wong L, Wood PL 1985 Calcium metabolism in aluminum encephalopathy. Exp Neurol 88:68–83

Johnson GVW, Jope RS 1987 Aluminum alters cyclic AMP and cyclic GMP levels but not presynaptic cholinergic markers in rat brain in vivo. Brain Res 403:1–6

Johnson GVW, Jope RS 1988 Phosphorylation of rat brain cytoskeletal proteins is increased after orally administered aluminum. Brain Res 456:95–103

Koenig ML, Jope RS 1987 Aluminum inhibits the fast phase of voltage-dependent calcium influx into synaptosomes. J Neurochem 49:316–320

Macdonald TL, Humphreys WG, Martin RB 1987 Promotion of tubulin assembly by aluminum ion in vitro. Science (Wash DC) 236:183–186

Strong MJ, Yanagihara R, Wolff AN, Shankar SK, Garruto RM 1990 Experimental neurofilamentous aggregates: acute and chronic models of aluminum-induced encephalomyelopathy in rabbits. In: Rose FC, Norris FB (eds) Amyotrophic lateral sclerosis: new advances in toxicology and epidemiology. Smith-Gordon, London, p 157–173

Aluminium and the neurofibrillary tangle: results of tissue microprobe studies

Daniel P. Perl and Paul F. Good

Neuropathology Division, Department of Pathology, and the Arthur M. Fishberg Research Center for Neurobiology, Mount Sinai School of Medicine, One Gustave L. Levy Place, New York, NY 10029, USA

Abstract. Despite the contradictory results of studies attempting to compare the bulk brain tissue aluminium content of specimens from Alzheimer's disease patients and controls, microprobe studies from our laboratory have consistently documented evidence of selective accumulation of the element within the neurofibrillary tangle-bearing cells associated with this condition. Laser microprobe mass analysis (a highly sensitive and precise technique for trace elemental microprobe analysis) has now demonstrated that the most prominent aluminium accumulations occur within the neurofibrillary tangle itself. Similar findings have been obtained from microprobe studies of the neurofibrillary tangles which are a characteristic feature of amyotrophic lateral sclerosis/parkinsonism-dementia complex of Guam. Although the intraneuronal localization of aluminium in the Guam-derived specimens is similar to that of Alzheimer's disease, the concentration of aluminium is considerably higher than is encountered in Alzheimer's disease specimens. We conclude that aluminium is an integral component of the neurofibrillary tangle and raise the possibility that the cross-linking properties of this highly reactive metal may stabilize the constituent cytoskeletal proteins which make up this pathological structure.

1992 Aluminium in biology and medicine. Wiley, Chichester (Ciba Foundation Symposium 169) p 217–236

Aluminium is a silver-white metallic element and is the third most abundant element in the earth's crust. Indeed, aluminium is the earth's most abundant metallic element. A highly charged and reactive element, aluminium is virtually always encountered in combined form. In fact, its ability to form strong chemical bonds initially frustrated man's attempts to obtain even the most minuscule of purified samples of the metal. This made it so rare that Napoleon Bonaparte is said to have dined using utensils made of aluminium while his guests were relegated to using forks and spoons made of gold. However, in 1886 the Hall–Heroult process for the electrolytic separation of the metal finally yielded commercially feasible quantities of aluminium and transformed what was an exceedingly rare substance into a widely employed component of numerous industrial and household products. Despite aluminium's abundance as a constituent in the earth's crust, it is not known to be used in any biological process and thus is not regarded as an essential component of the diet. Indeed, and probably as a consequence of its small ionic radius and capacity for covalent bonding, aluminium must be effectively excluded from the tissues of most biological species, and in man tissue concentrations tend to be exceedingly small.

Over the past 20 years, research on the role of aluminium in Alzheimer's disease has been characterized by a growing body of solid reproducible scientific data which has unfortunately been intertwined with contradiction and controversy. The idea of an association between aluminium and Alzheimer's disease began with the development of an animal model of the human neurofibrillary tangle (NFT) in rabbits after direct exposure of the cerebral cortex to aluminium-containing salts (Klatzo et al 1965, Terry & Peña 1965). Although animals exposed in this way show a marked deficit in short-term memory and other cognitive abnormalities (King et al 1975, Solomon et al 1988), the ultrastructural features of the experimentally induced tangles are different from those of the paired helical filaments which characterize man's neurofibrillary tangles (Wisniewski et al 1976). Nevertheless, the existence of the experimental model led to a series of analytical studies designed to determine the bulk aluminium content of brain tissue from Alzheimer's disease patients and controls. These studies produced conflicting data, with Crapper and his colleagues claiming a 2–3-fold increase in aluminium concentration in Alzheimer's disease tissues (Crapper et al 1973, 1976), while other studies showed no significant difference between Alzheimer's disease and age-matched controls (McDermott et al 1979, Markesbery et al 1981). These discrepancies remain unresolved, although differences in analytic techniques and sample sizes continue to be debated.

Microprobe approach

Amid this controversy, information has emerged from our laboratory which serves as a scientific basis for a linkage between aluminium and the pathogenesis of Alzheimer's disease. We were initially concerned that bulk tissue aluminium assays might fail to identify accumulations of the element associated with focally distributed cellular lesions, which would make up only a very small proportion of the total tissue sample under analysis. To deal with this concern, our approach has been to use microprobe techniques to investigate the possible association of aluminium and other trace elemental abnormalities with Alzheimer's disease.

In 1980 we published evidence, obtained using scanning electron microscopy in conjunction with X-ray spectrometry, of selective aluminium accumulation within the NFT-bearing neurons of patients with Alzheimer's disease (Perl &

Brody 1980). At the time this represented a considerable technical achievement, considering the nature of the tissues being examined and the amounts of the element being detected. Using these techniques, we then proceeded to examine the trace elemental content of neurons identified in brain specimens derived from native inhabitants of the Pacific island of Guam who had suffered from amyotrophic lateral sclerosis (ALS) or from parkinsonism with dementia. The ALS/parkinsonism-dementia complex of Guam represents an endemic focus of neurodegenerative disease which, on the basis of extensive epidemiological investigation, is related to unique, though poorly understood, local environmental factors present on the island. Brain specimens obtained at autopsy from patients with ALS/parkinsonism-dementia of Guam were particularly relevant to these studies, since both conditions characteristically show large numbers of neurofibrillary tangles which are virtually identical to those encountered in patients with Alzheimer's disease. In the Guam-derived specimens we demonstrated evidence of dramatic intraneuronal aluminium accumulation, again within the neurofibrillary tangle-bearing cells (Perl et al 1982). Despite the recent resurgence of interest in the role of certain plant-derived neurotoxins (Spencer et al 1987, Spencer 1987), aluminium is still considered by most investigators to play a major role in the pathogenesis of this intriguing endemic focus of neurodegenerative disease (Garruto & Yase 1986, Perl & Pendlebury 1986).

Laser microprobe mass analysis

In the past few years we have adapted a new instrumental technology of tissue microprobe elemental analysis to our research, namely laser microprobe mass analysis (LAMMA). This technology has enabled us to provide much more precise information on the location, concentration and nature of trace elemental abnormalities in central nervous system tissues (Heinen et al 1980, Perl et al 1986). The laser microprobe focuses a high energy pulsed laser beam through the objective of a light microscope to perforate and ionize a minute portion of a semi-thin histological section of brain tissue which has been embedded in plastic. The ions formed by the laser perforation are attracted by a charged ion-lens into the column of a time-of-flight mass spectrometer, where they are separated and identified by differences in their mass. This remarkable instrument can detect the constituent elements of biological tissue samples at the level of one part per million and has a spatial resolution of about 1 µm. The intensity of the signal strength for an individual element, such as aluminium, is proportional to the concentration of the element in the specimen. Using this property, we are beginning to obtain accurate elemental concentrations within specific cellular compartments identified in the brain tissue specimens that we investigate.

The LAMMA instrument in our laboratory is now providing incontrovertible evidence that aluminium concentrations in NFT-bearing neurons of Alzheimer's disease patients are many times greater than those found in the immediately adjacent non-tangled neurons or in the adjacent neuropil. Indeed, as will be discussed later, these studies indicate that most of the excess aluminium in the tangle-bearing neuron is localized specifically in association with the NFT itself.

LAMMA studies of hippocampal neurons in Alzheimer's disease

Our approach has involved the analysis of brain tissue specimens derived at autopsy from patients who have satisfied clinical diagnostic criteria for Alzheimer's disease and whose diagnosis has been confirmed at autopsy. Excluded from these studies are any individuals with neuropathological evidence of other significant neurological conditions, such as superimposed infarcts, lesions associated with idiopathic Parkinson's disease, or hypoxic damage. Control cases were age-matched individuals who were noted to be free of dementia or other significant neurological disorders in the months before their death and, at autopsy, were free of evidence of neurofibrillary tangles or large numbers of senile plaques within the hippocampus or neocortex.

Most of the LAMMA data that we have obtained so far have been derived from probing neurons in the CA_1 region of the posterior portion of the hippocampus (we sample the hippocampus at a coronal level, lying between the level of the lateral geniculate nucleus and the caudal pole of the pulvinar complex of the thalamus). Samples are taken from the formalin-fixed brain tissues and are dissected into blocks measuring about $2 \, mm^3$. The tissues are post-fixed in OsO_4 and embedded in Spurr's epoxy resin. Sections are then cut at $0.75 \, \mu m$ thickness and are mounted on 3 mm electron microscopy grids. The grids are then introduced into a Leyboldt-Heraeus LAMMA 500 and subjected to multipoint microprobe analysis.

Within the LAMMA specimen chamber, NFT-bearing and NFT-free neurons are readily identified. The cellular elements to be analysed are identified using high magnification light microscope optics integral to the instrument ($100 \times$ Zeiss Ultrafluar glycerine immersion objective). Probe sites are determined by aiming a continuous helium-neon (HeNe) laser coaxial with the pulsed high energy neodymium-YAG laser. Probe sites within the NFT-bearing neurons are directed to the nuclear heterochromatin, the NFT itself, the NFT-free cytoplasm, and the immediately adjacent neuropil. In a similar fashion, probe sites of the tangle-free neurons are directed to the nuclear heterochromatin, the cytoplasm and the immediately adjacent neuropil. A maximum of five laser shots are placed into the four probe areas of an NFT-bearing cell, followed by the three probe areas of an NFT-free cell. The neuronal types (NFT-bearing and NFT-free cells) are alternated sequentially until spectra from eight to 15 neurons of each type are collected for each case. The control cases are probed in a similar manner, using identical instrumental parameters.

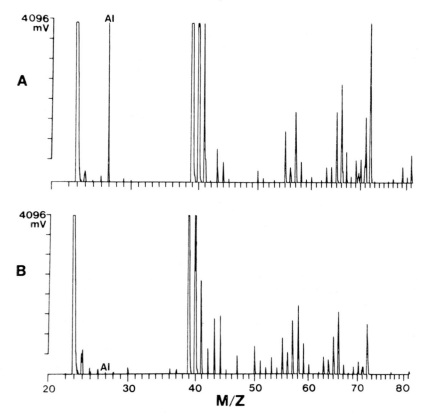

FIG. 1.   A. Partial mass spectrum taken from a probe site directed to the neurofibrillary tangle within a tangle-bearing neuron from the hippocampus of an Alzheimer's disease patient. Notice the prominent mass 27 peak related to the presence of aluminium. B. Partial mass spectrum taken from a probe site directed to the cytoplasm adjacent to the probe site displayed in A. This portion of cytoplasm is not involved in the neurofibrillary tangle within this cell. Notice the relative absence of a peak at mass 27.

Figure 1 (A and B) demonstrates representative partial mass spectra obtained from laser microprobe sites directed to the neurofibrillary tangle identified in one of the Alzheimer's disease-derived hippocampal samples (A) and the spectrum obtained from a probe site directed to the immediately adjacent uninvolved cytoplasm of the same neuron (B). Notice the prominent mass 27 signal present in the spectrum obtained from the neurofibrillary tangle, indicative of the presence of aluminium, whereas the spectrum obtained from the uninvolved cytoplasm of that cell shows almost no mass 27 signal. Not every laser probe site directed to a neurofibrillary tangle produces such a prominent

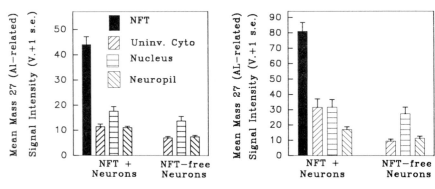

FIG. 2. The mean mass 27 (aluminium-related) signal intensity (V ± 1 SE) obtained from hippocampal neurofibrillary tangle-bearing and tangle-free neurons of two cases of Alzheimer's disease. The data are collected for the neurofibrillary tangle, uninvolved cytoplasm, nucleochromatin and adjacent neuropil.

aluminium-related signal; yet, when the mean spectral intensity obtained from each cellular compartment of hippocampal tissue from an Alzheimer's disease case is analysed, a clear association between an increased aluminium-related signal and the NFT is seen. Figure 2 shows the mean aluminium-related signal intensities from the seven probe sites of samples obtained from two Alzheimer's disease patients.

In a recent LAMMA microprobe study (Good et al 1992), hippocampal samples probed from 10 Alzheimer's disease patients revealed evidence of aluminium accumulation within the neurofibrillary tangles. Figure 3 shows the mean aluminium-related signal obtained from probe sites in the neurofibrillary tangles, uninvolved cytoplasm of the tangle-bearing cells, and the cytoplasm of the adjacent tangle-free cells from the 10 cases. The results show the consistency with which excess aluminium is detected within the neurofibrillary tangles of the tangle-bearing cells in the cases examined. We have evaluated the spectra for the presence of a consistent correlation of any of 60 elements of biological significance with neurofibrillary tangle formation. Within the limits of detection of the laser microprobe instrumentation (1–5 parts per million, minimum detection limit), no other trace element appears to accumulate at the intense level of aluminium in association with the neurofibrillary tangle-bearing neurons of this disorder; however, the signal for iron is also increased within the neurofibrillary tangle. It is of interest that the accumulation appears to be specific for aluminium and less so for iron, and that we could find no other element which was associated with this form of cytopathology. This statement must be qualified, because we are using brain tissues that have been obtained at autopsy and were fixed in water-based fixatives. We acknowledge that the procedures used in these studies do not enable us to

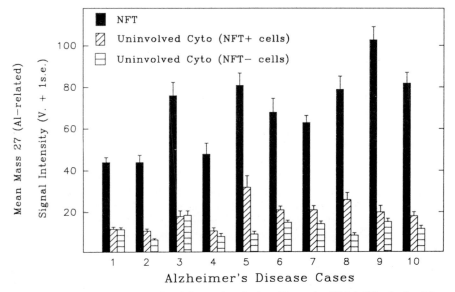

FIG. 3.   The mean mass 27 (aluminium-related) signal intensity (V ± 1 SE) obtained from the neurofibrillary tangle probe sites and the uninvolved cytoplasm within hippocampal neurofibrillary tangle-bearing (NFT +) and tangle-free (NFT −) neurons of 10 cases of Alzheimer's disease.

evaluate many of the loosely bound physiological ions, such as sodium, calcium and magnesium.

Discussion

On the basis of our recent LAMMA findings, it can no longer reasonably be argued that aluminium does not accumulate in the NFT-bearing neurons of Alzheimer's disease. The difficult problem remaining is to determine the functional significance of this finding. There can be no doubt that when aluminium gains access to the central nervous system of certain animal species, it acts as a potent neurotoxin. The origin of the aluminium we have detected and its route of access to the CNS remain unclear. Overt evidence of excessive or abnormal exposure to aluminium-containing products is not a prerequisite for the development of Alzheimer's disease (Heyman et al 1984), although some studies have shown an association between exposure to environmental sources of aluminium and the subsequent development of this disease (Martyn et al 1989, Graves et al 1990). However, all such studies must be viewed as inconclusive because they fail to take into account critical factors related to aluminium speciation and bioavailability. Furthermore, the ubiquity and abundance of this element in the environment limit our ability to take a reliable aluminium exposure history.

Although many species, including man, are heavily exposed to aluminium, little evidence of acute overt damage has been noted. However, as we mentioned earlier, the major part of the aluminium in the natural and commercial environment is present in the form of aluminium oxides or hydroxides and is thus not bioavailable at physiological pH (Martin 1986, Ganrot 1986, Macdonald & Martin 1988). It has been clear to those working on the biological effects of acid rain that shifts in aluminium speciation can markedly alter bioavailability, with dramatic pathological effects (Driscoll et al 1980, Muniz & Leivestad 1980). Most studies on the neurotoxicity of aluminium in mammals have been performed using inorganic aluminium salts by individuals who were not familiar with aluminium's complex chemistry (see Martin 1986, Macdonald & Martin 1988). More recently, aluminium linked to organic ligands has been introduced into this research, with evidence of dramatically increased bioavailability and consequently increased toxicity (Finnegan et al 1986, Corain et al 1988).

Biochemical research has provided considerable evidence for a number of mechanisms by which aluminium neurotoxicity may be relevant to pathogenetic events encountered in Alzheimer's disease. Now that we have demonstrated the presence of aluminium within intact neurons showing cytopathological changes in patients with Alzheimer's disease, we must consider what effects the presence of such an active ion might have on important biological processes.

Since aluminium has been consistently observed to be co-localized with the neurofibrillary tangle, we suggest that this represents the interaction of this highly charged metal with specific binding sites on integral constituents of this structure. The protein composition of the paired helical filaments of the neurofibrillary tangle is now beginning to yield to analysis. Recent studies suggest that the cores of these fibrils are composed of an abnormally phosphorylated form of the microtubule-associated protein, tau (Goedert et al 1988, Wischik et al 1988, Crowther et al 1989, Lee et al 1991). Himmler and colleagues (Himmler et al 1989, Himmler 1989) have noted that the tau molecule contains a metal-binding site which might account for aluminium binding to neurofibrillary tangles. Alternatively, aluminium might bind to the highly acidic tau-binding site of the tubulin molecule (Littauer et al 1986, Serrano et al 1985), thus preventing the binding of tau. This blockage would result in an abnormal accumulation of the tau molecule within the neuronal perikaryon. Aluminium can replace magnesium ions within the magnesium–phosphate complex at the GTP-binding sites of, for example, tubulin, and thus block normal GTP hydrolysis and GTP–GDP exchange (Macdonald et al 1987). Aluminium has long been known to bind covalently to proteins and to function as a cross-linking stabilizer.

The studies in our laboratory clearly demonstrate that aluminium selectively binds to the neurofibrillary tangle of Alzheimer's disease at some point in the biological progression of the disease. The nature of this binding and its consequences remain unknown. However, the presence of this highly reactive and potentially toxic metal must be taken into account in any investigation into the nature and mechanisms of formation of this important pathological structure.

Acknowledgements

The authors would like to thank Gina Jabbar, Amy Hsu and Essie Smith for their technical and secretarial assistance. Supported, in part, by grants AG-05138, AG-08802 and ES-00928 from the National Institutes of Health.

References

Corain B, Bombi GG, Zatta P 1988 Differential effects of covalent compounds in aluminum toxicology. Neurobiol Aging 9:413–414

Crapper DR, Krishnan SS, Dalton AJ 1973 Brain aluminum distribution in Alzheimer's disease and experimental neurofibrillary degeneration. Science (Wash DC) 180: 511–513

Crapper DR, Krishnan SS, Quittkat S 1976 Aluminium, neurofibrillary degeneration and Alzheimer's disease. Brain 99:67–80

Crowther T, Goedert M, Wischik CM 1989 The repeat region of microtubule-associated protein tau forms part of the core of the paired helical filament of Alzheimer's disease. Ann Med 21:127–132

Driscoll CT, Baker JP, Bisogni JJ, Schofield CL 1980 Effect of aluminium speciation on fish in dilute acidified waters. Nature (Lond) 284:161–164

Finnegan MM, Rettig ST, Orvig C 1986 A neutral water-soluble aluminum complex of neurological interest. J Am Chem Soc 108:5033–5035

Ganrot PO 1986 Metabolism and possible health effects of aluminum. Environ Health Perspect 65:363–441

Garruto RM, Yase Y 1986 Neurodegenerative disorders of the western Pacific: the search for mechanisms of pathogenesis. Trends Neurosci 9:368–374

Goedert M, Wischik CM, Crowther RA, Walker JE, Klug A 1988 Cloning and sequencing of the cDNA encoding a core protein of the paired helical filament of Alzheimer disease: identification as the microtubule-associated protein tau. Proc Natl Acad Sci USA 85:4051–4055

Good PF, Perl DP, Bierer LM, Schmeidler J 1992 Selective accumulation of aluminum and iron in the neurofibrillary tangles of Alzheimer's disease: a laser microprobe (LAMMA) study. Ann Neurol 31:286–292

Graves AB, White E, Koepsell TD, Reifler BV, van Belle G, Larson EB 1990 The association between aluminum-containing products and Alzheimer's disease. J Clin Epidemiol 43:35–44

Heinen HJ, Hillenkamp F, Kaufmann R, Schroder W, Wechsung R 1980 A new laser microprobe mass analyzer for biomedicine and biological materials analysis. In: Frigerio A, McCamish M (eds) Recent developments in mass spectrometry in biochemistry and medicine. Elsevier, Amsterdam, vol 6:435–451

Heyman A, Wilkinson WE, Stafford JA, Helms MJ, Sigmon AH, Weinberg T 1984 Alzheimer's disease: a study of epidemiological aspects. Ann Neurol 15:335–341

Himmler A 1989 Structure of the bovine tau gene: alternatively spliced transcripts generate a protein family. Mol Cell Biol 9:1389–1396

Himmler A, Drechsel D, Kirschner MW, Martin DW Jr 1989 Tau consists of a set of proteins with repeated C-terminal microtubule-binding domains and variable N-terminal domains. Mol Cell Biol 9:1381–1388

King GA, De Boni U, Crapper DR 1975 Effect of aluminum upon conditioned avoidance response acquisition in the absence of neurofibrillary degeneration. Pharmacol Biochem Behav 3:1003–1009

Klatzo I, Wisniewski H, Streicher E 1965 Experimental production of neurofibrillary pathology. I. Light microscopic observations. J Neuropathol & Exp Neurol 24:187–199

Lee VMY, Balin BJ, Otvos L Jr, Trojanowski JQ 1991 A major subunit of paired helical filaments and derivatized forms of normal tau. Science (Wash DC) 251: 675–678

Littauer UZ, Giveon D, Thierauf M, Ginzburg I, Ponstingl H 1986 Common and distinct tubulin binding sites for microtubule-associated proteins. Proc Natl Acad Sci USA 83:7162–7166

Macdonald TL, Martin RB 1988 Aluminum ion in biological systems. Trends Biochem Sci 13:15–19

Macdonald TL, Humphreys WG, Martin RB 1987 Promotion of tubulin assembly by aluminum ion in vitro. Science (Wash DC) 236:183–186

Markesbery WR, Ehmann WD, Hossain TI, Alauddin M, Goodin DT 1981 Instrumental neutron activation analysis of brain aluminum in Alzheimer disease and aging. Ann Neurol 10:511–516

Martin RB 1986 The chemistry of aluminum as related to biology and medicine. Clin Chem 32:1797–1806

Martyn CN, Barker DJP, Osmond C, Harris EC, Edwardson JA, Lacey RF 1989 Geographical relation between Alzheimer's disease and aluminium in drinking water. Lancet 1:59–62

McDermott JR, Smith AI, Iqbal K, Wisniewski HM 1979 Brain aluminum in aging and Alzheimer disease. Neurology 29:809–814

Muniz IP, Leivestad H 1980 In: Drablos D, Tollan A (eds) Ecological impact of acid precipitation. SNSF Project, Oslo, Norway, p 269

Perl DP, Brody AR 1980 Alzheimer's disease: x-ray spectrometric evidence of aluminum accumulation in neurofibrillary tangle-bearing neurons. Science (Wash DC) 208: 297–299

Perl DP, Pendlebury WW 1986 Aluminum neurotoxicity—potential role in the pathogenesis of neurofibrillary tangle formation. Can J Neurol Sci 13:441–445

Perl DP, Gajdusek DC, Garruto RM, Yanagihara RT, Gibbs CJ Jr 1982 Intraneuronal aluminum accumulation in amyotrophic lateral sclerosis and parkinsonism-dementia of Guam. Science (Wash DC) 217:1053–1055

Perl DP, Muñoz-Garcia D, Good P, Pendlebury WW 1986 Laser microprobe mass analyzer (LAMMA)—a new approach to the study of the association of aluminum and neurofibrillary tangle formation. In: Fisher A, Hanin I, Lachman C (eds) Alzheimer's and Parkinson's disease: strategies for research and development. Plenum Publishing Corporation, New York, p 241–248

Serrano L, Montejo de Garcini E, Hernandez MA, Avila J 1985 Localization of the tubulin binding site for tau protein. J Biochem 153:595–600

Solomon PR, Pingree TM, Baldwin D, Koota D, Perl DP, Pendlebury WW 1988 Disrupted retention of the classically conditioned nictitating membrane response in rabbits with aluminum-induced neurofibrillary degeneration. Neurotoxicology 9:209–221

Spencer PS 1987 Guam ALS/parkinsonism-dementia: a long-latency neurotoxic disorder caused by 'slow toxin(s)' in food? Can J Neurol Sci 14:347–357

Spencer PS, Nunn PB, Hugon J et al 1987 Guam amyotrophic lateral sclerosis–parkinsonism–dementia linked to a plant excitant neurotoxin. Science (Wash DC) 237:517–522

Terry RD, Peña C 1965 Experimental production of neurofibrillary degeneration. 2. Electron microscopy, phosphatase histochemistry and electron probe analysis. J Neuropathol & Exp Neurol 24:200–210

Wischik CM, Novak M, Thøgersen HC et al 1988 Isolation of a fragment of tau derived from the core of the paired helical filament of Alzheimer's disease. Proc Natl Acad Sci USA 85:4506–4510

Wisniewski HM, Narang HK, Terry RD 1976 Neurofibrillary tangles of paired helical filaments. J Neurol Sci 27:173–181

DISCUSSION

Fawell: Is there any way that you can visualize the very earliest stage of the neurofibril formation in the cell and measure whether Al is present? Obviously this is difficult, but if you are saying that Al is causal, then Al has to be there at the beginning of the formation of the tangle, which presumably develops gradually. Can you study that very early stage and see whether Al is present?

Perl: We think we can. There are a number of experiments that we can do for which we now have the tools available. We intend to look for cells that show cytoskeletal abnormalities (for example, Alz-50-positive cells), but don't yet contain tangles, to see what their Al content is. We also want to examine the hippocampi of Down's syndrome patients of intermediate age (20–30 years), as another approach to this. And ultimately we want to be able to trace out the anatomy, in terms of the spread of these lesions in the brain. That approach will be very much more difficult.

Other ingredients are obviously needed, apart from Al; it has to be a properly primed cell, in order for tangle formation to happen. The time when Al is added to the process, to finally 'seed' the mixture, leading to the binding up of the whole thing, we don't know. The timing of many of the events in Alzheimer's disease remains a very difficult problem.

Birchall: An important point is that the entry of a small (undetectable) concentration of aluminium into a cell may, if it becomes bound at a critical site, perhaps displacing Mg^{2+}, trigger off a cascade of events and damage amplification. Professor Petersen will describe an effect on Ca^{2+} mobilization of aluminium injected into a cell (p 237). This could be such a trigger event.

I find Dr Perl's results fascinating: the presence of aluminium in tangles may be most significant. Aluminium would be expected to bind to phosphorylated proteins as does iron. Fe^{3+} binds to phosvitin with high affinity (Log K = 18.0) at clustered phosphorylserine residues (Hegenauer et al 1979) and it appears that aluminium behaves similarly (Hegenauer et al 1977). Such binding reduces the electrostatic repulsion between the phosphate groups of phosvitin at neutral pH and induces a conformational change towards the β-structure. Perhaps this is a clue to the effect of aluminium in tangle formation.

Williams: You always detect Mg and Ca in LAMMA-studied systems, Dr Perl? How much Mg and Ca do you actually have in the neurofibrillary tangle region?

Perl: The detected Mg concentration is relatively low in tangles and we don't see any significant difference between the Mg signal in the tangle and in the other six cellular components. Mg is increased in nuclei, as you would expect, but there is no difference between these two cell types (tangle-bearing and non-tangled neurons).

The Ca data are difficult to interpret, only because the LAMMA is set to maximize the signal from the trace elements, and Ca saturates rather readily.

Williams: That would mean you have Ca levels above the level of trace elements?

Perl: Yes, definitely.

Williams: But Ca is supposed to be 10^{-7} M in cytoplasm.

Perl: The problem is that the ionization efficiency of each element varies. The instrument is more sensitive for detecting potassium, calcium and sodium than for some other elements. There is a significant difference, for instance, between Al and Si in terms of the ionization potential of the two elements.

Williams: It would be good to know those numbers.

Perl: Yes. We would have to re-do the collection of the data at different settings; I might expect to find increased Ca in tangled cells if we did that.

Wisniewski: In my opinion, the presence of Al in the tangle does not mean that Al is necessary for PHF formation. However, Dr Perl thinks that Al is important in PHF formation. Yet, Jim Edwardson has found that 15–16 years of dialysis increased the level of Al in the brain at least 100%, but there was no accelerated pace of PHF development. In necrotic brain tissue, particularly in children, we see accumulations of calcium, iron, silicon, and probably other trace atoms. It is possible, therefore, that the large NFT may act as a sink for Al and other atoms.

Another point: in tangles induced by Al, there is no Al present. Therefore, we need to be more open-minded about the presence of Al and its importance in the pathogenesis of the PHF. Again, the neuronal β-protein-positive immunoreactivity is probably in lipofuscin, because we see it often in neurons filled with lipofuscin. I would expect that neurons are affected by lipofuscin accumulation in patients with dialysis more than in age-matched controls. Therefore, my interpretation of Drs Candy and Edwardson's data would be that the immunoreactivity to APP which they see in neurons in dialysis patients is associated with lipofuscin.

Candy: I am fairly confident that the APP staining that we have observed is associated with the nucleus, but, as you know, in post mortem tissue, at the ultrastructural level, intracellular organelles are frequently disrupted!

Williams: We must not confuse the good part of the data with what *may be* the poor part of the interpretation! What Dr Perl has shown and established is the good part of the data; after that we can argue about the interpretation. He showed that Al is in the tangles in all his experimental cases, with a method (LAMMA) which is better than any other current method.

Candy: It's not better than SIMS, where we have, at present, an imaging sensitivity limit of approximately 3 p.p.m. In an initial study we have failed to find evidence of an association between Al and neurofibrillary tangles in the frontal cortex in Alzheimer's disease. If Al was present at 20–30 p.p.m, we would have clearly seen it.

Williams: So what is the difference between your Alzheimer's patients and his?

Candy: We are using unfixed, unstained material, which is critical for establishing the 'gold' standard for trace element analysis. In the dialysis patients that we have studied, the highest focal concentration of Al was between 300 and 500 p.p.m. This is what is claimed to be the focal Al concentration associated with neurofibrillary tangles in patients with the parkinsonism-dementia complex of Guam. I don't understand how in the renal dialysis patients we are only finding these levels, yet your Guam cases have very similar focal Al concentrations in the absence of renal impairment.

Williams: So you have to say that Dr Perl's data are artifacts?

Candy: I would like to see LAMMA done on unfixed, unstained material. Until you have established that Al is present in such material, you can't rule out analytical artifacts. We have developed a strategy using unfixed, unstained tissue sections and we have taken great care to ensure that the substrates on which we mount the sections do not have focal Al deposits.

Perl: But we have seen the same peaks related to the presence of aluminium in unfixed frozen sections that have not been osmicated, plastic-embedded or otherwise handled.

Candy: How did you visualize a tangle in an unfixed frozen section?

Perl: I can't visualize a tangle in these relatively thick air-dried frozen sections. In unfixed frozen sections we find the same aluminium-related peaks, but with the LAMMA you cannot specifically visualize a tangle; you can only examine an area in the hippocampus that is loaded with tangles, and find the same Al peaks (Good et al 1992).

Zatta: The presence of aluminium in features of Alzheimer's disease has been a matter of controversy for a very long time. These phenomenological discrepancies are producing an enormous confusion and a final answer has become an imperative task. I think it is time (and this is also the proper place) to activate an international interlaboratory project aimed at ending the controversy on the presence and compartmentation of Al in AD and other related pathologies. The involved laboratories should be requested to exchange biological material and standards suitable for more analytical methods. The final evaluation and reliability of the results should be left to a panel of experts belonging to international bodies such as IUPAC. The unambiguous, internationally controlled resolution of this controversy would represent a milestone in the understanding of the molecular bases of this challenging issue.

Perl: Dr Wisniewski raises the possibility that our findings are non-specific and that Al sticks to damaged tissues. If you fix Alzheimer's disease brain tissue in formalin to which you have added Al and then process the tissue and analyse it by LAMMA, you find Al to be increased in concentration throughout all the cellular components. It raises the level of Al in the tissue, when compared to an adjacent piece, fixed in Al-free formalin, not only in the tangles but in the cytoplasm and neuropil and in the non-involved cells; it doesn't go selectively to the tangle. The NFT is not a sink for Al, therefore. You cannot dismiss a finding of 100 p.p.m. of Al associated with this specific pathological structure. Nor can you dismiss the 300–600 p.p.m. of Al seen in the tangles of the ALS/parkinsonism-dementia cases of Guam. It is an abnormality that has to be explained. You may not like my interpretation of the data, but you cannot dispute the findings.

Williams: We would like to ask that you and Dr Candy exchange samples.

Zatta: I want to emphasize the importance of standardizing the analytical data, otherwise for the next decade or so we shall be here again saying the same controversial things!

Perl: We have done this for standards with the group in Newcastle.

Garruto: We have extensively discussed the role of Al in Alzheimer's disease, but in my opinion the role of Al in neurodegeneration can also be understood by reviewing some of the data on Guamanian ALS and parkinsonism-dementia (PD), two disorders found in high incidence in the Western Pacific (Garruto 1991). The neurofibrillary degeneration that occurs in ALS and PD of Guam is in many cases probably more extensive than what you see in Alzheimer's disease. The neurofibrillary pathology is dramatic and in some cases nearly all the remaining neurons contain tangles. These tangles are histochemically, immunocytochemically, ultrastructurally and molecularly indistinguishable from those of patients with Alzheimer's disease.

Using computer-controlled X-ray microanalysis and wavelength dispersive spectrometry we produced chemical maps of the hippocampus of the brain and subsequently of the spinal cord of ALS and PD patients from Guam. These elemental images demonstrated the striking co-localization of calcium (7200 p.p.m. dry weight), aluminium (500 p.p.m. dry weight), and later silicon (3000 p.p.m. dry weight) in neurofibrillary tangle-bearing neurons and in the walls of cerebral blood vessels (Garruto et al 1984, 1986). The co-localization of these elements has been repeatedly confirmed by other techniques, including secondary ion mass spectrometry (that also demonstrated the co-localization of these elements in oligodendrocytes) (Linton et al 1987). No other elements, with the occasional exception of iron, were found in elevated amounts in these tissues. Similarly, the abnormal deposition of calcium, aluminium and silicon occurs in neurologically normal Guamanians with NFT formation, but not in those free of such lesions (Piccardo et al 1988). I am not suggesting what the temporal order of elemental accumulation is in these neurons, nor am I suggesting that all neurofibrillary tangle-bearing neurons were intact at the time of the patient's death, but clearly some were, because we were able to visualize the deposition of these elements in neuritic process of some cells (Garruto et al 1984). Lastly, our order-of-magnitude estimates of the amount of Al in affected neurons in Guamanian ALS and PD are one to two orders higher than what has been reported for Alzheimer's disease. Every research group who has ever looked for aluminium in nervous tissue of these patients has found plenty of it!

Perl: In Guam-derived specimens, many of the tangle-bearing cells have died, leaving neuron-shaped structures called 'ghost tangles'. When that happens, we find by LAMMA that the Al level is much lower in these 'ghosts', or extracellular tangles.

Williams: Your analytical numbers are almost exactly those we gave for calcium, Dr Garruto, so the three analytical methods are giving the same range of values for calcium, whereas for Al, using the proton microprobe, we could only just detect the element; but Dr Perl (using LAMMA) and Dr Garruto (using X-ray microanalysis and wavelength dispersive spectrometry) find around 10^{-4} M. So the analytical methods so far are matching two ways, for aluminium. I trust this is correct?

Wisniewski: There are data on trace elements in necrotic tissue: when these data are compared with your data, Dr Perl, are they similar? Studies with GFAP (glial fibrillary acidic protein) antibody show many of the NFT-bearing neurons as positive. It looks like a 'petrified forest'. Many of the ghost profiles are already penetrated by astrocytic processes, suggesting that many of the NFT are already in 'extracellular' space.

Perl: The issue of 'ghost' tangles is important, because those structures contain much lower concentrations of Al by LAMMA analysis. One of the few neuropathological criteria that we have for including cases in our series is that if we find that significant number of neurons have dropped out, we do not use such a case. We have looked at damaged (infarcted) tissue as well, and don't find that the damaged tissue accumulates Al. We have also studied other forms of neuronal pathology and don't find Al accumulating, so we don't think Al is a non-specific marker of cell damage. This element is unlike Ca, which is present in large amounts in all tissue, and starts to move around after a cell dies.

Candy: The key to the problem of trace element microanalysis is technique dependent. Using a magnetic sector SIMS instrument it is possible to achieve a very high mass resolution, of 10 000 (Dr Perl has unit resolution using LAMMA). Figure 1 shows a high resolution mass spectrum of mass 27 in the frontal cortex of a renal dialysis patient. This mass spectrum demonstrates that

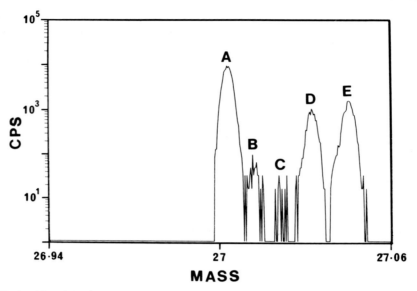

FIG. 1 (*Candy*) High resolution spectrum of mass 27 in the frontal cortex of a renal dialysis patient. The mass spectrum shows the presence of A, Al; B, ^{26}MgH; C, C^{15}N; D, CNH; and E, C_2H_3.

C_2H_3, CNH, C^{15}N and ^{26}MgH are mass interferences on the aluminium signal when recording Al ion images at low mass resolution (M/ΔM = 300, FWHM). I don't know to what extent you get mass interferences with LAMMA? So that at low mass 27 signals, one has to be careful of mass interferences, which may depend on the laser transmission.

Perl: In our spectra comparing the tangle and the adjacent cytoplasm (see Fig. 1, p 221), are you saying that this would be related to some extraneous organic fragment with mass 27 present in extremely high concentrations, yet virtually no appreciable accompanying signal would be present at mass 25, 26 or 28? That does not make sense. What we are detecting is Al.

Candy: You are not analysing a homogeneous material; you are using a relatively inconstant pulse of laser energy to evaporate and ionize a tangle of fibrillary protein and comparing that with the LAMMA signal from predominantly non-fibrillary material in the cytoplasm. You can't compare the LAMMA mass 27 signals in a tangle and cytoplasm in that way, because you are not sure that these matrices will evaporate and ionize in an identical way.

I was disappointed that you used epoxy resin-embedded, Al-loaded brain homogenate standards for quantification, because we don't think that these standards are satisfactory! In an unfixed, unembedded Al-loaded brain homogenate, we have shown using SIMS that Al is homogeneously distributed. However, when we fix and epoxy resin embed this standard, we see focal precipitates of aluminium. So I am not sure how, with LAMMA, you get such a very good linear relationship between the Al concentration of the brain homogenate and LAMMA mass 27 signal.

Perl: LAMMA has a very small 'viewing window'. By this I mean that the probe is only 1 μm in diameter. Yours is 400 μm. We are averaging 50–100 1 μm probe spots on a semi-thin section and we collect data on whatever is distributed in this tissue.

Candy: But that will be lower than the bulk estimate of the concentration of aluminium in the brain homogenate standard, because the aluminium is not homogeneously distributed in the standard.

Perl: It's homogeneous enough. We can't be sure of the precise concentrations, I agree.

Wischik: I would like to raise a point that bears directly on your central thesis, that Al is required for PHF formation. I understand that the extracellular tangles (i.e. 'ghosts' of former NFT-bearing cells) are negative for Al, or that the levels are much lower?

Perl: They tend to show a lower concentration of Al, but are not 'negative' for Al.

Wischik: This observation undermines the hypothesis that Al is the glue that holds the PHF together. We have demonstrated that we can visualize paired helical filaments that come from the extracellular tangles and that these PHFs are morphologically intact. The fundamental difference between intracellular and extracellular PHFs is the loss of the N-terminal 200 amino acids of tau in the extracellular ones. Now, the only phosphate site that has unequivocally been localized to the PHF is the Tau-1 site. The only firm evidence of abnormal phosphorylation is in the N-terminal part of the tau molecule, detected by the Tau-1 antibody, the only available marker which can be used to make that point.

We have shown that this region of the molecule is lost as tangles pass from intracellular to extracellular compartments (Bondareff et al 1990). Therefore, if you are saying that the aberrant phosphate site is the most likely Al-binding site, that would be entirely consistent: there may well be an Al-binding site which is phosphate, and is in the intracellular tangle; nevertheless, the paired helical filament can remain structurally intact without the N-terminal domain of tau (Wischik et al 1988), and in fact without (from your evidence) high levels of Al.

Perl: This brings us to another technical problem regarding the use of LAMMA for trace elemental analysis of tissues. A tangle in an intact neuron is a dense structure and the laser shot is thus an almost pure sample of PHF. The extracellular tangles are much more diffuse. In a sense, the extracellular tangle has been diluted out by extracellular matrix. So in a laser probe site to an extracellular tangle you will have relatively less PHF material in this admittedly very small sample size. I don't take the 'lower concentration of Al' concept as a given; I think we have to look at this further. We need to isolate purified samples of tangles and start to take apart the PHF, to see at what point

Al leaves the structure. That's another approach, to try to find what Al is binding to. Something in the tangles is binding Al selectively; the data suggest to me that Al represents an essential component of the PHF.

Wischik: It would be equally consistent with the data available from both you and Dr Garruto that there is a high affinity Al-binding site in PHFs; but as to *when* the Al comes in, it may enter during the technical processing of the specimen. It is not established that Al is there *in vivo.*

Perl: From our positive data on the untreated frozen sections, I don't see how that would happen.

Williams: That would be a problem for the Newcastle group to respond to. Using EDXA (electron dispersed X-ray analysis) we virtually just repeated the work done before the development of the new methods, which had claimed Al to be in these tangles. We tried to improve analysis using the proton microprobe. We found no Al, or only a trace. But Perl and Garruto now discuss one or two orders of magnitude better in the analyses, and at this stage showed Al was present. The Newcastle data, just presented, seem to show, from SIMS, that they don't see detectable Al in equivalent(?) samples. Until we straighten this out by exchanging samples, we shall not know who is wrong!

Wischik: Taking the value of 500 p.p.m. of Al (Dr Garruto's figure), what is the ratio of that to the quantity of carbon–nitrogen–oxygen in the PHF?

Williams: The phosphate that we all see in tangles is more than the Al and Ca in total. Thus phosphate is not saturated with respect to these metal ions, but it's not far off it. On a volume basis, if you consider the whole mass as wet protein, and there is 10^{-4} M Al present, it would mean that the protein was about 50 000 molecular weight. That's the right order of magnitude for tau, but it would be very near the saturation of the phosphorus on it, if there are two phosphorus per mole. With a higher number of phosphates it is more reasonable.

Wisniewski: Considering the presence of trace atoms in tangles, we should remember that all our best staining methods showing the tangles are based on silver techniques. This means that there is something in the property of the tangles to bind to metallic ions. Because NFT have binding sites to metals, it is possible that during tissue processing, the tangles pick up Al, as was discussed earlier.

Williams: What you see with silver staining depends critically on the organic accompanying bases used in the staining procedures. I believe heavy metal stain procedures cause problems with our type of analyses. Ms F. E. S. Murray has data in her graduate thesis (Part II of the Oxford Chemistry School) on the problems of different silver stains.

Dr Garruto, did you measure phosphorus in your SIMS? This would help.

Garruto: No; unfortunately, we could not measure phosphorus.

Martin: Would anyone care to comment on the paper (Jacobs et al 1989) denying an association of elevated Al levels with the brains of Alzheimer's disease patients?

Perl: They were using X-ray spectrometry. They claim a sensitivity of 25 p.p.m., but they were not aware of the serious limitations of this technology.

Williams: That technique won't measure 25 p.p.m. Al with any certainty.

Perl: Technically, their microprobe findings cannot be supported. In order for many of these X-ray techniques to work, you need a certain mass of atoms to be present in order to produce detectable X-rays. It's not merely the peak-to-background ratio; a certain amount of tissue is needed. Some groups have attempted to use scanning transmission electron microscopy (STEM) technology using semi-thin sections. You can't detect Al in tissues using this approach because not enough of the element is present. We took involved brain tissue from Guam ALS/parkinsonism-dementia patients, that we knew had very high Al concentrations (10 times those seen in Alzheimer's disease). I went to a world-class STEM laboratory with that material to try to get localization on semi-thin sections, and we couldn't detect any Al in these tissues. The technology one employs is very important, and unfortunately these techniques are difficult to learn. The facilities tend to be run by people not trained in biology, and the marriage between the two requires great dedication and much time and effort to make this kind of analytical procedure work.

Let me make an additional comment, on the concept that tangles may pick up elements like Al in a non-specific fashion. We did analysis of variance looking at every element you could think of, and found no significant association between tangles and any other element, apart from Al and iron. The structure is associated with these two elements in case after case, and no other element. That is not what I would expect for non-specific binding.

Williams: But you couldn't measure negative ions such as phosphate; and you can get swamped by Mg and Ca; so you can't say anything yet about these elements.

Perl: Mg is not saturated. The three elements that are saturated are Na, K and Ca, but looking at levels of these three in post mortem material is rather misleading in any case.

References

Bondareff W, Wischik CM, Novak M, Amos WB, Klug A, Roth M 1990 Molecular analysis of neurofibrillary degeneration in Alzheimer's disease: an immunohistochemical study. Am J Pathol 137:711–723

Garruto RM 1991 Pacific paradigms of environmentally-induced neurological disorders: clinical, epidemiological and molecular perspectives. Neurotoxicology 12:347–378

Garruto RM, Fukatsu R, Yanagihara R, Gajdusek DC, Hook G, Fiori CE 1984 Imaging of calcium and aluminum in neurofibrillary tangle-bearing neurons in parkinsonism-dementia of Guam. Proc Natl Acad Sci USA 81:1875–1879

Garruto RM, Swyt C, Yanagihara R, Fiori CE, Gajdusek DC 1986 Intraneuronal co-localization of silicon with calcium and aluminum in amyotrophic lateral sclerosis and parkinsonism with dementia of Guam. N Engl J Med 315:711–712

Good PF, Perl DP, Bierer LM, Schmeidler J 1992 Selective accumulation of aluminum and iron in the neurofibrillary tangles of Alzheimer's disease: a laser microprobe (LAMMA) study. Ann Neurol 31:286–292

Hegenauer J, Ripley L, Nace G 1977 Staining acid phosphoproteins (phosvitin) in electrophoretic gels. Anal Biochem 78:308–311

Hegenauer J, Saltman P, Nace G 1979 Iron(III)—phosphoprotein chelates: stoichiometric equilibrium constant for interaction of iron(III) and phosphorylserine residues of phosvitin and casein. Biochemistry 18:3865–3979

Jacobs RW, Duong T, Jones RE, Trapp GA, Scheibel AB 1989 A reexamination of aluminum in Alzheimer's disease: analysis by energy dispersive X-ray microprobe and flameless atomic absorption spectrophotometry. Can J Neurol Sci 16 (4 suppl):498–503

Linton RW, Bryan SR, Griffis DP, Shelburne JD, Fiori CE, Garruto RM 1987 Digital imaging studies of aluminum and calcium in neurofibrillary tangle-bearing neurons using secondary ion mass spectrometry. Trace Elem Med 4:99–104

Piccardo P, Yanagihara R, Garruto RM, Gibbs CJ Jr, Gajdusek DC 1988 Histochemical and X-ray microanalytical localization of aluminum in amyotrophic lateral sclerosis and parkinsonism-dementia of Guam. Acta Neuropathol 77:1–4

Wischik CM, Novak M, Edwards PC, Klug A, Tichelaar W, Crowther RA 1988 Structural characterization of the core of the paired helical filament of Alzheimer disease. Proc Natl Acad Sci USA 85:4884–4888

Intracellular effects of aluminium on receptor-activated cytoplasmic Ca^{2+} signals in pancreatic acinar cells

O. H. Petersen, M. Wakui* and C. C. H. Petersen

MRC Secretory Control Research Group, The Physiological Laboratory, University of Liverpool, PO Box 147, Liverpool, L69 3BX, UK

Abstract. The hypothesis that intracellular aluminium may interfere with cytoplasmic Ca^{2+} signals evoked by the activation of receptors linked to inositol lipid hydrolysis has been tested. Single mouse pancreatic acinar cells were used, because there is much information in this system on the mechanism by which acetylcholine (ACh) evokes cytoplasmic Ca^{2+} oscillations (spiking) and these spikes can be monitored in internally perfused cells by measuring the Ca^{2+}-dependent chloride current. ACh normally evokes repetitive Ca^{2+} spikes, but when aluminium (1 µM–1 mM) is present in the internal perfusion solution the responses are reduced or absent. When aluminium is acutely infused into the internal perfusion solution the ACh-evoked Ca^{2+} signals quickly disappear. Aluminium also inhibits Ca^{2+} signals evoked by the Ca^{2+} releasing agent caffeine. Preliminary results suggest that silicic acid may protect against the toxic effects of aluminium. Silicic acid and citrate, in the absence of added Al^{3+}, have the effect of enhancing the ACh-evoked Ca^{2+} signals. This could be due to binding of traces of Al^{3+} in the solutions. We conclude that aluminium can disrupt receptor-activated cytosolic Ca^{2+} signals when present inside cells.

1992 Aluminium in biology and medicine. Wiley, Chichester (Ciba Foundation Symposium 169) p 237–253

This article deals with the effects of intracellular aluminium on receptor-activated cytoplasmic Ca^{2+} signals. A Ca^{2+} signal is, for the purpose of this paper, defined as an increase in the cytoplasmic Ca^{2+} concentration ($[Ca^{2+}]_i$) and can be generated either by release of Ca^{2+} from intracellular stores or by opening of Ca^{2+} channels in the plasma membrane. The work described here was undertaken to test the hypothesis that intracellular aluminium interferes

---

Present address: Department of Physiology, Tohoku University School of Medicine, Sendai 980, Japan.

with cytoplasmic Ca^{2+} signals evoked by the activation of receptors linked to production of the Ca^{2+}-releasing messenger inositol 1,4,5-trisphosphate ($InsP_3$) (Birchall & Chappell 1988). Since we have a considerable body of information on the mechanisms underlying cytoplasmic Ca^{2+} signal generation in pancreatic cells it was decided in the first instance to investigate the possible effects of aluminium in these cells.

The first part of this article provides the necessary background explaining the mechanisms of calcium signal generation in electrically non-excitable cells and the second part presents the results so far obtained with intracellular infusions of aluminium into pancreatic cells.

Ca^{2+} signals evoked by activation of receptors linked to inositol lipid hydrolysis

Cytoplasmic Ca^{2+} signals can be initiated by the opening of voltage-sensitive Ca^{2+} channels in electrically excitable cells when receptor activation evokes depolarization of the plasma membrane (Jan & Jan 1989) or by generation of $InsP_3$, a messenger that releases Ca^{2+} from endoplasmic reticulum (ER) Ca^{2+} stores (Berridge & Irvine 1989). $InsP_3$ production can be induced by a variety of agonists interacting with their specific cell surface receptors that stimulate phosphoinositidase C (phospholipase C) (PIC) via functionally distinct guanine nucleotide-binding (G) proteins (Ashkenazi et al 1989). PIC hydrolyses phosphatidylinositol 4,5-bisphosphate ($PtdInsP_2$), thus generating $InsP_3$ as well as 1,2-diacylglycerol (Berridge & Irvine 1989).

Activation of receptors linked to $InsP_3$ formation generally evokes oscillating cytoplasmic Ca^{2+} signals (spikes) when submaximal agonist concentrations are used. The frequency of such cytoplasmic Ca^{2+} spikes is often dependent on agonist concentration and may provide the basis for particularly precise cellular control mechanisms (Berridge & Irvine 1989). It has been suggested that oscillating or spiking Ca^{2+} signals can be explained by fluctuating $InsP_3$ levels (Meyer & Stryer 1988), but intracellular infusion of either $InsP_3$, or the non-hydrolysable analogue inositol 1,4,5-trisphosphorothioate ($InsPS_3$) (Wakui et al 1989), can generate repetitive Ca^{2+} spikes. We have therefore concluded that pulsatile intracellular Ca^{2+} release does not depend on fluctuations in the $InsP_3$ concentration (Petersen & Wakui 1990).

A two-pool quantitative model for signal-induced Ca^{2+} oscillations has been proposed that predicts the occurrence of periodic cytoplasmic Ca^{2+} spikes in the absence of $InsP_3$ oscillations (Goldbeter et al 1990). In this model a steady $InsP_3$ level causes a steady flow of Ca^{2+} into the cytosol that evokes the emptying of separate intracellular Ca^{2+} pools via Ca^{2+}-induced Ca^{2+} release (Fig. 1). The emptying of these $InsP_3$-insensitive pools gives rise to a Ca^{2+} spike, and the time taken to refill the pools determines the interspike interval. According to this model, cytoplasmic Ca^{2+} spikes should be evoked by

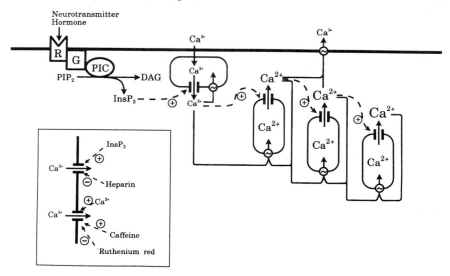

FIG. 1. Working hypothesis of mechanisms underlying cytoplasmic Ca^{2+} oscillations evoked by receptor activation. The concept is based on the model of Goldbeter et al (1990). This present version of the model emphasizes the different characteristics of the two types of Ca^{2+} channels in intracellular Ca^{2+}-storing organelles which are now known and also highlights the progression of the Ca^{2+} signal from the surface cell membrane to the cell interior. This idea is based on data from Osipchuk et al (1990). R, receptor; G, guanine nucleotide-binding protein; PIC, phosphoinositidase C; PIP$_2$, phosphatidylinositol 4,5-bisphosphate; DAG, 1,2-diacylglycerol; InsP$_3$, inositol 1,4,5-trisphosphate. (Adapted from Petersen & Wakui 1990.)

intracellular Ca^{2+} infusion. In experiments on single internally perfused pancreatic acinar cells such a phenomenon has been directly demonstrated (Osipchuk et al 1990).

Two different channels that mediate Ca^{2+} release from intracellular non-mitochondrial stores into the cytosol are known. The InsP$_3$-activated Ca^{2+} channel has been isolated and sequenced (Furuichi et al 1989), reconstituted and functionally investigated in lipid bilayers (Ehrlich & Watras 1988, Ferris et al 1989), and localized to the ER (Ross et al 1989). The Ca^{2+}-induced Ca^{2+}-release channel has been isolated from muscle sarcoplasmic reticulum (SR) (Lai et al 1988), sequenced (Takeshima et al 1989) and studied functionally by reconstitution in lipid bilayers (Lai et al 1988). The InsP$_3$-activated Ca^{2+} channel can be blocked specifically by heparin, a substance that has no effect on the SR Ca^{2+} channel (Ehrlich & Watras 1988), which can be activated by submicromolar to micromolar Ca^{2+} concentrations, specifically from the cytoplasmic side (Lai et al 1988). Caffeine, a well-known potentiator of Ca^{2+}-induced Ca^{2+} release, can open SR Ca^{2+} release channels at normal

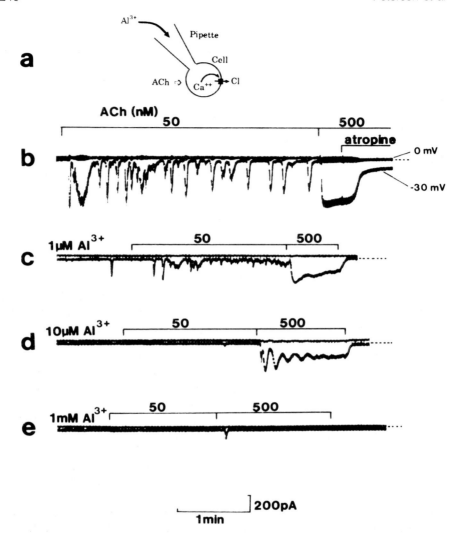

FIG. 2. Effect of acetylcholine (ACh) on Ca^{2+}-dependent Cl^- current in single internally perfused mouse pancreatic acinar cells in the absence and presence of $AlCl_3$ in the pipette solution. Acinar cells were voltage-clamped at a holding potential of -30 mV and repetitive depolarizing voltage jumps of 100 ms duration to 0 mV applied throughout the experiments. The Cl^- equilibrium potential $(E_{Cl^-}) = 0$. Because of compression of the pen-recording traces all records seem to show currents at -30 and 0 mV simultaneously. The downward deflections in the current traces obtained at -30 mV represent increases in the Cl^- current due to increases in $[Ca^{2+}]_i$. Dotted horizontal lines indicate zero current level. (Traces b and e are from Wakui et al 1990b by permission of Elsevier Science Publishers.)

cytoplasmic Ca^{2+} concentrations (Penner et al 1989), but has no effect on InsP$_3$-activated Ca^{2+} channels (Ehrlich & Watras 1988).

In pancreatic acinar cells we have shown that intracellular heparin infusion blocks cytoplasmic Ca^{2+} spikes evoked by external acetylcholine (ACh) or internal InsP$_3$ application but fails to inhibit Ca^{2+} spiking induced by intracellular Ca^{2+} infusion (Wakui et al 1990a). Caffeine potentiates the induction by Ca^{2+} of Ca^{2+} oscillations and can also, under certain circumstances, itself evoke such oscillations (Wakui et al 1990a). These results show that continued opening of InsP$_3$-activated Ca^{2+} channels is needed to sustain repetitive cytoplasmic Ca^{2+}-spiking evoked by agonist–receptor interaction, but they also demonstrate that Ca^{2+}-induced Ca^{2+} spikes can occur independently of InsP$_3$. It would appear that InsP$_3$ evokes a small steady outflow of Ca^{2+} into the cytosol from a heparin-sensitive channel, generating repetitive Ca^{2+} spikes from caffeine-sensitive channels (Fig. 1).

Effects of intracellular aluminium infusion on cytoplasmic Ca^{2+} signal generation

The intracellular Ca^{2+} signals can be monitored in single internally perfused pancreatic acinar cells by measuring the Ca^{2+}-dependent Cl$^-$ current, using the patch clamp whole-cell recording configuration (Fig. 2a). Figure 2b shows typical control responses to external application of ACh. As previously described (Wakui et al 1989), ACh evokes repetitive pulses of inward Cl$^-$ current at a membrane potential of $-30\,\text{mV}$, as a result of the repetitive release of Ca^{2+} from intracellular stores. When the membrane potential is clamped at zero there is hardly any effect, because this is close to the Cl$^-$ equilibrium potential (E$_{\text{Cl}^-}$). The ACh effect is blocked by the muscarinic antagonist atropine (Fig. 2b) and by a high intracellular concentration of the Ca^{2+} chelator, EGTA (Wakui et al 1989). We have previously shown that the reversal potential for the ACh-evoked current response varies with E$_{\text{Cl}^-}$ over a wide range of values. The ACh-evoked inward current is therefore due to the opening of Ca^{2+}-dependent Cl$^-$ channels and this current can be used to assess [Ca^{2+}]$_i$ close to the plasma membrane (Wakui et al 1989). When the pipette solution contained 1 mM AlCl$_3$ the ACh responses were abolished (Fig. 2e). At the lower concentrations of 10 and 1 μM, Al^{3+} caused marked reductions in ACh-evoked responses (Fig. 2c,d).

Figure 3(a,b) shows repetitive spikes of Ca^{2+}-dependent Cl$^-$ current due to repetitive [Ca^{2+}]$_i$ spikes evoked by the intracellular infusion of InsP$_3$. Such spikes could not be evoked when aluminium was present in the internal solution (Fig. 3c). This experiment provides direct evidence for the hypothesis proposed by Birchall & Chappell (1988). It is important to realize that InsP$_3$ is metabolized in the cell by the action of phosphatases and kinases (Berridge & Irvine 1989) and that it is therefore unlikely that the InsP$_3$ concentration at

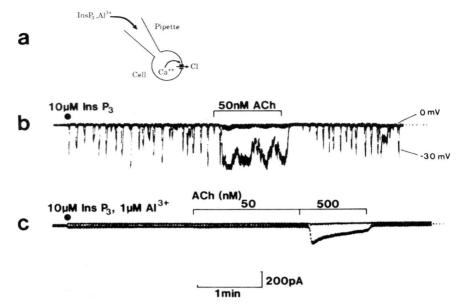

FIG. 3. InsP$_3$-evoked repetitive spikes of Ca^{2+}-dependent Cl$^-$ current and the abolition of this effect by aluminium. InsP$_3$ (10 µM) or InsP$_3$ plus 1 µM AlCl$_3$ were present in the pipette solution. At the times indicated by filled circles the whole-cell current recording configuration was attained by breaking through the cell membrane under the patch-clamp pipette and InsP$_3$ or InsP$_3$ + Al^{3+} started to flow into the cell interior (M. Wakui & O. H. Petersen, unpublished observation).

the intracellular receptor sites is as high as in the pipette solution. It is quite possible that the actual InsP$_3$ concentration at the relevant intracellular sites is less than 1 µM in these experiments (Fig. 3).

We have previously shown that caffeine, a well-known potentiator of Ca^{2+}-induced Ca^{2+} release through the ryanodine receptor channel in muscle sarcoplasmic reticulum, can markedly potentiate the intracellular Ca^{2+} release evoked by low concentrations of ACh and InsP$_3$ (Osipchuk et al 1990, Wakui et al 1990a). We have also demonstrated that when pancreatic acinar cells are perfused internally with a solution not containing the normally used low concentration of the Ca^{2+} chelator EGTA (0.25 mM), then caffeine alone evokes repetitive pulses of Ca^{2+}-dependent Cl$^-$ current (Wakui et al 1990a). Caffeine (1 mM) evoked responses of the type shown in Fig. 4b in a control series of experiments whether applied before or after a period of ACh stimulation. The effect of an acute intracellular AlCl$_3$ infusion via a tube inserted into the patch pipette is shown in Fig. 4c, which starts with a control response to ACh. Thereafter the intracellular tube is opened and pressure applied, enabling the AlCl$_3$ (10 µM)-containing solution to enter the pipette tip

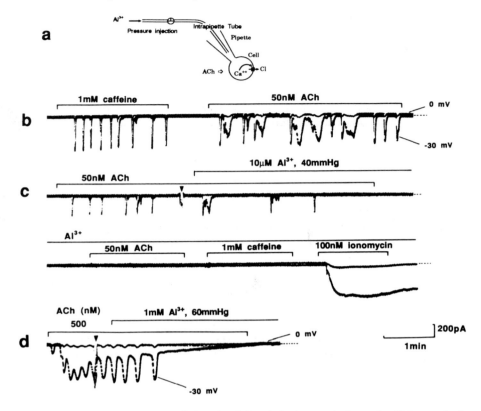

FIG. 4.   Effect of acute intracellular aluminium infusion on ACh- and caffeine-evoked
Ca²⁺ signals. (a) shows the arrangement of the infusion tube for aluminium, patch-
clamp pipette and whole-cell recording configuration. In (b), control responses
(in the absence of an AlCl₃-containing solution in the tube) to caffeine and ACh
(applied externally) are shown. The two traces in (c) are from the same cell and
show an uninterrupted experimental run starting with a control ACh response (in
all experiments with an AlCl₃-containing solution in the infusion tube the response
seemed weaker than in real control cells; we cannot exclude some leak of AlCl₃
into the cell even when the tube is closed). At the arrowhead the tube is opened
and thereafter pressure injection of intracellular solution containing 10 μM AlCl₃
occurs, which abolishes the ACh-evoked Cl⁻ current pulses within 2.5 min. ACh
stimulation is then discontinued. A renewed ACh challenge fails to elicit a response
and caffeine is also ineffective. Finally, 100 nM ionomycin is applied to the bath,
evoking a large sustained inward current response. The record shown in (d) is from
a separate experiment. ACh (500 nM) causes an oscillating response. Intrapipette infusion
of AlCl₃ (1 mM)-containing solution slows down the Cl⁻ current pulses and then
abolishes the response. In this experiment EGTA (0.25 mM) was present in the pipette
solution, but since Al³⁺ was present in excess of EGTA the chelator has relatively
little effect on the Al³⁺ concentration. (From Wakui et al 1990b by permission of
Elsevier Science Publishers.)

and therefore the cell interior. Within 2.5 min from the start of the aluminium infusion the ACh-evoked response has been abolished. The ACh stimulation is then discontinued; after an interval of about 2 min, ACh is reapplied but without any effect. Thereafter caffeine is applied, but also without effect. Finally it is shown that in the same cell the Ca^{2+} ionophore ionomycin (100 nM) evokes a large and sustained inward current. Figure 4d shows the results of an experiment in which intrapipette infusion of a solution containing 1 mM $AlCl_3$ first changed the quasi-sustained oscillating response to 500 nM ACh to a clearly pulsatile pattern and thereafter abolished the Ca^{2+} signal.

The results in Figs. 2, 3 and 4 indicate that intracellular application of aluminium can block ACh-evoked and $InsP_3$-mediated cytoplasmic Ca^{2+} signals. There are many steps from receptor activation by ACh to the cytoplasmic Ca^{2+} signal (Fig. 1), and aluminium could in principle act on several of these, but, in view of the finding that aluminium also inhibits caffeine-evoked intracellular Ca^{2+} release (Fig. 4c), the simplest hypothesis to account for our results is that aluminium inhibits opening of the caffeine-sensitive Ca^{2+}-induced Ca^{2+} release channel. However, other effects can certainly not be ruled out.

Derek Birchall has proposed that silicon and silicic acid play a major role in the way biological systems reject aluminium, since silicic acid reduces the toxicity of aluminium to fish and since silicon deficiency in experimental animals

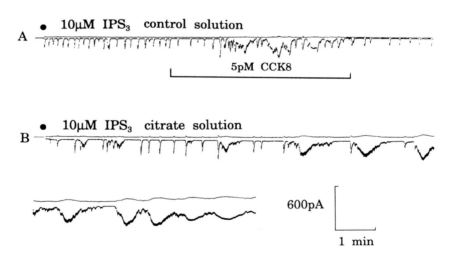

FIG. 5.   Comparison of the effects of 10 μM $InsPS_3$ in a control pipette solution (A) and a pipette solution containing 10 mM citrate (B). The experimental set-up was the same as in Fig. 2. In the control experiment (A), $InsPS_3$ (10 μM) evokes repetitive slim spikes, but the addition of 5 pM cholecystokinin (CCK8) externally evokes longer transients; whereas in the citrate experiment (B), the same concentration of $InsPS_3$ by itself is able to induce longer Ca^{2+} transients. The lower part of panel B is a direct continuation of the upper part. (From Petersen et al 1991b by permission of Elsevier Science Publishers.)

produces symptoms similar to those seen after excessive aluminium absorption. There is also a unique affinity between aluminium and silicic acid, resulting in the formation of hydroxyaluminosilicates (see review by J. D. Birchall 1992, this volume). In a preliminary series of experiments (C. C. H. Petersen, J. D. Birchall & O. H. Petersen, unpublished results) we have tested the effects of adding 100 µM silicic acid to the solution perfusing internally single mouse pancreatic acinar cells, using the approach already described in Fig. 2. In these experiments silicic acid itself enhanced the ACh-evoked Ca^{2+} signals by making them much broader and appeared to inhibit the aluminium-induced block so that, for example, 50 nM ACh would evoke normal signals in the combined presence of 10 µM aluminium chloride and 100 µM silicic acid. Silicic acid in the cytoplasm may therefore protect against the toxic effects of Al^{3+}.

The effects of silicic acid alone (100 µM) were unexpected, but could perhaps be explained by the binding of trace amounts of aluminium that may always be present in our solutions. In this context it is interesting that citrate (in the higher concentration of 10 mM) has much the same effect on ACh-evoked Ca^{2+} spikes as silicic acid by making the individual $[Ca^{2+}]_i$ transients (spikes) considerably longer (Petersen et al 1991a). Such an effect has also recently been observed on the Ca^{2+} spikes evoked directly by the non-metabolizable InsP$_3$ analogue, InsPS$_3$ (Fig. 5). Although the citrate effects may be explained without reference to aluminium (Petersen et al 1991a,b), it is interesting to note that the effects of silicic acid and citrate seem to be very similar and that the one thing these two agents have in common is the ability to bind aluminium.

Summary

Our results indicate that intracellular aluminium can disrupt receptor-activated cytosolic Ca^{2+} signals that are mediated by InsP$_3$. Preliminary results indicate that silicic acid may protect against this effect, but much work remains to be done before a clear picture emerges of the mechanism of action of aluminium at the cellular level. So far only a limited number of experiments on pancreatic acinar cells have been carried out. It is clearly desirable to extend these studies to many other cell types.

Acknowledgements

We thank Derek Birchall for many valuable suggestions and helpful discussions. This work was supported by an MRC programme grant and by ICI.

References

Ashkenazi A, Peralta EG, Winslow JW, Ramachandran J, Capon DJ 1989 Functionally distinct G proteins selectively couple different receptors to PI hydrolysis in the same cell. Cell 56:487–493

Berridge MJ, Irvine RF 1989 Inositol trisphosphate and cell signalling. Nature (Lond) 341:197–205

Birchall JD 1992 The interrelationship between silicon and aluminium in the biological effects of aluminium. In: Aluminium in biology and medicine. Wiley, Chichester (Ciba Found Symp 169) p 50–68

Birchall JD, Chappell JS 1988 Aluminium, chemical physiology and Alzheimer's disease. Lancet 1:1008–1010

Ehrlich BE, Watras J 1988 Inositol 1,4,5-trisphosphate activates a channel from smooth muscle sarcoplasmic reticulum. Nature (Lond) 336:583–586

Ferris CD, Huganir RL, Supattapone R, Snyder SH 1989 Purified inositol 1,4,5-trisphosphate receptor mediates calcium flux in reconstituted lipid vesicles. Nature (Lond) 342:87–89

Furuichi T, Yoshikawa S, Miyawaki A, Wada K, Maeda N, Mikoshiba K 1989 Primary structure and functional expression of inositol 1,4,5-trisphosphate binding protein P_{400}. Nature (Lond) 342:32–38

Goldbeter A, Dupont G, Berridge MJ 1990 Minimal model for signal-induced Ca^{2+} oscillations and for their frequency encoding through protein phosphorylation. Proc Natl Acad Sci USA 87:1461–1465

Jan LY, Jan YN 1989 Voltage-sensitive ion channels. Cell 56:13–25

Lai FA, Erickson HP, Rousseau E, Liu AY, Meissner G 1988 Purification and reconstitution of the calcium release channel from skeletal muscle. Nature (Lond) 331:315–319

Meyer T, Stryer L 1988 Molecular model for receptor-stimulated calcium spiking. Proc Natl Acad Sci USA 85:5051–5055

Osipchuk YV, Wakui M, Yule DI, Gallacher DV, Petersen OH 1990 Cytoplasmic Ca^{2+} oscillations evoked by receptor stimulation, G-protein activation, internal application of inositol trisphosphate or Ca^{2+}: simultaneous microfluorimetry and Ca^{2+} dependent Cl^- current recording in single pancreatic acinar cells. EMBO (Eur Mol Biol Organ) J 9:697–704

Penner R, Neher E, Takeshima H, Nishimura S, Numa S 1989 Functional expression of the calcium release channel from skeletal muscle ryanodine receptor cDNA. FEBS (Fed Eur Biochem Soc) Lett 259:217–221

Petersen OH, Wakui M 1990 Oscillating intracellular Ca^{2+} signals evoked by activation of receptors linked to inositol lipid hydrolysis: mechanism of generation. J Membr Biol 118:93–105

Petersen CCH, Toescu EC, Petersen OH 1991a Different patterns of receptor-activated cytoplasmic Ca^{2+} oscillations in single pancreatic acinar cells: dependence on receptor type, agonist concentration and intracellular Ca^{2+} buffering. EMBO (Eur Mol Biol Organ) J 10:527–533

Petersen CCH, Toescu EC, Potter BVL, Petersen OH 1991b Inositol trisphosphate produces different patterns of cytoplasmic Ca^{2+} spiking depending on its concentration. FEBS (Fed Eur Biochem Soc) Lett 293:179–182

Ross CA, Meldolesi J, Milner TA, Satoh T, Supattapone S, Snyder SH 1989 Inositol 1,4,5-trisphosphate receptor localized to endoplasmic reticulum in cerebellar Purkinje neurons. Nature (Lond) 339:468–470

Takeshima H, Nishimura S, Matsumoto T et al 1989 Primary structure and expression from complementary DNA of skeletal muscle ryanodine receptor. Nature (Lond) 339:439–445

Wakui M, Potter BVL, Petersen OH 1989 Pulsatile intracellular calcium release does not depend on fluctuations in inositol trisphosphate concentration. Nature (Lond) 339:317–320

Wakui M, Osipchuk YV, Petersen OH 1990a Receptor activated cytoplasmic Ca^{2+} spiking mediated by inositol trisphosphate is due to Ca^{2+}-induced Ca^{2+} release. Cell 63:1025–1032
Wakui M, Itaya K, Birchall JD, Petersen OH 1990b Intracellular aluminium inhibits acetylcholine- and caffeine-evoked Ca^{2+} mobilization. FEBS (Fed Eur Biochem Soc) Lett 267:301–304

DISCUSSION

Williams: We have asked the question whether InsP$_3$ is a better binder of cations such as Mg^{2+} than ATP, and it turns out not to be (White et al 1991). The InsP$_3$ species is protonated on one phosphate group at pH 7.0, so of the two phosphates that come together, one is protonated. This reduces the cation binding relative to ATP, which is not protonated at this pH. We then compared Mg binding to ATP, to see whether Mg would interfere with the InsP$_3$ reaction mechanisms. Mg binding is weaker to InsP$_3$ than to ATP. We didn't test Al, but I suspect Al binding to InsP$_3$ is also weaker than to ATP. Therefore, if you have a big buffer of ATP in a cell system, there seems unlikely to be a direct effect of Al on InsP$_3$. It is more likely to be an indirect effect of Al on some other part of the cell system, as you have described it. However, we must always beware of special effects in proteins.

Petersen: The hypothesis in our original paper (Wakui et al 1990) was that Al acted on the Ca^{2+}-activated Ca^{2+} release channel, because that is the most parsimonious explanation of all the data.

Birchall: The binding of cations (Ca^{2+}, Mg^{2+}) by inositol 1,4,5-trisphosphate is thought to be strongest across adjacent 4,5 phosphate groups (Hendrickson & Reinertsen 1969). A computation of the energy of binding of aluminium by ATP and InsP$_3$ suggested a three-fold advantage for InsP$_3$ (Birchall & Chappell 1988). We were looking for a site of aluminium binding such as would produce a cascade of events and amplification from a small concentration of aluminium. It was this thinking that drew attention to second messenger systems and Ca^{2+} manipulation. There is indeed an effect, the mechanism of which remains unclear.

Martin: We studied Al binding of 2,3-diphosphoglycerate, as I mentioned in my paper. We concluded that Al was bound to one phosphate and the carboxylate, not to two phosphates.

Wischik: Have you tested Li in this system, and do you know what happens?

Petersen: We have not tested Li in pancreatic acinar cells. Lithium inhibits inositol 1-phosphate phosphatase (Berridge 1987). Li would therefore inhibit the resynthesis of phosphatidylinositol 4,5-bisphosphate. It would not directly interfere with InsP$_3$-induced calcium signalling.

Williams: Li is teratogenic; is Al?

Edwardson: There's a large literature on the subject. Most studies have found *no* evidence of teratogenic effects of Al in rats or mice, and epidemiological data in man are also negative. In a smaller number of studies, some evidence of retardation of motor function was reported in the offspring of rats and mice after very large doses of aluminium had been administered during gestation (e.g. Muller et al 1990). More recently it has been suggested (R. Clayton, personal communication) that exposure to Al *in utero* produces persistent cognitive and other behavioural changes in mice, without any evidence of morphological abnormalities.

Kerr: Many of the Al-overloaded patients have had kidney transplants and some have gone on to have children. There is no striking increase in congenital defects in the offspring of transplant recipients generally, but I am not aware of studies that have compared the offspring of aluminium-overloaded mothers with control transplant recipients.

Ward: Successful pregnancy leads to intellectually normal children, in such cases.

Edwardson: As I understand it from Derek Birchall, one would expect $Si(OH)_4$ to cross cell membranes freely. Silicic acid is available in the circulation, and presumably present in tissue culture media. Do we know the pre-existing intracellular levels of silicic acid, and what reactions might be expected between it and Al at intracellular pH?

Birchall: I don't know how much silicic acid there is in cells. As a neutral molecule, it may be able to cross cell membranes freely, but it has not been measured.

Petersen: We shall test that, but so far we have only done experiments with silicic acid in the pipette solution, putting it internally. Another way to investigate its action would be to put silicic acid into the extracellular solution and see whether we get the same effects.

Birchall: You can find Si in cells—in osteoblasts, for instance.

Edwardson: Your interpretation of the effects of silicic acid alone suggests that, although this molecule is abundant in extracellular fluid normally and is freely diffusible across the cell membrane, there is not sufficient in the cell to mop up all the available Al.

Birchall: Professor Petersen is studying the isolated cell, so it's in a rather artificial environment in which there may not be enough silicic acid to neutralize injected Al. We find that when you isolate a tissue or cell, there's no such thing as a blank. You immediately introduce aluminium, since all reagents contain some Al. There may not have been much silicic acid in the medium but Al would be present. It is then necessary to add silicic acid, to go back to normal.

Petersen: But not necessarily at 100 µM?

Birchall: No; maybe only 5 µM.

Petersen: We are introducing large concentrations of silicic acid (100 µM).

Birchall: Yes. That's not abnormal, however. Dialysis patients can reach plasma levels of that order, for example.

Day: Any laboratory reagent might contain silicic acid.

Birchall: Reagents may contain some Si in some form, but only silicic acid will be active.

Day: You were suggesting that an underlying concentration of Al may have been shown up by your experiment using silicic acid alone, Professor Petersen. Could you test this hypothesis by specifically binding the Al, perhaps by introducing desferrioxamine into the system?

Petersen: These are experiments we would certainly like to do.

Birchall: We are doing experiments on the hexokinase reaction, its inhibition when Al-contaminated ATP is involved, and the removal of inhibition by citrate, which withdraws Al from ATP and hence allows Mg^{2+} to operate (Viola et al 1980). We are repeating that experiment to discover if silicic acid will similarly activate the system. The earliest evidence is that it will do so.

Williams: What is the concentration of citrate in these pancreatic cells?

Petersen: I don't know. In this particular study, however, citrate would be subject to wash-out into the pipette solution. There would have to be some citrate in the cells, but probably not 10 mM!

Williams: Maybe 1 mM?

Petersen: Yes, perhaps; so it could have some effects in this situation. There is evidence, although it is still not clear-cut, that cells may rapidly lose calcium-buffering components when you break into the cell in the whole-cell recording configuration. Part of that may be citrate.

Candy: The key experiment to confirm the relevance of this model in relation to the physiological effects of Al is to attempt to introduce Al into the cells using Al-loaded transferrin. This is because you don't know whether Al which is internalized bound to transferrin reaches the intracellular compartment into which you are injecting Al.

Petersen: This is more difficult to do. You can put small molecules quite easily into the cells, but with larger molecules, equilibration is not very rapid. You cannot be sure that the substance in practice gets into the cell.

Candy: You almost need two sets of experiments, one which uses cells preloaded with Al and one where the cells are loaded during the experiment with Al-transferrin.

Petersen: I agree, and it would be interesting to look at other types of Al loading.

Candy: I think it's essential, if you want to confirm that Al is *potentially* toxic to Ca signalling and show that it is physiologically relevant.

Birchall: The importance of the experiments described by Professor Petersen is that they help us to get down to some fundamental mechanism of the cellular toxicity of aluminium. An effect on Ca^{2+} mobilization is a good start.

McLachlan: To me, the mystery has always been that when lethal amounts of Al are slowly infused over 5–15 minutes into awake cats or rabbits through a chronically implanted, stereotopically placed cannula, surface and depth EEG activity, evoked potentials, learning and memory performance, fine motor control and sleep states do not change for several days (Crapper & Dalton 1973a,b, Crapper 1973). It would appear that injected soluble aluminium salts are rapidly sequestrated and functionally inactive for several days; then slowly the toxicity is expressed. This may be because aluminium, at the doses used, inactivates a protein or a biochemical process but toxicity does not occur until the reserve of this inhibited product is used up; or aluminium slowly changes ligands and eventually causes toxicity.

Petersen: That result does not contradict the kind of results that I have presented, because in our experiments Al enters cells in a very different way from the *in vivo* situation. I imagine it is a question of the route of Al entry that is important in this case.

McLachlan: There seems to be some shifting of compartments, or of ligands, over time.

Petersen: That could well be. To some extent, one sees what one is looking for, and in our experiments, we are looking for acute effects, so we can only see such effects; we cannot look at effects in this type of system over several days.

Birchall: Professor McLachlan once made the very perceptive comment (Kruck & McLachlan 1988) that aluminium is involved in the disruption of Ca^{2+}-dependent electrophysiological functioning and of intracellular Ca^{2+} regulation. I agree. The fate of aluminium in a biological milieu is probably to pass from one ligand to another with higher binding strength—a hierarchy of binding sites—for example, from citrate to transferrin and thence to a series of intracellular sites. At some critical site, damage amplification is produced. Ultimately, aluminium will become 'locked' in insoluble form in bone, as phosphate in lysosomes, bound on phosphorylated proteins, or as aluminosilicate. Then, unless it is re-mobilized, I suspect it has done its damage.

Martin: In what form is Al being administered in those infusion studies?

McLachlan: We have used the salts of chloride, phosphate and lactate. Provided the infusion is 'tissue friendly', we do not observe behavioural or electrophysiological changes for several days, even after the administration of supra-lethal doses which eventually go on to kill 100% of animals. Sensitive electrophysiological changes such as long-term potentiation, which is a marker for conditioning, are delayed in onset. Later changes in EPSPs and spike initiation occur in a hierarchical sequence of changes associated with a rise in total tissue calcium. Perhaps, as Dr Birchall has described, Al shifts from ligand to ligand. One of our operating hypotheses is that applied Al reaches the cell nucleus very rapidly, within hours of delivery. Subsequently, there is a slow loss of Al from the nucleus, as though the nucleus is one of the reservoirs from which Al shifts to other neurotoxic targets.

Petersen: We cannot, from our experiments, say whether there is any correspondence with these *in vivo* studies, because you cannot make comparisons about the kind of Al levels and in what form Al exists inside the cells. We were trying to make conceptually simple experiments and to test a particular hypothesis; only time will show whether these circumstances correspond to those that can be created in other types of experiments or in pathological situations. It is conceivable that they have no counterpart in disease states, but it still seems sensible to try to create clear-cut experimental situations in which to test a particular hypothesis.

Birchall: The suggestion that aluminium binds to InsP$_3$ and would influence Ca^{2+} manipulation by that means (Birchall & Chappell 1988) may or may not be correct. However, the idea did attract attention to Ca^{2+} mobilization as a way by which small, intracellular concentrations of aluminium could provoke a cascade effect. Aluminium does indeed affect the inositol phosphate system in several ways (see, for example, McDonald & Mamrack 1988). Oral aluminium alters protein phosphorylation in rat brain (Johnson et al 1990) and would be expected to become bound to phosphate groups, perhaps inducing conformational change as it does when it binds to phosvitin. We may need to look for two types of effect of aluminium—a 'catalytic' effect, and 'conformational' effects.

Williams: I don't think phosvitin carries iron naturally, so it probably wouldn't carry Al very well?

Birchall: Phosvitin does bind Fe.

Williams: Yes, but it's not a strong binder, because the phosphates are quite spread out.

Birchall: A stoichiometric equilibrium contant of 10^{18} has been calculated for the binding of iron by the di-*O*-phosphorylserine residues of phosvitin, and the relative binding strengths of various ligands for iron have been shown to be EDTA > phosvitin > citrate > nitrilotriacetate. Aluminium behaves similarly (Hegenauer et al 1979).

Williams: The presence of ATP always worries me, in all these competitive binding studies. It is so powerful as a ligand. Does phosvitin bind with ATP present?

Birchall: Since citrate binds aluminium more strongly than ATP (Viola et al 1980) and phosvitin binds more strongly than citrate, I suspect that phosphorylated proteins could bind in the presence of ATP.

Kerr: I am intrigued by your hierarchical scheme of binding as a possible explanation for the phenomenon in bone where Al is deposited at the ossification surface and then ossification stops. If you start removing Al from bone with desferrioxamine infusions, ossification recommences after only a small amount of Al has been removed. How do you envisage that taking place? Is there an insoluble deposit of Al phosphate or silicate covered by an active thin layer of some other aluminium compound which has to be removed before ossification can recommence?

Birchall: In healthy bone, Si appears at the mineralization front and is found within the osteoblast (Carlisle 1986). In dialysis osteodystrophy, aluminium is found at the cement line. Is it also within the osteoblast? As regards its effect on bone, aluminium affects both the formation of the collagen matrix and the nucleation of the calcium hydroxyapatite mineral phase (Birchall & Espie 1986).

Williams: And, at this bone front, are found the phosphoproteins that are not quite like phosvitin. These mop up ions. It may be that Al interferes with them. They are essential for the synthesis of bone.

Edwardson: Because of the potential importance of the binding between silicic acid and Al, Professor Birchall, can you speculate on the possibilities of the binding of silicic acid to other ions in the cell?

Birchall: Silicic acid is a very weak acid (pK_1 of 9.8). It interacts only with metals when they are basic. The only relevant metals basic at physiological pH are Al and iron as Fe^{3+}. Calcium and magnesium are not, and silicic acid does not interact with them. The interaction with Fe^{3+} is weak. The strong and specific interaction of silicic acid with aluminium is what drew my attention to this as a 'protective' role for Si.

Williams: Ferric iron would bind more strongly than Al if it were not for the precipitation of ferric hydroxide. Al hydroxide, being more soluble than ferric hydroxide, allows the possibility of its interaction with silicic acid around pH 7.0.

Birchall: That is so, and also iron is very compartmentalized in biology. At neutral pH, Fe^{3+} polymerizes so fast (= precipitation) that it can't react with silicic acid. The polymerization of basic aluminium species is relatively slow (see this volume: Martin 1992).

Edwardson: So you don't envisage any kind of ligand hopping for silicic acid? What about other potential non-ionic ligands within the cell could there be interactions of silicic acid with other intracellular constituents, which would impair its ability to interact with Al?

Birchall: At near-neutral pH, I cannot envisage any strong interaction between silicic acid and other cellular constituents. Maybe weak hydrogen bonding, but none such as would inhibit reactions with aluminium.

Williams: We have talked about using gallium as a following probe for Al. It would be nice to know if gallium behaves with silicic acid more like iron or more like Al. I suspect that it is more like iron.

Birchall: It's more like iron, so it's not a good model for alluminium in this respect.

Williams: So silicic acid as a reagent against Al is a very selective agent, but not so as far as gallium is concerned.

References

Berridge MJ 1987 Inositol trisphosphate and diacylglycerol: two interacting second messengers. Annu Rev Biochem 56:159–193

Birchall JD, Chappell JS 1988 Aluminium, chemical physiology and Alzheimer's disease. Lancet 1:1008–1010

Birchall JD, Espie AW 1986 Biological implications of the interaction (via silanol groups) of silicon with metal ions. In: Silicon biochemistry. Wiley, Chichester (Ciba Found Symp 121) p 140–159

Carlisle EM 1986 Silicon as an essential trace element in animal nutrition. In: Silicon biochemistry. Wiley, Chichester (Ciba Found Symp 121) p 123–139

Crapper DR 1973 Experimental neurofibrillary degeneration and altered electrical activity. Electroencephalogr Clin Neurophysiol 35:575–588

Crapper DR, Dalton AJ 1973a Alterations in short-term retention, conditioned avoidance response acquisition and motivation following aluminium induced neurofibrillary degeneration. Physiol Behav 10:925–933

Crapper DR, Dalton AJ 1973b Aluminum induced neurofibrillary degeneration, brain electrical activity and alterations in acquisition and retention. Physiol Behav 10:935–945

Hegenauer J, Saltman P, Nace G 1979 Iron(III)—phosphoprotein chelates: stoichiometric equilibrium constant for interaction of iron(III) and phosphorylserine residues of phosvitin and casein. Biochemistry 18:3865–3879

Hendrickson HS, Reinertsen JL 1969 Comparison of metal binding properties of trans-1,2-cyclohexanediol diphosphate and deacylated phosphoinositides. Biochemistry 8:4855–4858

Johnson GVW, Cogdill KW, Jope RS 1990 Orally administered aluminum alters in vitro protein phosphorylation and protein kinase activities in rat brain. Neurobiol Aging 11:209–216

Kruck TPA, McLachlan DR 1988 Mechanisms of aluminium neurotoxicity— relevance to human disease. In: Sigel H, Sigel A (eds) Metal ions in biological systems, vol 24: Aluminum and its role in biology. Marcel Dekker, New York, p 285–314

Martin RB 1992 Aluminium speciation in biology. In: Aluminium in biology and medicine. Wiley, Chichester (Ciba Found Symp 169) p 5–25

McDonald LJ, Mamrack MD 1988 Aluminium affects phosphoinositide hydrolysis by phosphoinositidase C. Biochem Biophys Res Commun 155:203–208

Muller G, Bernuzzi V, Desor D, Hutin MF, Burnel D, Lehr PR 1990 Developmental alterations in offspring of female rats orally intoxicated by aluminium lactate at different gestation periods. Teratology 12:253–261

Viola RE, Morrison JF, Cleland WW 1980 Interaction of metal(III)–adenosine 5′-triphosphate complexes with yeast hexokinase. Biochemistry 19:3131–3137

Wakui M, Itaya K, Birchall JD, Petersen OH 1990 Intracellular aluminium inhibits acetylcholine- and caffeine-evoked Ca^{2+} mobilization. FEBS (Fed Eur Biochem Soc) Lett 267:301–304

White AM, Varney MA, Watson SP et al 1991 Influence of Mg^{2+} binding and pH on n.m.r. spectra and radio ligand binding to inositol 1,4,5-trisphosphate. Biochem J 278:759–764

Neurotoxic effects of dietary aluminium

Richard S. Jope and Gail V. W. Johnson

Department of Psychiatry and Behavioral Neurobiology, University of Alabama at Birmingham, Birmingham, AL 35294-0017, USA

Abstract. Neurochemical responses to chronic oral aluminium administration have been studied in rats. Aluminium (0.3%) was added to drinking water of adult rats for four weeks or longer and weanling rats were given aluminium for eight weeks. Selective cognitive impairment was demonstrated in the adult rats. Aluminium inhibited calcium flux and phosphoinositide metabolism, one product of which (inositol 1,4,5-trisphosphate) modulates intracellular calcium levels. In weanling rats aluminium decreased the *in vivo* concentration of inositol 1,4,5-trisphosphate in the hippocampus. An increase in cyclic AMP concentrations by 30–70% in various brain regions in adult and weanling rats was found. Aluminium enhanced agonist-stimulated but not basal cyclic AMP production *in vitro*. It was postulated that aluminium inhibits the GTPase activity of the stimulatory G protein, G_s, leading to prolonged activation of G_s after receptor stimulation and increased cyclic AMP production. Aluminium treatment also increased the phosphorylation of microtubule-associated protein 2 (MAP-2) and the 200 kDa neurofilament protein (NF-H) but several other phosphoproteins were unaffected. Concentrations of seven structural proteins—MAP-2, tau, NF-H, NF-M (150 kDa), NF-L (68 kDa), tubulin and spectrin—were measured in rat brain regions by immunoblot methods. MAP-2 was most consistently decreased.

These studies show that chronic oral aluminium administration to rats has significant neurochemical consequences. Three sites of action are implicated: altered calcium homeostasis, enhanced cyclic AMP production, and changes in cytoskeletal protein phosphorylation states and concentrations.

1992 Aluminium in biology and medicine. Wiley, Chichester (Ciba Foundation Symposium 169) p 254–267

Aluminium neurotoxicity has been associated with a number of neurodegenerative disorders, as discussed by others in this symposium, but the causative mechanisms are not known. Discoveries made in a number of model systems have however contributed towards identifying the mechanisms underlying the neurotoxic consequences of exposure to aluminium. The most widely used experimental approaches employ the direct application to brain tissue of aluminium, either by *in vitro* incubation with aluminium or by its central administration (reviewed by McLachlan 1986). We felt that it would be informative to develop an additional model system in which chronic exposure to ingested aluminium could be investigated. After a number of trials we settled on adding aluminium sulphate to the drinking water (0.3%) given to rats over a period of at least four weeks. Such a treatment avoids acute exposure of the brain to excessively high levels of aluminium, and would perhaps allow us to identify the processes most susceptible to longer-term elevated aluminium concentrations. As we shall discuss, this treatment increases the serum aluminium concentration eight-fold, causes subtle cognitive impairments, and results in large changes in signal transduction mechanisms and cytoskeletal proteins in the brain without producing generalized toxicity or altering mortality. These findings suggest that ingestion of aluminium over a period of several weeks causes specific alterations in the brain. It is possible that longer exposures to lower levels of aluminium, or disruption of barriers limiting the access of aluminium to the brain, may result in similar neurotoxic consequences.

Aluminium administration

Most of these studies utilized aluminium sulphate (0.3% Al) in the drinking water given to adult rats for four weeks. Longer times of administration were used in later investigations. Treated rats (initially weighing 120 g) increased in body weight at a reduced rate during the first week or two of aluminium administration and thereafter gained weight at a rate similar to that of controls. The initial lag resulted in body weights approximately 10% below controls after four weeks. The initial studies were limited to four weeks of treatment because of the size constraints of the microwave irradiation device used to kill the rats. This method of sacrifice is necessary to limit postmortem changes of rapidly turning over metabolites, such as second messengers, but its capacity is limited to rats weighing less than 275 g. Treated rats consumed 18 ± 0.8 ml of drinking solution per day each (2 mmol aluminium/day per rat) and serum aluminium concentrations were increased from 1.8 ± 0.5 µg/l in controls to 15.8 ± 0.5 µg/l after four weeks of aluminium administration. In a recent developmental study, weanling rats were given 0.1% Al for two weeks, followed by 0.2% Al for two weeks and 0.3% Al for four weeks.

Cognition

Behavioural tests showed that four weeks of aluminium treatment induced a specific cognitive impairment in adult rats. Although, locomotor activity and performance on a radial eight-arm maze or an active avoidance task were not different in aluminium-treated and control rats (Connor et al 1988, 1989), retention of a learned passive avoidance task was maintained only half as long in the treated rats as in controls. Thus, chronic oral administration of aluminium caused specific cognitive deficits without resulting in generalized toxicity.

Calcium and phosphoinositide hydrolysis

Signal transduction mechanisms in the brain have been one of the primary focuses of investigation in this laboratory. Aluminium was found to inhibit Ca^{2+} flux in synaptosomes (Koenig & Jope 1987), and other inhibitory effects of aluminium on Ca^{2+}-dependent processes have since been described (Schöfl et al 1990, Wakui et al 1990), but further studies with brain tissue are required to clarify the interactions of aluminium with such processes.

The effect of aluminium on the phosphoinositide second messenger-producing system, which modulates intracellular calcium concentrations, has been investigated. Aluminium was reported in 1986 to inhibit agonist-stimulated phosphoinositide hydrolysis in brain tissue (Johnson & Jope 1986a). This finding has been extended in later studies (Jope 1988, McDonald & Mamrack 1988), all of which used *in vitro* exposure of brain tissue to aluminium. With the recent development of a method of measuring the *in vivo* concentration of the primary second messenger, inositol 1,4,5-trisphosphate ($Ins(1,4,5)P_3$) (Challiss et al 1988), we have extended those studies. In the rats treated with aluminium during development, the hippocampal $Ins(1,4,5)P_3$ concentration was significantly lower than in controls (21.1 ± 0.8 and 27.7 ± 0.4 pmol/mg, $P < 0.05$), indicating that aluminium exposure either directly or indirectly impairs the activity of this important second messenger-producing system in the rat hippocampus. Thus, there is sufficient evidence to support the hypothesis that aluminium influences calcium homeostasis and calcium-dependent processes in the brain (Farnell et al 1985). However, the most susceptible sites and the consequences for neuronal function have yet to be clarified. One intriguing possibility raised recently is that aluminium may modulate the activity of calcium-activated neutral proteinases (calpains) (Nixon et al 1990), which may alter the distribution and concentration of their substrates, such as cytoskeletal proteins.

Cyclic AMP

The most startling response of a second messenger system to aluminium that we have observed came when the *in vivo* concentrations of cyclic AMP were measured in rat brain regions after four weeks of aluminium ingestion. In these experiments, cyclic AMP was elevated by 30–70% in all five brain regions that we examined after aluminium treatment (Johnson & Jope 1986b, 1987). The largest increases in cyclic AMP were in the hippocampus and cerebral cortex. In the same tissues the concentration of cyclic GMP was similar to that in controls except for an elevation in the cerebellum. We recently found that cyclic AMP was increased by 33% in the hippocampus and cortex of the rats treated during development. Aluminium treatment has also been reported to increase cyclic AMP in neuroblastoma × glioma hybridoma cells (NG108–15) (Singer et al 1990). Such large, chronically elevated concentrations of this major second messenger suggest that aluminium has profound effects on neuronal processes associated with cyclic AMP.

The increased cyclic AMP concentrations found after aluminium administration led us to attempt to reproduce this response *in vitro* in order to investigate the mechanism of the interaction. Agonist-stimulated (2-chloroadenosine or isoproterenol) cyclic AMP production in rat brain cortical slices was enhanced by 100 µM AlCl$_3$ when the slices were preincubated with aluminium (suggesting an intracellular site of action) and when the agonists were present at submaximal concentrations (suggesting facilitation of the coupling mechanisms) (Johnson et al 1989). Direct effects of aluminium on phosphodiesterase and adenylate cyclase were shown not to account for this response. Because aluminium had previously been reported to inhibit the GTPase activity associated with transducin and with tubulin (Kanaho et al 1985, Macdonald et al 1987), we suggested that a potential explanation of the elevated cyclic AMP production induced by aluminium would be the inhibition of the GTPase activity of the stimulatory G protein, G$_s$ (Johnson et al 1989). This would result in the prolonged action of G$_s$ on adenylate cyclase, causing increased cyclic AMP production. These findings suggest that cells containing receptors coupled to cyclic AMP production are likely to be particularly susceptible to exposure to aluminium and that protein phosphorylation systems and genomic cyclic AMP response elements may both be influenced by this response to aluminium.

Protein phosphorylation

Since the function of cyclic AMP is to activate specific kinases, and because alterations in cytoskeletal proteins are associated with many neurodegenerative disorders, we investigated the effects of four weeks of oral aluminium administration to rats on brain cytoskeletal protein phosphorylation. *In vivo* phosphorylation was measured by the intraventricular infusion of ^{32}P, the rats being killed four hours later. Aluminium administration significantly increased the phosphorylation of microtubule-associated protein 2 (MAP-2) and the 200 kDa neurofilament protein (NF-H), while the phosphorylation of several other proteins was not different from that in control rats (Johnson & Jope 1988, Johnson et al 1990). These cytoskeletal proteins are of particular interest because abnormalities have been noted in a number of neurodegenerative disorders (Kosik et al 1984, Sternberger et al 1985, Forno et al 1986), as well as after the central administration of aluminium to rabbits (Bizzi & Gambetti 1986, Troncoso et al 1986). Concomitant with these increases in the phosphorylation of MAP-2 and NF-H, aluminium induced a significant increase in the activity of a cofactor-independent protein kinase that co-purifies with microtubules

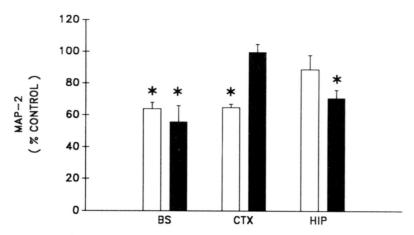

FIG. 1.   Effects of aluminium ingestion on the concentration of MAP-2 in rat brain regions. Adult rats were given drinking water containing 0.3% Al for three months, or weanling rats were given water containing 0.1% Al for two weeks, 0.2% Al for two weeks, and 0.3% Al for four weeks. The concentration of MAP-2 was measured in the brainstem (BS), cerebral cortex (CTX) and hippocampus (HIP), as described in the text. $n = 3$ for adults; $n = 4$ for weanlings. *$P < 0.05$ compared with age-matched controls.

and can catalyse the phosphorylation of both MAP-2 and NF-H (Caputo et al 1989). It is possible that abnormal phosphorylation will impair the axonal transport of these proteins, leading to abnormal accumulations and depletions in the cell soma and nerve endings, respectively.

Cytoskeletal protein concentrations

Our finding of the altered phosphorylation of cytoskeletal proteins after aluminium ingestion led to a study of the effects of aluminium treatment on the concentrations of seven structural proteins, including the three neurofilament proteins, NF-H (200 kDa), NF-M (150 kDa) and NF-L (68 kDa), tubulin, MAP-2, tau and spectrin, in rat cortex, hippocampus and brainstem. These measurements were made after three months of 0.3% aluminium administration in the adult rat study and after two months of such treatment in the developmental study. MAP-2 was the most sensitive of these proteins to aluminium; it was reduced by 29% and 44% in the hippocampus and brainstem after three months of aluminium treatment, and by 35% in both the cortex and brainstem of developing rats given aluminium (Fig. 1). All the other proteins were unaffected by both aluminium treatment protocols, except for a 43%

decrease in spectrin in the adult hippocampus after three months of treatment. The decrease in MAP-2 in the hippocampus after three months of aluminium treatment observed by Western blot analysis was further confirmed by immunohistochemistry.

Two findings from these experiments are especially notable. First is the selective sensitivity of MAP-2 to aluminium ingestion. The loss of MAP-2 was exceptionally high, considering both the resistance of the other proteins to aluminium administration and the fact that many neurochemical and behavioural processes are unchanged by these treatments. It is interesting that others have concluded that MAP-2 is the most sensitive of several structural proteins to the deleterious effects of ischaemia (Kitagawa et al 1989, Kudo et al 1990). Since there is no evidence for ischaemia in aluminium-treated rats, it is possible that MAP-2 is generally sensitive to neurotoxic insults and may play a role in the initial events associated with neurodegeneration.

The second finding of note is the resistance of tau (a microtubule-associated protein) to aluminium administration. Abnormalities of tau in Alzheimer's disease are clearly documented in many studies (Wood et al 1986, Goedert et al 1989, Ksiezak-Redding et al 1990). Thus, the aluminium administration protocols used in our studies do not model Alzheimer's disease insofar as tau is concerned. Several possible explanations arise—for example, there may be species differences, or longer treatment with aluminium might be required, because tau abnormalities might follow the loss of MAP-2. But at this point it must be concluded that aluminium administration as utilized here does not model the major changes in tau that are associated with Alzheimer's disease.

Summary

More questions remain than are answered by our current understanding of the neurotoxic consequences of exposure to aluminium. The investigations summarized here demonstrate that in the rat the ingestion of large quantities of aluminium alters central nervous system processes without causing generalized toxicity. These findings indicate that long-term exposure to aluminium can have deleterious effects on the brain. The primary sites of action of aluminium are difficult to discern from the many effects that have been reported using a wide variety of *in vivo* and *in vitro* model systems. It may be useful to study models in which the effects of aluminium are fairly subtle, as in our studies summarized here, rather than models that are acutely toxic and/or lethal. With any model system, however, species differences in the absorption, distribution, and sites of action of aluminium must be considered.

The effects of aluminium that have been identified in our and other laboratories may be connected in serial or parallel pathways leading to the ultimate manifestations of neurotoxicity (Fig. 2). The cognitive deficits associated with aluminium exposure are likely to result from impaired neuronal function.

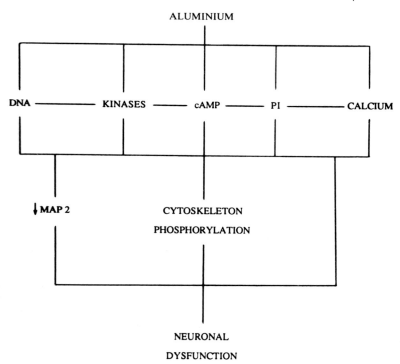

FIG. 2. Summary of some of the sites of neurotoxic responses to aluminium administration. Many potential sites of action of aluminium have been identified (including others not shown here but discussed elsewhere in this book). These may be connected in multiple parallel pathways, with mutual interactive influences, or as fewer primary responses that are linked in series. Altered cytoskeletal proteins appears to be an important consequence of the ingestion of aluminium, probably reflecting indirect effects following the actions of aluminium at the initial sites depicted here. PI, phosphoinositide metabolism.

The commonality of alterations of cytoskeletal proteins associated with aluminium treatment and with several neurodegenerative disorders suggests that the cause of abnormal neuronal function may reside in changes in the cytoskeletal protein functions in affected cells. The changes in cytoskeletal proteins may be the result of several processes, including altered phosphorylation, proteolysis, transport and synthesis. Interactions among these processes undoubtedly contribute to the final outcome, and each can be influenced by kinases, some of which are activated by second messengers. Since aluminium alters the activities of second messenger systems, possibly by direct effects on G proteins, this may be an important early site of action of aluminium which ultimately contributes substantially to the neurotoxicity. Also, genomic interactions with either aluminium, or second messenger systems affected by aluminium, must be considered as important potential sites of action. Whether the neurotoxicity occurs as a series of directly connected events, or represents the culmination of parallel processes following multiple effects of aluminium, remains to be clarified.

Acknowledgements

The authors thank their many colleagues who have made important contributions to these investigations. This work was supported by National Institutes of Health grants AG06569 and NS27538.

References

Bizzi I, Gambetti PC 1986 Phosphorylation of neurofilaments is altered in aluminum intoxication. Acta Neuropathol 71:154–158

Caputo CB, Sygowski LA, Scott CW, Johnson GVW, Brunner WF, Salama AI 1989 Changes in protein kinase activities following aluminum administration to aged rats. Brain Dysfunct 2:297–309

Challiss RAJ, Batty IA, Nahorski SR 1988 Mass measurements of inositol (1, 4, 5) trisphosphate in rat cerebral cortex slices using a radioreceptor assay: effects of neurotransmitters and depolarization. Biochem Biophys Res Commun 157:684–694

Connor DJ, Jope RS, Harrell LE 1988 Chronic oral aluminum administration to rats: cognition and cholinergic parameters. Pharmacol Biochem Behav 31:467–474

Connor DJ, Harrell LE, Jope RS 1989 Reversal of an aluminum-induced behavioral deficit by administration of deferoxamine. Behav Neurosci 103:779–783

Farnell BJ, Crapper McLachlan DR, Baimbridge K, De Boni U, Wong L, Wood PL 1985 Calcium metabolism in aluminum encephalopathy. Exp Neurol 88:68–83

Forno LS, Sternberger LA, Sternberger NH, Strefling AM, Swansen K, Eng LF 1986 Reaction of Lewy bodies with antibodies to phosphorylated and non-phosphorylated neurofilaments. Neurosci Lett 64:253–258

Goedert M, Spillantini MG, Jakes R, Rutherford D, Crowther RA 1989 Multiple isoforms of human microtubule-associated protein tau: sequences and localization in neuro-fibrillary tangles of Alzheimer's disease. Neuron 3:519–526

Johnson GVW, Jope RS 1986a Aluminum impairs glucose utilization and cholinergic activity in rat brain in vitro. Toxicology 40:93–102

Johnson GVW, Jope RS 1986b Aluminum increases cyclic AMP in rat cerebral cortex in vivo. Life Sci 39:1301–1305

Johnson GVW, Jope RS 1987 Aluminum alters cyclic AMP and cyclic GMP levels but not presynaptic cholinergic markers in rat brain in vivo. Brain Res 403:1–6

Johnson GVW, Jope RS 1988 Phosphorylation of rat brain cytoskeletal proteins is increased after orally administered aluminum. Brain Res 456:95–103

Johnson GVW, Li X, Jope RS 1989 Aluminum increases agonist-stimulated cyclic AMP production in rat cerebral cortical slices. J Neurochem 53:258–263

Johnson GVW, Cogdill KW, Jope RS 1990 Orally administered aluminum alters in vitro protein phosphorylation and protein kinase activities in rat brain. Neurobiol Aging 11:209–216

Jope RS 1988 Modulation of phosphoinositide hydrolysis by NaF and aluminum in rat cortical slices. J Neurochem 51:1731–1736

Kanaho Y, Moss J, Vaughan M 1985 Mechanism of inhibition of transducin GTPase activity by fluoride and aluminum. J Biol Chem 260:11493–11497

Kitagawa K, Matsumoto M, Niinobe M et al 1989 Microtubule-associated protein 2 as a sensitive marker for cerebral ischemic damage—immunohistochemical investigation of dendritic damage. Neuroscience 31:401–411

Koenig ML, Jope RS 1987 Aluminum inhibits the fast phase of voltage-dependent calcium influx into synaptosomes. J Neurochem 49:316–320

Kosik KS, Duffy LK, Dawling MM, Abraham C, McCluskey A, Selkoe DJ 1984 Microtubule-associated protein 2: monoclonal antibodies demonstrate selective incorporation of certain epitopes into Alzheimer neurofibrillary tangles. Proc Natl Acad Sci USA 81:7941–7945

Ksiezak-Redding H, Binder LI, Yen SH 1990 Alzheimer's disease proteins (A68) share epitopes with tau but show distinct biochemical properties. J Neurosci Res 25:420–430

Kudo T, Tada K, Takeda M, Nishimura T 1990 Learning impairment and microtubule-associated protein 2 decrease in gerbils under chronic cerebral hypoperfusion. Stroke 21:1205–1209

Macdonald TL, Humphreys WG, Martin RB 1987 Promotion of tubulin assembly by aluminum ion in vitro. Science (Wash DC) 236:183–186

McDonald LJ, Mamrack MD 1988 Aluminum affects phosphoinositide hydrolysis by phosphoinositidase C. Biochem Biophys Res Commun 155:203–208

McLachlan DRC 1986 Aluminum and Alzheimer's disease. Neurobiol Aging 7:525–531

Nixon RA, Clarke JF, Logvinenko KB, Tan MKH, Hoult M, Grynspan F 1990 Aluminum inhibits calpain-mediated proteolysis and induces human neurofilament proteins to form protease-resistant high molecular weight complexes. J Neurochem 55:1950–1959

Schöfl C, Sanchez-Bueno A, Dixon J et al 1990 Aluminum perturbs oscillatory phosphoinositide-mediated calcium signalling in hormone-stimulated hepatocytes. Biochem J 269:547–550

Singer HS, Searles CD, Hahn I-H, March JL, Troncoso JC 1990 The effect of aluminum on markers for synaptic neurotransmission, cyclic AMP, and neurofilaments in a neuroblastoma × glioma hybridoma (NG108 − 15). Brain Res 528:73–79

Sternberger NH, Sternberger LA, Ulrich J 1985 Aberrant neurofilament phosphorylation in Alzheimer's disease. Proc Natl Acad Sci USA 82:4274–4276

Troncoso JC, Sternberger NH, Sternberger LA, Hoffman DN, Price DL 1986 Immunocytochemical studies of neurofilament antigens in neurofibrillary pathology induced by aluminum. Brain Res 364:295–300

Wakui M, Itaya K, Birchall JD, Petersen OH 1990 Intracellular aluminum inhibits acetylcholine- and caffeine-evoked Ca^{2+} mobilization. FEBS (Fed Eur Biochem Soc) Lett 267:301–304

Wood JG, Mirra SS, Pollock NJ, Binder LI 1986 Neurofibrillary tangles of Alzheimer's disease share antigenic determinants with the axonal microtubule-associated protein tau. Proc Natl Acad Sci USA 83:4040–4043

DISCUSSION

Petersen: Dr Jope, did your assay discriminate between the different forms of $InsP_3$?

Jope: Yes; we used the binding assay for inositol 1,4,5-trisphosphate, so it is the active $InsP_3$ that is decreased *in vivo* by Al administration.

Wischik: You demonstrated that the cyclic AMP produced was increased by Al ingestion much more in cortex and hippocampus than in the brainstem, but the changes in phosphorylated cytoskeletal proteins appeared to be greater in brainstem than in cortex. Can you explain this apparent difference?

Johnson: The *in vivo* phosphorylation of MAP-2 and NF-H in brainstem was greater than in the cortex, but the difference between control and Al-treated rats was similiar in both regions.

Wischik: In relation to the decreases in MAP-2 concentration after Al treatment, you suggest that this could be due to increased proteolysis, rather than reduced synthesis.

Jope: We don't know what causes the decrease in MAP-2 after Al administration. Dr Johnson has shown that phosphorylation can change proteolysis, but we don't know if synthesis of MAP-2 is altered, so we don't know the mechanism for the decrease.

Wischik: It would be a remarkable connection if you could show increased proteolysis. MAP-2 is the predominant dendritic microtubule-associated protein. In Alzheimer's disease, tau protein, which is normally axonal, becomes abnormally sequestered within the somato-dendritic compartment. The loss that you demonstrate is a very striking effect; if you could demonstrate proteolytic cleavage of MAP-2 occurring, this could shed light on an intracellular amyloidogenic cascade.

Jope: MAP-2 is not like spectrin, where one can measure the accumulation of breakdown products. The decrease in MAP-2 concentration is surprisingly large, considering that it is a chronic Al treatment. These are recent results, and we haven't gone into the mechanisms as yet.

Edwardson: These observations on cytoskeletal protein changes are very important. However, going back to your behavioural studies, using changes in a passive avoidance task as evidence of cognitive impairment is skating on very thin ice. Why don't you use a paradigm such as delayed matching to sample, or another more sophisticated memory test which would provide real evidence of cognitive impairment? There are numerous examples of animals performing differently in passive avoidance tasks where it's clear that motivational parameters or arousal are being affected, and not cognitive processes *per se.*

Jope: The rats learn the task; it is the retention that is impaired after Al administration. Also, they learn the other tests used.

Edwardson: But it's not a test of cognitive performance; it is a test which is frequently confounded by motivational and other non-cognitive factors.

Martyn: I wonder whether you have more details about how long this effect persisted after you treated the animal with Al? Was this a permanent effect, or did it diminish if the animals were tested some time after Al treatment?

Jope: The effect on this behaviour is reversible. We have done a second study in which we treated rats for a month with Al, followed by a month off Al, or with administration of desferrioxamine; we found that their impaired performance was reversible (Connor et al 1989).

Martyn: There are two aspects of the passive avoidance test; there is the question of how long rats retain the memory of the shock, but you could also measure how long it takes them to learn the task—that is, to avoid the area in which they are shocked. Is there any difference in the time which Al-treated and not Al-treated rats take to learn this?

Jope: There is no significant difference. Some Al-treated rats had a little more difficulty in learning the task, but it was not significant. The Al-treated and control rats seem to learn in a similar number of trials, but the Al-treated rats don't retain the memory as well.

Williams: I gather, Dr Jope, that you see G proteins as a potential site of action of Al. Can you see any biochemical reason why the GTP system should be more sensitive to Al than an ATP system?

Jope: No!

Williams: I suspect that if we just take GTP and ATP, we will hardly see any difference at all.

Jope: It is GTP bound to a protein (i.e., a G protein) which is relevant when discussing the effects of Al on second messenger systems.

Williams: Yes. I have studied one simple, ATP, kinase system, so I know *why* the affinity at the site of binding on the protein could go up for a metal ion such as Al^{3+}, because in those kinases, magnesium also interacts with a carboxylate which is on the protein. If you have increased the local anion density, Mg binding and Al binding will increase. This has been shown to be essential for the activity of several kinases. Now, the crystal structure of the GTP-binding site of the Ras21 protein, a G protein, is known (Pai et al 1990), so one can investigate whether the chemistry in the binding site is different for a GTP type protein than for an ATP. Obviously, there must be some differences, because the adenosine is recognizing a very specific loop system. The difference may not lie in the ligand atoms to the metal ion, but the protein fold energies and the GTP binding. Mg^{2+} is very strongly retained by Ras21 at the GTP site. (See the parallel discussion of transferrin on p 47, and of fluoride binding on p 103–107.)

Jope: It would be very interesting if somebody would look at the effects of Al on ATPases and GTPases.

Williams: I agree. I don't know whether the data exist. I have looked specifically at other trivalent ions in this situation, like lanthanum, using them as probes for enzyme activity. Such ions block the enzymes very neatly; we think that they bind both to the phosphate and to the carboxylate. Biology does not like trivalent cations!

Birchall: We know that Al inhibits hexokinase activity by replacing Mg^{2+} on ATP (Viola et al 1980).

Williams: It also binds across the phosphates and to the protein.

Jope: It would be logical to examine the effects of Al on several ATP- and GTP-binding proteins, including ATPases, GTPases (which includes the second messenger-coupled G proteins), and kinases.

Williams: We ought to find a small GTPase, such as Ras21, where we can look at the chemistry of Al at the GTP-binding site.

Birchall: Is yours a system where you have to watch fluorine levels as well? You could get activation because of AlF_4^-, perhaps.

Jope: You would need to have enormous endogenous concentrations of fluorine for it to have significant effects in these assays, so I doubt that it plays a role in the results we have reported.

Blair: In all our concern with features of Alzheimer's disease such as the $\beta/A4$ protein, or the deposition of Al in particular brain regions, we may forget that we are looking at a problem where Al is not the cause of Alzheimer's disease. Dr Jope has reminded us that the experimental animal is very useful in this connection, and the results may well be independent of the intermediate steps you would care to take.

We developed a system like his in which we gave rats Al in their drinking water over a period of three months. We found no changes in weight, and no marked changes in behaviour patterns. At the end of the three months we made brain preparations for analysis for the absence or presence of various enzyme activities, and matched these with deficiencies we had found in human Alzheimer's disease brain tissue by similar methods. We hoped to produce a biochemical mapping of enzyme activities in an experimental animal against that of human subjects.

For example, we found no change in the activity of dihydropteridine reductase in the Al-treated rat or in human AD tissue, or in two other enzymes (GTP cyclohydrolase and sepiapterin reductase), whereas tyrosine hydroxylase was inhibited in both. Choline acetyltransferase was also inhibited in rat and in AD tissue. This is a key observation, because the diminution of choline acetyltransferase activity is thought to be a touchstone of Alzheimer's disease biochemistry. We were able to show that the difference was at least specific between Al and lead; we could not reproduce the inhibition of choline acetyltransferase by administering lead.

Glutamate decarboxylase was also inhibited; amino acid decarboxylase showed no change. Dopamine β-hydroxylase was inhibited. Monoamine oxidase activity showed no change, but MAO B was elevated in AD brain and has since been shown to be elevated in the rat model. Hexokinase was inhibited in the model. Cyclic AMP levels were increased significantly, as Dr Jope has found. We did not measure AMP levels in the Al model (Cowburn 1989).

So you can draw up a biochemical map showing the consequences of Al exposure over a long time which produces no obvious behavioural changes, but produce the biochemical changes characteristic of Alzheimer's disease.

We can modify this system further. Some of the effects of Al or other neurotoxins on the brain are lethal; there's no possibility of reversing the biochemical changes. Other changes simply represent inhibition of a biochemical parameter which could be reversed. We measured tetrahydrobiopterin (BH_4) biosynthesis and biopterin levels in the brain. This is a key molecule which affects human cognitive behaviour: we know that total destruction of BH_4 leads to total damage in the brain; partial destruction leads to reduced IQ scores. We treated rats for three months with Al; the level of synthesis of the enzyme was reduced to about 60% of the previous level. If we treated these Al-treated rat brains with transferrin, we could restore some of the original BH_4 biosynthetic activity. The effect was not just due to the addition of transferrin, as shown by the controls (Cowburn & Blair 1989). So in this system we are able to demonstrate that you can reverse the effect of Al on BH_4 synthesis. That this reversal happens in man has been supported by work on Alzheimer's disease patients by Dr S. Milstein at the National Institutes of Health (personal communication).

Jope: This work might demonstrate that there are many common mechanisms underlying different neurotoxic or neurodegenerative disorders; whether or not Al plays a role in initiating neurodegeneration, or whether after the initial toxic insult there's a common sequence of events, either way it's an interesting study.

Blair: Lead had no effect on choline acetyltransferase but it mimics the effects on BH_4 metabolism in its suppression of the biosynthesis. We didn't test the effect of any chelating agent in removing the toxic metal. And transferrin has quite a high level of metal specificity.

Williams: If Al were to affect standing levels (homeostatic levels) of Ca and cyclic AMP in a cell, so many things would be affected that consequences would develop right across the activity of the cell.

Blair: My view is that Al produces an enormous range of damage to the cell, and some of that would be lethal, so you would no longer be able to restore the biochemical function, or any cognitive function relying on it. In other areas, damage would be reversible. In the NIH studies, they administered BH_4 to the Alzheimer's disease patients and showed the appearance of its metabolites in the CSF, but they could show no significant clinical improvement in the patients.

Perl: We have talked about the protective barriers to Al exposure—the gut barrier, serum binding, urinary excretion, and finally the blood–brain barrier. Yet in your model, Dr Jope, these are apparently overcome. I assume that brain Al levels are increased in the animals?

Jope: We haven't measured it. If it's in focal accumulations, we might have the same problem as with Alzheimer's disease, that we would not observe significant accumulations of Al if large areas of the brain were used.

Perl: But certainly you are seeing a CNS effect, and I wonder what the implications are; they may be very important, in terms of day-to-day Al exposure.

Jope: With a chronic elevation of Al in the plasma, there is clearly a detrimental effect in the CNS.

Perl: Seemingly, in your model, this mechanism was not protective, and the general implications of that are considerable.

Blair: We didn't look at Al levels in plasma or brain in the work I just mentioned. But in another situation we have looked at speciation in the plasma of gallium, as an Al surrogate (Farrar et al 1988). As the gallium levels rise, we see a switch from a high molecular mass species to a lower one. So, effectively, with these long-term exposures, you are probably changing the ease of entry of the species into the brain. And if you administer gallium to animals, you can show that amounts of radioactivity entering the brain increase in a fashion consistent with the enhanced appearance of another low molecular mass species which can penetrate the brain more readily.

References

Connor DJ, Harrell LE, Jope RS 1989 Reversal of an aluminum-induced behavioral deficit by administration of deferoxamine. Behav Neurosci 103:779–783

Cowburn JD 1989 Tetrahydrobiopterin metabolism in mental disorders. PhD thesis, Aston University, Birmingham, UK

Cowburn JD, Blair JA 1989 Aluminium chelator (transferrin) reverses biochemical deficiency in Alzheimer brain preparations. Lancet 1:99

Farrar G, Morton AP, Blair JA 1988 The intestinal speciation of gallium: possible models to describe the bioavailability of aluminium. In: Brätter P, Schramel P (eds) Trace element analytical chemistry in medicine and biology. Walter de Gruyter, Berlin, vol 5:342–347

Pai EF, Krengel U, Petsko GA, Goody RS, Kabsch W, Wittinghofer A 1990 Refined crystal structure of the triphosphate conformation of H-ras21 at 1.3Å resolution: implications for the mechanism of GTP hydrolysis. EMBO (Eur Mol Biol Organ) J 9:2351–2359

Viola RE, Morrison JF, Cleland WW 1980 Interaction of metal(III)–adenosine 5'-triphosphate complexes with yeast hexokinase. Biochemistry 19:3131–3137

Molecular characterization and measurement of Alzheimer's disease pathology: implications for genetic and environmental aetiology

C. M. Wischik*,**, C. R. Harrington*,**, E. B. Mukaetova-Ladinska*, M Novak**,†, P. C. Edwards** and F. K. McArthur‡

*Cambridge Brain Bank Laboratory, University of Cambridge Clinical School, Department of Psychiatry, MRC Centre, Hills Road, Cambridge CB2 2QQ, UK, **MRC Laboratory of Molecular Biology, MRC Centre, Hills Road, Cambridge CB2 2QQ, UK, †Institute of Virology, 84246 Bratislava, Czechoslovakia, and ‡MRC Neurochemical Pathology Unit, Newcastle General Hospital, Westgate Road, Newcastle upon Tyne NE4 6BE, UK

Abstract. The neuropathological changes seen in Alzheimer's disease represent an interaction between the ageing process in which normal intellectual function is retained, and changes which are specifically associated with severe cognitive deterioration. Molecular analysis of these changes has tended to emphasize the distinction between neurofibrillary pathology, which is intracellular and highly correlated with cognitive deterioration, and the changes associated with the deposition of extracellular amyloid, which appears to be widespread in normal ageing. Extracellular amyloid deposits consist of fibrils composed of a short 42 amino acid peptide (β/A4) derived by abnormal proteolysis from a much larger precursor molecule (APP). The recent demonstration of a mutation associated with APP in rare cases with familial dementia, neurofibrillary pathology in the hippocampus and atypical cortical Lewy body pathology raises the possibility that abnormal processing of APP could be linked directly with neurofibrillary pathology. Neurofibrillary tangles and neuritic plaques are sites of dense accumulation of pathological paired helical filaments (PHFs) which are composed in part of an antigenically modified form of the microtubule-associated protein tau. The average brain tissue content of PHFs measured biochemically does not increase in the course of normal ageing but increases 10-fold relative to age-matched controls in patients with Alzheimer's disease. There is also a substantial (three-fold) disease-related decline in normal soluble tau protein relative to age-matched controls. This intracellular redistribution of a protein essential for microtubule stability in cortico-cortical association circuits may play an important part in the molecular pathogenesis of dementia in Alzheimer's disease. The role of abnormal proteolysis of APP in this process remains to be elucidated. Immunohistochemical studies on renal dialysis cases have failed to detect evidence of neurofibrillary pathology related to aluminium accumulation in brain tissue. Nevertheless it needs to be seen whether more sensitive biochemical assays of neurofibrillary pathology can demonstrate evidence of an association with aluminium.

1992 Aluminium in biology and medicine. Wiley, Chichester (Ciba Foundation Symposium 169) p 268–302

Alzheimer's disease is currently being transformed from a condition conceptualized largely in terms of classical histopathology into a condition in which the essential questions now concern molecular pathogenesis. The work of careful clinico-histopathological correlation still provides a necessary framework for evaluating which molecular constituents are key players in pathogenesis and in the mental deterioration seen clinically. This paper reviews the progress in identification of the molecular constituents of the neuropathology of Alzheimer's disease. These developments, by permitting the direct biochemical measurement of pathological change in brain tissues, shed new light on our understanding of the molecular pathogenesis of Alzheimer's disease.

β-Amyloid and pathobiology of the amyloid precursor protein (APP)

The discovery of plaques predates the discovery of neurofibrillary tangles. Blocq & Marinesco had described plaques in 1892 as a common neuropathological feature of ageing. In the classical description, the plaque consists of a dense congophilic amyloid core surrounded by a halo of less densely packed amyloid interspersed with swollen dystrophic neurites (Fig. 1). The pathologist's conception of amyloid is of a dense proteinaceous deposit consisting of arrays of fibrils of approximate diameter 10 nm, which are stained red by the dye congo red and, because of the paracrystalline molecular ordering of deposit, give rise to green birefringence in crossed-polarized light. Amyloid is thus a generic concept covering a range of different proteins in different disease states. A common theme is the abnormal accumulation of a partial degradation product originating from precursor over-production.

In Alzheimer's disease, the amyloid deposits found in plaques have similar tinctorial and ultrastructural properties to those found in the walls of cerebral and meningeal arterioles, referred to as cerebrovascular amyloid. Glenner & Wong (1984) identified a similar peptide fragment both in Alzheimer's disease and in aged Down's syndrome meninges. Similar peptides were also found in brain tissue extracts from Alzheimer's disease and Down's syndrome patients in the form of multimers of a basic 4 kDa peptide, which was termed A4 (Masters et al 1985). The peptide has been found to be 39–43 amino acid residues in length, depending on the source (Prelli et al 1988). Antibodies raised against parts or the whole of the synthetic A4 peptide label plaques and cerebrovascular β-amyloid deposits in tissue sections (Wong et al 1985, Masters et al 1985, Allsop

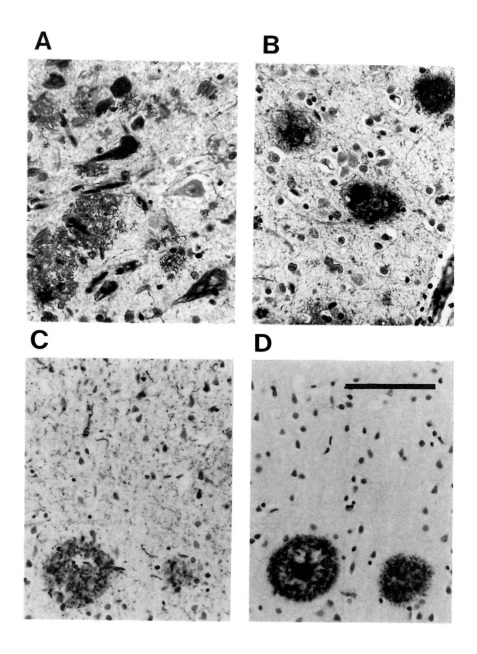

et al 1986). Synthetic peptides corresponding to A4 polymerize *in vitro* to form fibrils very similar to those found in β-amyloid deposits *in situ* (Gorevic et al 1987, Hilbich et al 1991). Thus the β-amyloid deposits of Alzheimer's disease are now generally accepted as being composed of the 'β/A4' protein.

This protein is derived from a much longer precursor termed 'amyloid precursor protein' (APP) (Kang et al 1987). Multiple alternatively spliced isoforms have been identified, all encoded on the long arm of chromosome 21 (reviewed by Selkoe 1991). A corresponding group of 110–135 kDa membrane-associated proteins have been identified in brain, in non-neural cells and in cell culture extracts. APP695 expression is closely associated with neuronal differentiation (Yoshikawa et al 1990) and neuronal maturation during embryogenesis (Ohyagi et al 1990), and it is the predominant mRNA isoform in brain (Neve et al 1988). Virtually all cells in the nervous system, including astrocytes, microglia and endothelial cells, as well as non-neural cells, including submandibular gland, epithelium and muscle (Catteruccia et al 1990), express some form of APP.

The simple hypothesis that an alteration in the absolute quantity or ratio of APP subtype mRNA causes Alzheimer's disease has been examined in many laboratories. Although there is an increase in mRNA for all APP isoforms in senescent cells in tissue culture (Adler et al 1991), neither an increase in the APP751/APP695 ratio (Johnson et al 1988), nor a specific increase in APP751 (Tanaka et al 1989), is correlated with β-amyloid pathology (Koo et al 1990, Oyama et al 1991), nor is there any correlation between particular mRNA levels and the neuroanatomical distribution of β-amyloid pathology (Chou et al 1990). Nevertheless, Down's syndrome cases, in which there is over-expression of APP (Delabar et al 1987), develop the neuropathology typical of Alzheimer's disease in later life.

FIG. 1.   Typical appearance of Alzheimer's disease neuropathology, using a variety of techniques to visualize the lesions. In A and B the methenamine silver method is used. This permits the visualization not only of neurofibrillary tangles and the dystrophic neurites of the plaque (A) but also the staining of plaques. (B). Section A was from hippocampus and section B from temporal cortex. In addition to labelling neurofibrillary tangles, anti-tau antibodies label the dystrophic neurites seen in the plaque periphery (C) and also extensive 'dystrophic neuropil threads' (C). These are all known to be sites of paired helical filament (PHF) accumulation from ultrastructural studies. By contrast, anti-β/A4 antibodies label plaque cores and produce denser labelling in the plaque periphery on an adjacent section (D). Anti-β/A4 antibodies do not label the dystrophic neuropil threads. The sections shown in C and D are from the occipital lobe in a case with very extensive PHF accumulation detected by biochemical analysis, but virtually no neurofibrillary tangles. When computed across all brain regions in Alzheimer's disease, PHF accumulation correlates best with neuritic plaques and dystrophic neuropil threads, and relatively poorly with neurofibrillary tangles. Scale bar, 100 μm.

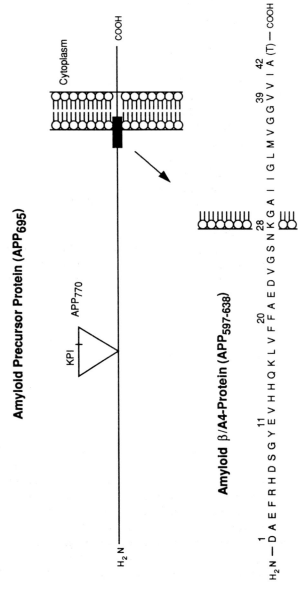

FIG. 2. Schematic representation of the shortest form of the amyloid precursor protein (APP695). Larger isoforms (APP751 and APP770) are produced by alternative splicing to include a Kunitz type II serine protease inhibitor domain (KPI). The segment of the molecule which is now known to accumulate as β-amyloid deposits in Alzheimer's disease includes the short stretch which spans the outer part of the cell membrane. Normal secretory cleavage, which releases the long N-terminal portion of APP, occurs within the N-terminal half of β/A4, and within a portion of the molecule which is thought to be critical for the polymerization of β/A4 fibrils. However, the C-terminal end of β/A4 is also thought to be important for aberrant polymerization.

There is genetic evidence suggesting that APP dysmetabolism can, in some cases, trigger the typical human neuropathology of Alzheimer's disease. Numerous epidemiological studies have identified family history as an important risk factor for developing Alzheimer's disease (Heston et al 1981). There are rare kindreds in which Alzheimer's disease is inherited as an autosomal dominant trait. Molecular genetic studies have shown that these are genetically heterogeneous (St George-Hyslop et al 1990), with some kindreds showing linkage to a marker from the proximal region of the long arm of chromosome 21 (St George-Hyslop et al 1987) but others having no linkage to any chromosome 21 markers (Schellenberg et al 1988). Despite earlier reports that the APP locus itself could be excluded as a linkage marker (van Broeckhoven et al 1987), several groups have now confirmed that a mutation at residue 642 of APP695 can be linked in certain kindreds with the Alzheimer's disease phenotype (Goate et al 1991, Naruse et al 1991, Chartier-Harlin et al 1991, Murrell et al 1991). On the other hand, a nearby mutation at residue 618 is associated only with cerebrovascular amyloidosis, without any evidence of neurofibrillary pathology (Levy et al 1990).

Attempts to understand how abnormal processing of APP could give rise to β-amyloid deposits have had to take account of the complexity of normal APP processing. The essential problem is to explain how the 39–43 amino acid β/A4 peptide is cleaved from APP since, once this fragment is released from the parent molecule, it will polymerize spontaneously. The C-terminal cleavage site at or near residue 638 would have to occur, according to secondary structure prediction (Kang et al 1987) and direct biochemical evidence (Dyrks et al 1988), within a segment of the molecule which is thought to span the membrane (Fig. 2). How this might occur is not known.

There is a normal proteolytic processing of APP which releases the large N-terminal extracellular portion of the molecule (Esch et al 1990, Wang et al 1991). The extracellular function of this region of the APP molecule is not known precisely, but that portion derived from the larger isoforms (APP751, APP770) may include activity as a Kunitz type II serine protease inhibitor which forms a stable covalent ester bond with the catalytic site serine of the protease, prior to internalization of the protease/inhibitor complex and lysosomal degradation (Van Nostrand et al 1989). This mechanism of protease regulation at or near the cell surface appears to be important for local regulation of the extracellular matrix and neurite outgrowth (Monard 1988). The proteolytic cleavage that releases the N-terminal segment of APP occurs C-terminal to either Lys612 or Glu611. This normal secretory cleavage would occur within the 13 residue minimal peptide thought to be required for the *in vitro* polymerization of β/A4 fibrils (Hilbich et al 1991).

Neurofibrillary pathology

It is for his discovery of the neurofibrillary tangle (Fig. 1) that Alzheimer's name is associated with this neurodegenerative disorder. Like β-amyloid deposits, the tangle also consists of a dense mass of pathological fibres. These are larger in diameter than amyloid fibrils, and have a characteristic wide-to-narrow modulation of diameter every 80 nm (Kidd 1963). They are called paired helical filaments (PHFs) and consist of two strands of repeating C-shaped structural subunits of about 100 kDa each (Wischik et al 1985, 1988a, Crowther & Wischik 1985). PHFs are also found in the dystrophic neurites of neuritic plaques, and diffusely distributed in the neuropil in the form of 'dystrophic neuropil threads' (Braak & Braak 1988, Yamaguchi et al 1990).

Cytoskeletal proteins which have been implicated in neurofibrillary pathology on immunohistochemical grounds include the microtubule-associated proteins MAP-2 and tau, neurofilament proteins, tubulin, vimentin, actin and tropomyosin (reviewed in Wischik 1989a). Of these, only tau protein has been identified as a definite constituent of the PHF. MAP-2 and tau proteins are members of a large group of proteins which co-purify with brain tubulin during repetitive cycles of temperature-dependent assembly and disassembly (Goedert et al 1991). MAP-2 and tau segregate into complementary somato-dendritic and axonal compartments in the mature neuron (Binder et al 1985). This segregation is developmentally controlled and coincides with the appearance of a specialized axonal neurite (Dotti et al 1987, Couchie et al 1990). Tau and MAP-2 share antigenic determinants, and have close sequence homology in the C-terminal domain, which consists of three or four tandem repeats of 31 amino acids each (Goedert et al 1991). These repeats contain shorter 18 residue segments which bind to C-terminal residues of tubulin and are separated by flexible linker sequences (Butner & Kirschner 1991).

The evidence for tau protein as an integral structural constituent of the PHF is based on two studies using complementary criteria to distinguish integral from associated PHF proteins. Kondo et al (1988) considered sodium dodecyl sulphate (SDS)-resistant association as a defining characteristic of integral PHF proteins. Although several studies have established that epitopes which span the entire tau molecule are associated with PHFs by histological (Kosik et al 1988), ultra-structural (Brion et al 1991a) and immunobiochemical (Caputo et al 1992) criteria, only epitopes associated with the C-terminal one-third of the molecule have an SDS-resistant association with PHFs (Ihara et al 1990, Brion et al 1991a). This also agrees with the results of Nieto et al (1991), who identified a 33 kDa tau protein, beginning at Ser129, which co-purifies with PHFs after SDS treatment. By this criterion, MAP-2 (Brion et al 1990) and N-terminally intact tau protein, which both co-purify with PHFs, would be excluded as integral PHF constituents.

Wischik et al (1988b) used protease-resistant association as a means of excluding associated proteins, and to define the domains of proteins integral to

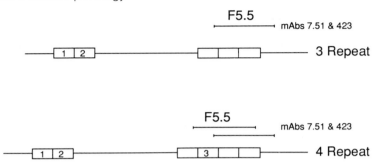

FIG. 3. Schematic representation of the various isoforms of tau protein, which arise by alternative splicing to include segments shown as 1, 2 in the N-terminal half of the molecule, and 3 in the tandem repeat region, giving rise to six possible isoforms. The tandem repeat region is known to function as the microtubule-binding domain of the molecule. Tau is required to maintain tubulin in the polymerized state as microtubules. The segments denoted F5.5, which span three tandem repeats of either 3- or 4-repeat isoforms, correspond to the fragments released from the protease-resistant core of the PHF by formic acid treatment. This fragment appears as a 12 kDa gel band, but contains various N-termini, depending on the isoform. The repeat region of tau is thus the segment which is embedded within the protease-resistant core of the PHF.

the protease-resistant core structure of the PHF. A 12 kDa tau fragment, beginning in the vicinity of His268, which encompasses only the repeat region, and excludes the C-terminal tail of tau, was identified as integral to the core PHF (Fig. 3). Since PHFs remain morphologically intact after proteolytic removal of the N-terminal two-thirds of tau (Wischik et al 1988a), this region of the molecule is not required for the structural integrity of the PHF. On the other hand, procedures which release the 12 kDa tau fragment from the repeat region are also associated with loss of PHF structure. The protease-protected core tau fragment contains several N-terminally distinct peptides which co-migrate as a single gel band. N-terminal sequence analysis of these species has shown that they originate from 3- and 4-repeat isoforms, and that their N-termini can be aligned according to the tandem repeat structure of the tau molecule (Jakes et al 1991). The PHF-core binding domain of tau is restricted to three repeats, regardless of tau isoform. The N-terminal cleavage sites align inside the 18 residue tubulin-binding segments of tau, which suggests that the residues required for tau/PHF association differ from those required for the normal tau/tubulin association. The procedures which release tau from tubulin are much milder than those required to release tau from the core PHF. Once assembled into the PHF, tau is sequestered irreversibly within a protease-resistant/SDS-resistant polymeric complex (Fig. 4).

Abnormal phosphorylation of tau protein is now widely regarded as an early event required for PHF assembly. The initial finding was that dephosphorylation with alkaline phosphatase enhanced immunostaining of neurofibrillary tangles

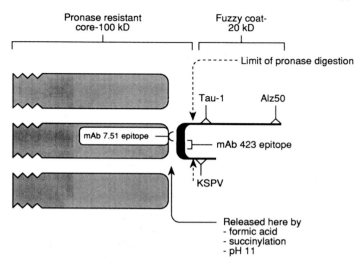

FIG. 4.   Schematic representation of the PHF as a stack of repeating subunits. The major part of the protease-resistant core (shown as stippled) has not been characterized fully. Studies of protease-resistant PHF preparations suggest that tau is tightly associated with the core structure of the PHF via the tandem repeat region, shown as the thicker part of the hairpin, which corresponds to a 12 kDa tau fragment that can be released after protease digestion by the various severe treatments shown. The Tau-1 epitope is phosphorylated both in Alzheimer's disease and in normal tau found in the somato-dendritic compartment of the cell, but not in tau found in the axonal compartment of the cell. A68 tau released from PHF preparations in Alzheimer's disease by SDS treatment is phosphorylated at the Tau-1 site. The epitope is contained in the N-terminal half of the molecule, located in the fuzzy coat of the PHF, which is lost after protease digestion of PHFs. Likewise, the Alz50 epitope, which was initially used to identify the A68 tau species of Alzheimer's disease, is located at the N-terminus of the molecule and is lost after protease digestion. The KSPV segment, thought to be another aberrant phosphorylation site, is located in the C-terminal tail of tau. This segment is probably not contained in the 12 kDa tau fragment which is protected from protease digestion. The epitopes recognized by mAbs 7.51 and 423 are both located in the repeat region of tau. The mAb 7.51 epitope is available in preparations of normal soluble tau protein, but is not available in PHF preparations, unless the tau is first released by formic acid treatment. Thus mAb 7.51 can be used to measure normal tau in brain homogenate without interference from PHF-tau. mAb 423 recognizes an epitope which is characteristic of PHF-associated tau, but is not found in normal tau preparations. This permits the measurement of PHF-tau without interference from normal tau in brain homogenate.

by the monoclonal antibody Tau-1, generated by Binder et al (1985), and shown by Grundke-Iqbal et al (1986) to recognize 55–62 kDa tau proteins isolated in a microtubule preparation from Alzheimer's disease brain tissues, but not from normal tissues. The Tau-1 epitope, mapped by Kosik et al (1988), is unavailable when tau is phosphorylated. Three bands with apparent gel mobilities 58 kDa,

63 kDa and 68 kDa, termed collectively 'A68 proteins' or 'Alzheimer's disease-associated proteins', have been identified in Alzheimer's disease brain tissues but not in controls (Flament & Delacourte 1989, Hanger et al 1991, Delacourte et al 1990, Flament et al 1990). These bands revert to the normal six-banded tau pattern after dephosphorylation (Brion et al 1991b, Goedert et al 1992). A68 proteins have been claimed to be major constituents of PHFs, based on the apparent complete solubilization of a preparation of dispersed PHFs, called 'A68 PHFs' (Greenberg & Davies 1990, Lee et al 1991). Brion et al (1991b) have demonstrated that the A68 proteins released from PHFs are recognized by Tau-1, although Lee et al (1991) have claimed the abnormal phosphorylation site to be at Ser396 in the KSPV segment that is shared with neurofilament proteins. However, since A68 proteins are intact at the N-terminus, it is not clear how they would meet the criterion of SDS-resistant association. Although abnormally phosphorylated tau species are present in Alzheimer's disease brain tissues, and co-purify with PHFs, further quantitative studies in post mortem brain tissues are needed to establish the nature of the relationship between PHF accumulation and the abnormal phosphorylation of tau in Alzheimer's disease. It remains possible that phosphorylation at the Tau-1 site occurs as part of a neuronal stress response, that can occur in the absence of PHFs (Papasozomenos & Su 1991), and may be a consequence rather than a cause of PHF assembly.

Since there are no changes in message for the various tau isoforms in Alzheimer's disease (Goedert et al 1988, 1989), it is likely that the incorporation of tau protein into PHFs is a consequence of some post-translational modification. The nature of this is at present unknown. Indeed, it is not even clear that tau is the only protein which is incorporated into the core structure of the PHF. There is some evidence that ubiquitin is associated with PHFs (Mori et al 1987), although its presence in the protease-resistant core structure has been questioned (Brion et al 1991a). The presence of conjugated ubiquitin in neurofibrillary pathology implies recognition by the cell of an aberrant structure and the activation of a non-lysosomal degradative pathway (Finley & Chou 1991).

Clinico-pathological correlation in Alzheimer's disease

Although it is generally agreed that cases with Alzheimer's disease can be distinguished from age-matched controls on the basis of the quantity of plaques (Tomlinson et al 1968) and tangles (Wilcock & Esiri 1982), neuronal loss (Mann et al 1988), and the quantity of granulo-vacuolar degeneration (Ball & Lo 1977), there is disagreement about the threshold values required to make a neuropathological diagnosis (Duyckaerts et al 1990). Cases with clinical dementia of two to five years duration, with numerous tangles and plaques in the neocortex, are not generally in doubt. The problems arise with cases that are in some way atypical. For example, as many as 30% of cases with dementia have no neocortical tangles, and have only plaques (Terry et al 1987). The general rate of clinico-pathological agreement is surprisingly low, estimated to be in the range of 43–87% (Boller et al 1989, Jellinger et al 1990). On the clinical side, an important ingredient for producing a higher rate of successful clinical prediction is informant-based assessment (Roth et al 1986, O'Connor et al 1989, Morris et al 1991).

The correlation between mental deterioration and the density of lesions is generally better for neurofibrillary tangles than for senile plaques (Wilcock & Esiri 1982, Tierney et al 1988, Delaère et al 1989, Duyckaerts et al 1990, Braak & Braak 1991), even for cases with very mild impairment (Morris et al 1991). Nevertheless, the evidence for cases with dementia but without tangles has led to the development of criteria based solely on an age-related scale of plaque counts (Khachaturian 1985), which have been incorporated into the CERAD criteria (Mirra et al 1991). Unfortunately, these criteria can give rise to a neuropathological diagnosis of Alzheimer's disease in cases without mental deterioration (Crystal et al 1988). To call such cases pre-clinical Alzheimer's disease begs the question of the biological difference between normal and pathological ageing.

The addition of immunohistochemical techniques to quantitative neuropathological studies has failed to provide a way of resolving the present diagnostic difficulties in neuropathology. The lesions detected by antibodies directed against tau protein correspond largely to those detected by conventional silver methods (Duyckaerts et al 1987). The presence of tau immunoreactivity, in the form of tangles, dystrophic neurites or neuritic plaques, is highly correlated with intellectual status (Delaère et al 1989) and is a distinctive discriminatory feature of Alzheimer's disease with respect to normal ageing (Ryong-Woon et al 1989, Barcikowska et al 1989, Tabaton et al 1989). Antibodies directed against the β/A4 peptide detect a broader range of deposits than conventional silver stains, including diffuse deposits (Yamaguchi et al 1988, 1990). Extensive deposits of the diffuse type appear to be compatible with normal intellectual function well into late life (Delaère et al 1990, 1991, Mann et al 1990, Braak & Braak 1991), and only certain types of β-amyloid deposit, specifically denser deposits often associated with neuritic (i.e. tau-reactive) pathology, correlate with dementia (Delaère et al 1991). Tau-reactive dystrophic neurites in the neuropil are also important discriminators between normal ageing and Alzheimer's disease (Ihara 1988, Braak & Braak 1988, Ryong-Woon et al 1989, Barcikowska et al 1989). Braak & Braak (1991) have recently shown that Alzheimer's disease can be staged on the basis of the spread of neurofibrillary pathology in CA1 of hippocampus, subiculum, transitional entorhinal cortex and temporal cortex. Neither diagnosis nor staging, therefore, is possible solely on the basis of β-amyloid pathology, but requires some additional assessment of neurofibrillary pathology.

Biochemical measurement of neurofibrillary pathology

Roth & Wischik (1985) first advanced the view some time ago that direct biochemical measurement of the PHF content in brain tissues might provide a more accurate means of quantifying the changes at the molecular level which underlie dementia, and which are expressed histologically as neurofibrillary pathology. It has taken some time to develop a methodology which would permit this hypothesis to be tested. It was necessary to identify a constituent of the inner core of the PHF (Wischik et al 1988b), to develop a reliable preparative protocol for the selective enrichment of PHFs (Wischik 1989b, Wischik et al 1989), to develop immunochemical reagents which could discriminate between PHF core constituents and normal brain proteins (Novak et al 1989, 1991), and finally to develop and validate assays by which these constituents could be measured reliably in brain homogenates (Harrington et al 1990, 1991). Although independent of classical neuropathology, the determination of PHF content ought in principle to measure the same process which is detected neuro-pathologically in the form of tangles and neuritic plaques, since both of these are simply indirect histological measures of PHF content.

Tau protein which is incorporated into the core of the PHF can be distinguished from normal tau protein by several immunochemical and biochemical criteria. This makes it possible to assay for PHF-bound tau independently of normal tau in brain homogenates. Since the amount of tau bound irreversibly within PHFs can be expected to be in a constant stoichiometric relationship with the quantity of PHFs, the measurement of PHF-tau provides a direct biochemical measure of the PHF content in brain homogenates, and hence a direct biochemical measure of neurofibrillary pathology in brain tissues. We have developed two independent assays for PHF-bound tau based on two monoclonal antibodies (mAbs). The first depends on immunoreactivity detected with mAb 423 in a brain tissue extract highly enriched in PHFs. This mAb is selective for PHF-tau, and does not recognize normal soluble tau protein. A second mAb, 7.51, recognizes an epitope in the repeat region of tau which is unavailable when tau is bound within PHFs, but becomes available after release from PHFs using formic acid (Fig. 4). These two independent immunochemical measures are correlated with one another, and provide comparable measures of PHF content. In the absence of formic acid extraction, mAb 7.51 detects only that tau protein which is not bound to PHFs. This includes normal free tau, or tubulin-bound tau, or tau rendered 'insoluble' by being non-specifically trapped with PHFs. We routinely use the mAb 7.51 immunoreactivity detected in the first soluble fraction from brain homogenate as a measure of normal tau protein content in brain tissue extracts.

We have applied these assays in several studies comparing cases of Alzheimer's disease with groups of young and old controls (Mukaetova-Ladinska et al 1992a,b). The overall PHF content in brain tissue measured by either of the immunochemical

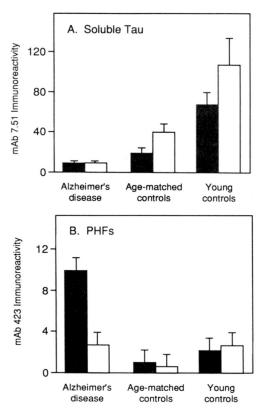

FIG. 5. Biochemical measurements (mean ± SEM) of normal soluble tau (A) and PHFs (B) in post mortem brain tissues dissected into grey and white matter in six brain regions per case, from 10 cases with Alzheimer's disease, five aged controls (>65 yr) and five younger controls (<65 yr). Black bars represent grey matter; white bars represent white matter. In the control cases, the normal white matter predominance of tau protein can be seen. Normal ageing is associated with a decline in soluble tau content, which is further accentuated in Alzheimer's disease. PHF accumulation clearly discriminates Alzheimer's disease cases from all controls, but is not a feature of normal ageing. In Alzheimer's disease, PHFs accumulate predominantly in grey matter.

methods described above distinguishes unambiguously between cases with and without dementia of the Alzheimer type. There is overall a 10-fold difference in PHF content relative to aged controls. Thus, irrespective of the manner in which PHF accumulation presents histologically—whether in neurofibrillary tangles, in neuritic plaques or in dystrophic neuropil threads—it is the overall content of PHFs in brain tissues which distinguishes controls from cases with dementia. The main site of PHF accumulation is, as expected from histological studies, in the grey matter (Fig. 5).

It has also been possible to address the question: which neuropathological parameter best predicts PHF content? We have examined this question in 18 cases of Alzheimer's diseases, examining brain tissues both histologically and biochemically in each of eight brain regions. The neuropathological parameter which correlates most highly with PHF content is the number of neuritic plaques ($P<0.005$), followed by the number of dystrophic neuropil threads ($P<0.005$). The number of neurofibrillary tangles correlates poorly with PHF content ($P<0.05$). We found no statistically significant correlation between PHF content and the number of β/A4-reactive amyloid plaques in brain tissues, nor indeed between the numbers of β-amyloid plaques and any of the three histological measures of neurofibrillary pathology (Mukaetova-Ladinska et al 1992b).

These findings confirm that a fundamental parameter which distinguishes normal ageing from Alzheimer's disease is PHF accumulation. The distribution of PHFs in the largely dendritic terminals found in neuritic plaques, in the dispersed dendritic neuropil threads, and in cell bodies as neurofibrillary tangles probably represents the different stages of somato-dendritic accumulation. It is likely, therefore, that the apparent dichotomy between plaques and tangles as major neuropathological correlates of dementia has more to do with the way PHFs are distributed between the somato-dendritic arborization and the cell body of affected cells, than with being a reflection of any fundamental biological difference. According to this view, the concept of 'plaque only' Alzheimer's disease (Terry et al 1987) is not essentially different from Alzheimer's disease in which neurofibrillary tangles are more prominent.

The distribution of PHFs by brain region tends to follow the pattern identified in early neuropathological studies. In the majority of cases, the regions worst affected are the temporal and parietal neocortices, the entorhinal cortex and the hippocampus (Fig. 6). The frontal and occipital neocortices have relatively lower levels of PHFs. This pattern also corresponds to the characteristic patterns of symptomatology seen in Alzheimer's disease, with more severe deficits in functions associated with the temporal and parietal brain regions.

This is not the pattern seen in all cases. We have recently identified a subset of cases with extensive PHF accumulation in the occipital cortex, and higher overall levels of PHFs (Mukaetova-Ladinska et al 1992a). Extensive PHF accumulation in the occipital neocortex is reflected neuropathologically neither in the tangle counts nor in β-amyloid plaques. Neither of the latter two parameters distinguish in the occipital cortex between cases with high and low PHF levels. The only significant differences are evident in the abundance of neuritic plaques and dystrophic neuropil threads (see Fig. 1C). This provides a striking instance of the histological correlations quoted above, namely that the main site of PHF accumulation tends to be in the somato-dendritic arborization rather than in the cell body. Cases with severe occipital lobe

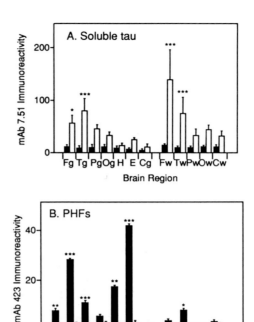

FIG. 6. Biochemical measurements (mean ± SEM) of normal soluble tau (A) and PHFs (B) in post mortem brain tissues dissected from 12 cases with Alzheimer's disease, 10 controls (five aged > 65 yr and five aged < 65 yr), shown by brain region (F, frontal; T, temporal; P, parietal; O, occipital; H, hippocampus; E, entorhinal; C, cerebellum) in grey matter (g) and white matter (w). Black bars show data from Alzheimer's disease cases; white bars show controls. The statistical significance of the difference between Alzheimer's disease and controls, tested by two-way analysis of variance, is indicated for each region (*, $P < 0.05$; **, $P < 0.01$; ***, $P < 0.001$). The main regions where PHF accumulation is found are the entorhinal, temporal and parietal neocortices, and hippocampus (B). By contrast, the very low levels of normal soluble tau are remarkably uniform throughout all brain regions (A). In controls, the highest levels of tau are found in the frontal and temporal association neocortices, and these suffer the greatest relative loss in Alzheimer's disease. (Adapted from Mukaetova-Ladinska et al 1992a, with permission from S. Karger AG, Basel.)

involvement have not yet been studied prospectively. In one of these variant cases, early onset of dyslexia had been noted clinically.

Redistribution of tau protein in Alzheimer's disease and normal ageing

A further parameter which distinguishes cases with and without Alzheimer's disease is the amount of normal soluble tau protein. Although tau levels are higher in white matter than in grey matter, the amount found in grey matter reflects the substantial afferent innervation of grey matter from corticopetal and cortico-cortical pathways. The normal distribution of soluble tau reflects frontal predominance, which is presumably due to the density of cortico-cortical axonal innervation of frontal cortex. In Alzheimer's disease, there is a substantial loss of normal soluble tau protein which affects all brain regions uniformly (Fig. 6). The loss of soluble tau is not well correlated with PHF content by brain region. Equally low levels of normal tau protein are found in regions with high and low PHF content. The molecular basis of this uniform decline in soluble tau protein in Alzheimer's disease is not known at present.

These results demonstrate that there is a major redistribution of available tau protein in Alzheimer's disease. In controls and in normal ageing, tau protein is mainly found in white matter, and in a form that is readily solubilized. In Alzheimer's disease, very little soluble tau is left. Most of the tau protein found in the brain in Alzheimer's disease is in the form of an irreversibly bound complex within the PHF.

The redistribution of available tau from soluble to PHF-bound compartments is likely to have important functional consequences in affected cortico-cortical association circuits. Tau is known to be essential for the stability of axonal microtubules, which are thought to be important both for the viability of axons and for fast axonal transport. It would be expected that this redistribution of tau from exchangeable to irreversibly bound states might have major consequences for microtubule-mediated functions in affected circuits. Neuronal death is probably not essential for these effects, since cases with and without extensive tangle-mediated cell death appear equally demented (Mountjoy et al 1983). In other words, the dementia of Alzheimer's disease may be visualized as a cortico-cortical disconnection syndrome, in which the disconnection could be functional rather than anatomical.

Neurofibrillary tangles are commonly seen in the hippocampus in non-demented elderly subjects. We have compared PHF levels in aged control cases with and without neurofibrillary tangles in the hippocampus. These did not differ significantly. PHF levels were, in both cases, substantially lower than those found in the hippocampus in cases with Alzheimer's disease. Thus, the presence of occasional tangles in the hippocampus, as indeed in the occipital neocortex in the cases described above, does not necessarily signal extensive PHF accumulation. PHF accumulation and extensive neurofibrillary degeneration in the hippocampus are part of a more widespread phenomenon that also affects entorhinal, temporal and parietal cortices in Alzheimer's disease. Even in hippocampus, therefore, and more particularly in the other affected regions, it is possible to distinguish normal ageing from dementia of the Alzheimer type simply by the quantity of PHFs.

Loss of soluble tau protein, on the other hand, is found both in normal ageing and in Alzheimer's disease. The extent of tau loss is significantly greater in

Alzheimer's disease subjects than in age-matched controls, but some loss of soluble tau protein appears to be part of the normal ageing process. Young and old controls differ significantly in their levels of soluble tau protein, but aged control subjects showed no significant increase in PHF content relative to younger controls.

These findings suggest that there are both quantitative and qualitative differences between normal ageing and Alzheimer's disease. Although it has been argued on psychometric grounds that cases with Alzheimer's disease are on a continuum with normal ageing (Brayne & Calloway 1988), this view is supported in our data only in respect of changes in soluble tau protein. PHF accumulation, on the other hand, which is highly correlated with neuropathological parameters of neurofibrillary degeneration, appears to provide a means of discriminating between normal and pathological ageing of the brain.

Relationship to various aetiological factors

As noted above, a genetic mutation at the locus encoding APP has been identified as a genetic cause of Alzheimer's disease in certain kindreds. However, the extensive 'APP dysmetabolism' (Hardy & Allsop 1991) which results in the appearance of β-amyloid deposits in brain tissues appears to be possible without any mental deterioration (Barcikowska et al 1989, Tabaton et al 1989, Delaère et al 1990, Braak & Braak 1991). Indeed, there are cases with extensive β-amyloid deposits caused by a genetic mutation at the APP locus, with neither dementia of the Alzheimer type nor neurofibrillary pathology (Levy et al 1990). By contrast, the cases of Alzheimer's disease that we have studied are sporadic, and typical of the majority of cases of Alzheimer's disease normally seen in clinical practice. No genetic mutation has been found in the vast majority of such cases, let alone any mutation at the APP locus (Schellenberg et al 1991). The main distinguishing characteristic of these cases at the molecular level is the accumulation of PHFs and loss of soluble tau protein. Although neither of these phenomena is correlated with the extent of β-amyloid deposition detected histologically in brain tissues (Mukaetova-Ladinska et al 1992b), it is important to develop an assay which will make it possible to test the biochemical relationship between β-amyloid and PHF accumulation in post mortem tissues.

On the available genetic data, if APP dysmetabolism is a cause of dementia of the Alzheimer type, it would appear to act as one of the possible triggers of PHF accumulation. It need not be the only, nor indeed aetiologically or epidemiologically the most significant, cause of PHF accumulation. The accumulation of PHFs, on the other hand, is intimately linked to clinical dementia, irrespective of its neuropathological presentation as plaques or tangles or largely invisible histologically in the form of extensive dystrophic neuropil threads. We would regard APP dysmetabolism as a cause of Alzheimer's disease only insofar as it is a cause of PHF formation. The nature of the molecular link between APP processing and PHF formation remains to be established. The data currently available leave open the possibility that PHF accumulation and dementia of the Alzheimer type could be the consequence of factors, either environmental or genetic, that act otherwise than through APP dysmetabolism (see, for example, evidence that a mutation in a mitochondrial protein is linked with Alzheimer's disease: Lin et al 1992).

Aluminium and neurofibrillary pathology

Brun & Dictor (1981) reported the presence of senile plaques and tangles in the dialysis encephalopathy syndrome, which had been found to be associated with increased brain aluminium content. Several species of animals (rabbits, cats, ferrets) have been shown to develop neurofibrillary changes in spinal cord, brainstem and hippocampus after intracerebral, subarachnoid, intravenous or subcutaneous injections of aluminium salts (see Wisniewski & Wen 1992: this volume). The ultrastructure of these neurofibrillary deposits differs from that of the PHFs found in Alzheimer's disease. The aluminium-induced tangles are composed of 10 nm filaments which have the morphological appearance of the neurofilaments which contribute to the normal neuronal cytoskeleton (Troncoso et al 1982). Biochemical studies of enriched preparations of aluminium-induced tangles showed that the major constituent was indeed the characteristic triplet of proteins known to make up neurofilaments (Selkoe et al 1979), and belonging to the broader class of cytoskeletal proteins known as intermediate filaments. Similar conclusions were reached in studies of aluminium-induced tangles in cultured rat neurons (Langui et al 1990).

At the time that neurofilament proteins were thought to be major constituents of Alzheimer neurofibrillary tangles (Anderton et al 1982), the well-known tissue culture phenomenon of 'cytoskeletal-collapse' was thought to account for the origin of the Alzheimer tangle (reviewed by Wischik & Crowther 1986). Deposits which resemble neurofibrillary tangles in histological morphology, and which are made of normal intermediate filament cytoskeletal proteins, can be induced in a wide variety of cell types by agents which inhibit microtubule polymerization (Wisniewski & Terry 1967, Seil & Lampert 1968), including aluminium salts (Wisniewski et al 1982, Langui et al 1990). However, structural studies of PHFs showed that these were more likely to arise by a process of aberrant *de novo* polymerization of an abnormal structural subunit than by the reorganization of an existing neuronal cytoskeleton (Wischik et al 1985, Crowther & Wischik 1985, Wischik & Crowther 1986). Furthermore, as outlined earlier, it is tau protein which is incorporated into the PHF, and not neurofilament protein, although there is clearly a disruption of the neuronal cytoskeleton in tangle-bearing cells.

We have attempted to examine the link between aluminium and PHF formation in several ongoing studies. In view of some of the data showing

TABLE 1   Immunohistochemistry of renal dialysis cases

Case	Age/Sex	Aluminium (µg/g)	β/A4 deposits	PHFs
104.90	68/m	14.3	+	
768.5	49/f	14.1		
938	47/f	13.1		
1004	59/f	10.0		
929	56/f	8.7		
902	59/m	8.0		
831	69/m	7.9		
812	64/f	6.3		
867	65/m	6.0		
782	54/m	5.6	+	
903	66/f	5.5	+	+
832	55/m	3.9	+	
885	38/f	3.7		
880	67/m	3.5		
936	60/m	2.7	+	
933	73/m	2.3	+	
934	37/f	1.2		
344.90 (AD)	72/f	3.2	+ +	+ +
309.90 (AD)	80/f	2.3	+ +	+ +
362.90 (C)	85	1.5		
402.90 (C)	67	3.1		

Frontal cortex was examined for the presence (+ , + +) of β/A4 protein deposits and paired helical filaments (PHFs) in 17 renal dialysis cases, two cases with Alzheimer's disease (AD) and two control subjects (C). Aluminium content is expressed as µg/g dry wt of tissue.

increased levels of aluminium in neurofibrillary tangles, we used microprobe X-ray spectroscopy to examine the aluminium content of isolated PHFs. We were unable to detect an aluminium peak by this methodology, although a caesium peak originating from caesium used in one of the preparative gradient centrifugations could be detected (A. Bourdillon & C. M. Wischik, unpublished observation).

We are also currently studying cases exposed to high therapeutic levels of aluminium hydroxide used as adjunctive therapy during renal dialysis. These cases were found to span a range of elevated brain aluminium levels, up to 14.3 µg/g (dry weight), compared with controls in the range 1.5–3.1 µg/g. We have so far examined these cases histologically with the PHF-tau-specific mAb 423, using a very sensitive immunohistochemical method for detecting PHF-containing dystrophic neuropil threads. Only one of the 17 cases examined, a 66-year-old female with a brain aluminium level of 5.6 µg/g, showed any evidence of mAb 423-reactive neurofibrillary pathology, which took the form of occasional dystrophic neuropil threads. This histological study does not provide support for aluminium as a risk factor for developing neurofibrillary pathology of the Alzheimer-type (Table 1). It may be that the more sensitive biochemical methods we have developed for measuring neurofibrillary pathology may indeed provide evidence for an association which has not been available from the application of purely histological methods.

Acknowledgement

The research was supported by the Medical Research Council; we are indebted to ICI Americas Inc and ICI plc for financial support. C. M. W. is a Lister Research Fellow.

References

Adler MJ, Coronel C, Shelton E, Seegmiller JE, Dewji NN 1991 Increased gene expression of Alzheimer's disease β-amyloid precursor protein in senescent cultured fibroblasts. Proc Natl Acad Sci USA 88:16–20

Allsop D, Landon M, Kidd M, Lowe JS, Reynolds GP, Gardner A 1986 Monoclonal antibodies raised against a subsequence of senile plaque core protein react with plaque cores, plaque periphery and cerebrovascular amyloid in Alzheimer's disease. Neurosci Lett 68:252–256

Anderton BH, Breinburg D, Downes MJ et al 1982 Monoclonal antibodies show that neurofibrillary tangles and neurofilaments share antigenic determinants. Nature (Lond) 298:84–86

Ball MJ Lo P 1977 Granulovacuolar degeneration in the aging brain and in dementia. J Pathol 106:165–185

Barcikowska M, Wisniewski HM, Bancher C, Grundke-Iqbal I 1989 About the presence of paired helical filaments in dystrophic neurites participating in the plaque formation. Acta Neuropathol 78:225–231

Binder LI, Frankfurther A, Rebhun LI 1985 The distribution of tau in the mammalian central nervous system. J Cell Biol 101:1371–1378

Blocq P, Marinesco G 1892 Sur les lésions et la pathogénie de l'épilepsie des dite essentielle. Semin Med 12:445–446

Boller F, Lopez OL, Moossy J 1989 Diagnosis of dementia: clinicopathologic correlations. Neurology 39:76–79

Braak H, Braak E 1988 Neuropil threads occur in dendrites of tangle-bearing nerve cells. Neuropathol Appl Neurobiol 14:39–44

Braak H, Braak E 1991 Neuropathological stageing of Alzheimer-related changes. Acta Neuropathol 82:239–259

Brayne C, Calloway P 1988 Normal aging, impaired cognitive function, and senile dementia of the Alzheimer's type: a continuum? Lancet 2:1265–1266

Brion J-P, Cheetham ME, Couck AM, Flament-Durand J, Hanger DP, Anderton BH 1990 Characterization of a partial cDNA specific for the high molecular weight microtubule-associated protein MAP2 that encodes epitopes shared with paired helical filaments of Alzheimer's disease. Dementia 1:304–315

Brion J-P, Hanger DP, Bruce MT, Couck A-M, Flament-Durant J, Anderton BH 1991a Tau in Alzheimer neurofibrillary tangles. N- and C-terminal regions are differently associated with paired helical filaments and the location of a putative abnormal phosphorylation site. Biochem J 273:127–133

Brion J-P, Hanger DP, Couck A-M, Anderton BM 1991b A68 proteins in Alzheimer's disease are composed of several tau isoforms in a phosphorylated state which affects their electrophoretic mobilities. Biochem J 279:831–836

Butner KA, Kirschner MW 1991 Tau protein binds to microtubules through a flexible array of distributed weak sites. J Cell Biol 115:717–730

Brun A, Dictor M 1981 Senile plaques and tangles in dialysis dementia. Acta Pathol Microbiol Scand A89:193–198

Caputo CB, Wischik C, Novak M et al 1992 Immunological characterization of the region of tau protein that is bound to the Alzheimer paired helical filament. Neurobiol Aging 13:267–274

Catteruccia N, Willingale-Theune J, Bunke D et al 1990 Ultrastructural localization of the putative precursors of the A4 amyloid protein associated with Alzheimer's disease. Am J Pathol 137:19–26

Chartier-Harlin M-C, Crawford F, Houlden H et al 1991 Early-onset Alzheimer's disease caused by mutations at codon 717 of the β-amyloid precursor protein gene. Nature (Lond) 353:844–846

Chou W-G, Zain SB, Rehman S et al 1990 Alzheimer cortical neurons containing abundant amyloid mRNA. Relationship to amyloid deposition and senile plaques. J Psychiatr Res 24:37–50

Couchie D, Legay F, Guilleminot J, Lebargy F, Brion J-P, Nunez J 1990 Expression of tau protein and tau mRNA in the cerebellum during axonal outgrowth. Exp Brain Res 82:589–596

Crowther RA, Wischik CM 1985 Image reconstruction of the Alzheimer paired helical filament. EMBO (Eur Mol Biol Organ) J 4:3661–3665

Crystal H, Dickson D, Fuld P et al 1988 Clinico-pathologic studies in dementia: non-demented subjects with pathologically confirmed Alzheimer's disease. Neurology 38:1682–1687

Delabar JM, Goldgaber D, Lamour Y et al 1987 β-Amyloid gene duplication in Alzheimer's disease and karyotypically normal Down's syndrome. Science (Wash DC) 235:1390–1392

Delacourte A, Flament S, Dibe EM et al 1990 Pathological proteins Tau 64 and 69 are specifically expressed in the somatodendritic domain of the degenerating cortical neurons during Alzheimer's disease. Demonstration with a panel of antibodies against tau proteins. Acta Neuropathol 80:111–117

Delaère P, Duyckaerts C, Brion JP, Poulain V, Hauw J-J 1989 Tau, paired helical filaments and amyloid in the neocortex: a morphometric study of 15 cases with graded intellectual status in aging and senile dementia of the Alzheimer type. Acta Neuropathol 77:645–653

Delaère P, Duyckaerts C, Masters C, Beyreuther K, Piette F, Hauw J-J 1990 Large amounts of neocortical βA4 deposits without neuritic plaques nor tangles in a psychometrically assessed, non-demented person. Neurosci Lett 116:87–93

Delaère P, Duyckaerts C, He Y, Piette F, Hauw JJ 1991 Subtypes and differential laminar distributions of βA4 deposits in Alzheimer's disease: relationship with the intellectual status of 26 cases. Acta Neuropathol 81:328–335

Dotti CG, Banker GA, Binder LI 1987 The expression and distribution of the microtubule-associated proteins tau and microtubule-associated protein 2 in hippocampal neurons in the rat *in situ* and in cell culture. Neuroscience 23:121–130

Duyckaerts C, Brion JP, Hauw J-J, Flament-Durant J 1987 Quantitative assessment of the density of neurofibrillary tangles and senile plaques in senile dementia of the Alzheimer type. Comparison of immunocytochemistry with a specific antibody and Bodian's protargol method. Acta Neuropathol 73:167–170

Duyckaerts C, Delaère P, Hauw J-J et al 1990 Rating of the lesions in senile dementia of the Alzheimer type: concordance between laboratories. A European multicenter study under the auspices of EURAGE. J Neurol Sci 97:295–323

Dyrks T, Weidemann A, Multhaup G et al 1988 Identification, transmembrane orientation and biogenesis of the amyloid A4 precursor of Alzheimer's disease. EMBO (Eur Mol Biol Organ) J 7:949–957

Esch FS, Keim PS, Beattie EC et al 1990 Cleavage of amyloid β peptide during constitutive processing of its precursor. Science (Wash DC) 248:1122–1124

Finley D, Chou V 1991 Ubiquitination. Annu Rev Cell Biol 7:25–69

Flament S, Delacourte A 1989 Abnormal tau species are produced during Alzheimer's disease neurodegenerating process. FEBS (Fed Eur Biochem Soc) Lett 247:213–216

Flament S, Delacourte A, Delaère P, Duyckaerts C, Hauw J-J 1990 Correlation between microscopical changes and tau 64 and 69 biochemical detection in senile dementia of the Alzheimer type. Tau 64 and 69 are reliable markers of the neurofibrillary degeneration. Acta Neuropathol 80:212–215

Glenner GG, Wong CW 1984 Alzheimer's disease: initial report of the purification and characterization of a novel cerebrovascular amyloid protein. Biochem Biophys Res Commun 120:885–890

Goate A, Chartier-Harlin M-C, Mullan M et al 1991 Segregation of a missense mutation in the amyloid precursor protein gene with familial Alzheimer's disease. Nature (Lond) 349:704–706

Goedert M, Wischik CM, Crowther RA, Walker JE, Klug A 1988 Cloning and sequencing of the cDNA encoding a core protein of the paired helical filament of Alzheimer disease: identification as the microtubule-associated protein tau. Proc Natl Acad Sci USA 85:4051–4055

Goedert M, Spillantini MG, Potier MC, Ulrich J, Crowther RA 1989 Cloning and sequencing of the cDNA encoding an isoform of microtubule-associated protein tau containing four tandem repeats: differential expression of tau protein mRNAs in human brain. EMBO (Eur Mol Biol Organ) J 8:393–399

Goedert M, Crowther RA, Garner CC 1991 Molecular characterization of microtubule-associated proteins tau and MAP2. Trends Neurosci 14:193–199

Goedert M, Spillantini MG, Cairns NJ, Crowther RA 1992 Tau proteins of Alzheimer paired helical filaments: abnormal phosphorylation of all six brain isoforms. Neuron 8:159–168

Gorevic PD, Castano EM, Sarma R, Frangione B 1987 Ten to fourteen residue peptides of Alzheimer's disease protein are sufficient for amyloid fibril formation and its characteristic X-ray diffraction pattern. Biochem Biophys Res Commun 147: 854–862

Greenberg SG, Davies P 1990 A preparation of Alzheimer paired helical filaments that displays distinct tau proteins by polyacrylamide gel electrophoresis. Proc Natl Acad Sci USA 87:5827–5831

Grundke-Iqbal I, Iqbal K, Tung YC et al 1986 Abnormal phosphorylation of the microtubule-associated protein tau in Alzheimer cytoskeletal pathology. Proc Natl Acad Sci USA 83:4913–4917

Hanger DP, Brion J-P, Gallo J-M, Cairns NJ, Luthert PJ, Anderton BH 1991 Tau in Alzheimer's disease and Down's syndrome is insoluble and abnormally phosphorylated. Biochem J 275:99–104

Hardy J, Allsop D 1991 Amyloid depositions as the central event in the aetiology of Alzheimer's disease. Trends Pharmacol Sci 12:383–388

Harrington CR, Edwards PC, Wischik CM 1990 Competitive ELISA for measurement of tau proteins in Alzheimer's disease. J Immunol Methods 134:261–271

Harrington CR, Mukaetova-Ladinska EB, Hills R et al 1991 Measurement of distinct immunochemical presentations of tau protein in Alzheimer disease. Proc Natl Acad Sci USA 88:5842–5846

Heston LL, Mastri AR, Anderson VE, White J 1981 Dementia of the Alzheimer type: clinical genetics, natural history and associated conditions. Arch Gen Psychiatry 38:1085–1090

Hilbich C, Kisters-Woike B, Reed J, Masters CL, Beyreuther K 1991 Aggregation and secondary structure of synthetic amyloid βA4 peptides of Alzheimer's disease. J Mol Biol 218:149–163

Ihara Y 1988 Massive somatodendritic sprouting of cortical neurons in Alzheimer's disease. Brain Res 459:138–144

Ihara Y, Kondo J, Miura R, Nakagawa Y, Mori H, Honda T 1990 Characterization of antisera to paired helical filaments and tau. Implication for the extent of tau tightly bound to paired helical filaments. Gerontology 36:15–24

Jakes R, Novak M, Davison M, Wischik CM 1991 Identification of 3- and 4-repeat tau isoforms within the PHF in Alzheimer's disease. EMBO (Eur Mol Biol Organ) J 10:2725–2729

Jellinger K, Danielczyk W, Fischer P, Gabriel E 1990 Clinico-pathological analysis of dementia disorders in the elderly. J Neurol Sci 95:239–258

Johnson SA, Pasinetti GM, May PC, Ponte PA, Cordell B, Finch CE 1988 Selective reduction of mRNA for the β-amyloid precursor protein that lacks a Kunitz-type protease inhibitor motif in cortex from Alzheimer brains. Exp Neurol 102: 264–268

Kang J, Lemaire H-G, Unterbeck A et al 1987 The precursor of Alzheimer's disease amyloid A4 protein resembles a cell-surface receptor. Nature (Lond) 325:733–736

Khachaturian ZS 1985 Diagnosis of Alzheimer's disease. Arch Neurol 42:1097–1105

Kidd M 1963 Paired helical filaments in electron microscopy in Alzheimer's disease. Nature (Lond) 197:192–193

Kondo J, Honda T, Mori H et al 1988 The carboxyl third of tau is tightly bound to paired helical filaments. Neuron 1:827–834

Koo EH, Sisodia SS, Cork LC, Unterbeck A, Bayney RM, Price DL 1990 Differential expression of amyloid precursor protein mRNAs in cases of Alzheimer's disease and in aged nonhuman primates. Neuron 2:97–104

Kosik KS, Orecchio LD, Binder L, Trojanowski JQ, Lee VM-Y, Lee G 1988 Epitopes that span the tau molecule are shared with paired helical filaments. Neuron 1:817–825

Langui D, Probst A, Anderton B, Brion J-P, Ulrich J 1990 Aluminium-induced tangles in cultured rat neurones. Enhanced effect of aluminium by addition of maltol. Acta Neuropathol 80:649–655

Lee VM-Y, Balin BJ, Otvos LJ Jr, Trojanowski JQ 1991 A68: a major subunit of paired helical filaments and derivatized forms of normal tau. Science (Wash DC) 251:675–678

Levy E, Carman MD, Fernandez-Madrid IJ et al 1990 Mutation of the Alzheimer's disease amyloid gene in hereditary cerebral haemorrhage, Dutch type. Science (Wash DC) 248:1124–1126

Lin F-H, Lin R, Wisniewski HM, Hwang YW, Grundke-Iqbal I, Healy-Louie G, Iqbal K 1992 Detection of point mutations in codon 331 of mitochondrial NADH dehydrogenase subunit 2 in Alzheimer brains. Biochem Biophys Res Commun 182:238–246

Mann DMA, Marcyniuk B, Yates PO, Neary D, Snowdon JS 1988 The progression of the pathological changes of Alzheimer's disease in frontal and temporal neocortex examined both at biopsy and at autopsy. Neuropathol Appl Neurobiol 14:177–195

Mann DMA, Jones D, Prinja D, Purkiss MS 1990 The prevalence of amyloid (A4) protein deposits within the cerebral and cerebellar cortex in Down's syndrome and Alzheimer's disease. Acta Neuropathol 80:318–327

Masters CL, Simms G, Weinman NA, Multhaup G, McDonald BL, Beyreuther K 1985 Amyloid plaque core protein in Alzheimer disease and Down's syndrome. Proc Natl Acad Sci USA 82:4245–4249

Mirra SS, Heyman A, McKeel D et al 1991 The consortium to establish a registry for Alzheimer's disease (CERAD). Part II. Standardization of the neuropathological assessment of Alzheimer's disease. Neurology 41:497–486

Monard D 1988 Cell-derived proteases and protease inhibitors as regulators of neurite outgrowth. Trends Neurosci 11:541–544

Mori H, Kondo J, Ihara Y 1987 Ubiquitin is a component of paired helical filaments in Alzheimer's disease. Science (Wash DC) 235:1641–1644

Morris JC, McKeel DW, Storandt M et al 1991 Very early Alzheimer's disease: informant-based clinical, psychometric and pathological distinction from normal aging. Neurology 41:469–478

Mountjoy CQ, Roth M, Evans NJR, Evans HM 1983 Cortical neuronal counts in normal elderly controls and demented patients. Neurobiol Aging 4:1–11

Mukaetova-Ladinska EB, Harrington CR, Hills R, Wischik CM 1992a Regional distribution of paired helical filaments and normal tau proteins in aging and in Alzheimer's disease with and without occipital lobe involvement. Dementia 3:61–69

Mukaetova-Ladinska EB, Harrington CR, Wischik CM 1992b Biochemical measurement of neurofibrillary pathology in Alzheimer's disease and normal aging. Am J Pathol, submitted

Murrell J, Farlow M, Ghetti B, Benson MD 1991 A mutation in the amyloid precursor protein associated with hereditary Alzheimer's disease. Science (Wash DC) 253: 97–98

Naruse S, Igarashi S, Aoki K et al 1991 Mis-sense mutation val→ile in exon 17 of amyloid precursor protein gene in Japanese familial Alzheimer's disease. Lancet 337:978–979

Neve RL, Finch FA, Dawes LR 1988 Expression of the Alzheimer amyloid precursor gene transcripts in the human brain. Neuron 1:669–677

Nieto A, Correas I, Lopez-Otin C, Avila J 1991 Tau related protein present in paired helical filaments has a decreased tubulin binding capacity as compared with microtubule-associated protein tau. Biochim Biophys Acta 1096:197–204

Novak M, Wischik CM, Edwards P, Pannell R, Milstein C 1989 Characterisation of the first monoclonal antibody against the pronase resistant core of the Alzheimer PHF. Prog Clin Biol Res 317:755–761

Novak M, Jakes R, Edwards PC, Milstein C, Wischik CM 1991 Difference between the tau protein of Alzheimer paired helical filament core and normal tau revealed by epitope analysis of mAbs 423 and 7.51. Proc Natl Acad Sci USA 88:5837–5841

O'Connor DW, Pollitt PA, Hyde JB et al 1989 The prevalence of dementia as measured by the Cambridge Mental Disorders of the Elderly examination. Acta Psychiatr Scand 79:190–198

Ohyagi Y, Takahashi K, Kamegai M, Tabira T 1990 Developmental and differential expression of beta amyloid protein precursor mRNAs in mouse brain. Biochem Biophys Res Commun 167:54–60

Oyama F, Shimada H, Oyama R, Titani K, Ihara Y 1991 Differential expression of β amyloid protein precursor (APP) and tau mRNA in the aged human brain: individual variability and correlation between APP-751 and four-repeat tau. J Neuropathol & Exp Neurol 50:560–578

Papasozomenos SCh, Su Y 1991 Altered phosphorylation of tau protein in heat-shocked rats and patients with Alzheimer's disease. Proc Natl Acad Sci USA 88:4543–4547

Prelli F, Castaño E, Glenner GG, Frangione B 1988 Differences between vascular and plaque core amyloid in Alzheimer's disease. J Neurochem 51:648–651

Roth M, Wischik CM 1985 The heterogeneity of Alzheimer's disease and its implications for scientific investigations of the disorder. In Arie T (ed) Recent advances in psychogeriatrics. Churchill Livingstone, Edinburgh, p 71–92

Roth M, Tym E, Mountjoy CQ, Huppert FA, Hendrie H, Verma S, Goddard R 1986 CAMDEX. A standardised instrument for the diagnosis of mental disorder in the elderly with special reference to the early detection of dementia. Br J Psychiatry 149:698–709

Ryong-Woon S, Ogomori K, Tetsuyuki K, Tateishi J 1989 Increased tau accumulation in senile plaques as a hallmark of Alzheimer's disease. Am J Pathol 134:1365–1371

Schellenberg GD, Bird TD, Wijsman EM et al 1988 Absence of linkage of chromosome 21q21 markers to familial Alzheimer's disease. Science (Wash DC) 241:1507–1510

Schellenberg GD, Anderson L, O'Dahl S et al 1991 APP_{717}, APP_{693}, and PRIP gene mutations are rare in Alzheimer's disease. Am J Hum Genet 49:511–517

Seil FJ, Lampert J 1968 Neurofibrillary tangles induced by vincristine and vinblastine sulfate in central and peripheral neurons in vitro. Exp Neurol 21:219–230

Selkoe DJ 1991 The molecular pathology of Alzheimer's disease. Neuron 6:487–498

Selkoe DJ, Liem RKH, Yen SH, Shelanski ML 1979 Biochemical and immunological characterisation of neurofilaments in experimental neurofibrillary degeneration induced by aluminium. Brain Res 163:235–252

St George-Hyslop PH, Haines JL, Farrar LA et al 1990 Genetic linkage studies suggest that Alzheimer's disease is not a single homogeneous disorder. Nature (Lond) 347:194–197

St George-Hyslop PH, Tanzi RE et al 1987 The genetic defect causing familial Alzheimer's disease maps on chromosome 21. Science (Wash DC) 235:885–890

Tabaton M, Mandybur TI, Perry G, Onorato M, Autilio-Gambetti L, Gambetti P 1989 The widespread alteration of neurites in Alzheimer's disease may be unrelated to amyloid deposition. Ann Neurol 26:771–778

Tanaka S, Shiojiri S, Takahashi Y et al 1989 Tissue-specific expression of three types of β-protein precursor mRNA: enhancement of protease inhibitor-harboring types in Alzheimer's disease brain. Biochem Biophys Res Commun 165:1406–1414

Terry RD, Hansen LA, DeTeresa R, Davies P, Tobias H, Katzman R 1987 Senile dementia of the Alzheimer type without neocortical neurofibrillary tangles. J Neuropathol & Exp Neurol 46:262–268

Tierney MC, Fisher RH, Lewis AJ et al 1988 The NINCDS-ADRDA work group criteria for the clinical diagnosis of probable Alzheimer's disease: a clinicopathologic study of 57 cases. Neurology 38:359–364

Tomlinson BE, Blessed G, Roth M 1968 Observations on the brains of nondemented old people. J Neurol Sci 7:331–356

Troncoso JC, Price DL, Griffen JW, Parhad IM 1982 Neurofibrillary axonal pathology in aluminium intoxication. Ann Neurol 12:278–283

Van Broeckhoven C, Genthe AM, Vandenberghe A et al 1987 Failure of familial Alzheimer's disease to segregate with the A4 amyloid gene in several European families. Nature (Lond) 329:153–155

Van Nostrand WE, Wagner SL, Suzuki M et al 1989 Protease nexin-II, a potent antichymotrypsin, shows identity to amyloid β-protein precursor. Nature (Lond) 341:546–549

Wang R, Meschia JF, Cotter RJ, Sisodia SS 1991 Secretion of the β/A4 amyloid precursor protein. Identification of a cleavage site in cultured mammalian cells. J Biol Chem 266:16960–16964

Wilcock GK, Esiri MM 1982 Plaques, tangles and dementia: a quantitative study. J Neurol Sci 56:407–417

Wischik CM 1989a Cell biology of the Alzheimer tangle. Curr Opin Cell Biol 1:115–122

Wischik CM 1989b Structure and biochemistry of paired helical filaments in Alzheimer's disease. PhD Thesis, University of Cambridge

Wischik CM, Crowther RA 1985 Subunit structure of paired helical filaments in Alzheimer's disease. J Cell Biol 100:1905–1912

Wischik CM, Crowther RA 1986 Subunit structure of the Alzheimer tangle. Br Med Bull 42:51–56

Wischik CM, Crowther RA, Stewart M, Roth M 1985 Subunit structure of paired helical filaments in Alzheimer's disease. J Cell Biol 100:1905–1912

Wischik CM, Novak M, Edwards PC, Klug A, Tichelaar W, Crowther RA 1988a Structural characterization of the core of the paired helical filament of Alzheimer disease. Proc Natl Acad Sci USA 85:4884–4888

Wischik CM, Novak M, Thøgersen HC et al 1988b Isolation of a fragment of tau derived from the core of the paired helical filament of Alzheimer's disease. Proc Natl Acad Sci USA 85:4506–4510

Wischik CM, Klug A, Milstein C 1989 Paired helical filament and antibody thereto useful in diagnosing Alzheimer's disease. British Patent No. 8724412

Wisniewski HM, Terry RD 1967 Experimental colchicine encephalopathy. I. Induction of neurofibrillary degeneration. Lab Invest 17:577–587

Wisniewski HM, Wen GY 1992 Aluminium and Alzheimer's disease. In: Aluminium in biology and medicine. Wiley, Chichester (Ciba Found Symp 169) p 142–164

Wisniewski HM, Sturman JA, Shek JW 1982 Chronic model of neurofibrillary changes induced in mature rabbits by metallic aluminum. Neurobiol Aging 3:11–22

Wong CW, Quaranta V, Glenner GG 1985 Neuritic plaques and cerebrovascular amyloid in Alzheimer disease are antigenically related. Proc Natl Acad Sci USA 82:8729–8732

Yamaguchi H, Hirai S, Morimatsu M, Shoji M, Ihara Y 1988 A variety of cerebral amyloid in the brains of the Alzheimer-type dementia by β-protein immunostaining. Acta Neuropathol 76:541–549

Yamaguchi H, Nakazato Y, Shoji M, Ihara Y, Hirai S 1990 Ultrastructure of the neuropil threads in the Alzheimer brain: their dendritic origin and accumulation in senile plaques. Acta Neuropathol 80:368–374

Yoshikawa K, Aizawa T, Maruyama K 1990 Neural differentiation increases expression of Alzheimer amyloid protein precursor gene in murine embryonal carcinoma cells. Biochem Biophys Res Commun 171:204–209

DISCUSSION

Kerr: Your clinical definition is an interesting contrast to Henry Wisniewski's demonstration of a continuum between old age and Alzheimer's disease. To what extent are your apraxic patients characteristic of the disease, rather than showing just an odd feature or two?

Wischik: I think the most general change is actually a change in practical function, which the relatives of the patients and the occupational therapists pick up best: the subtle loss of the ability to know where you left the flour in the sequence required to make a cake, or the proper sequence required to brush your teeth. Those changes are important for maintaining independence of life, and their loss represents the major cost to the health services in Alzheimer's disease. These changes can be manifest in various ways, depending on where the PHF accumulation may be predominant. For example, an occipito-parietal accumulation of PHFs is associated with visuo-spatial defects, though very little in the way of tangles may be present. The precise clinical presentation depends on where in the brain the cortico-cortical association pathway is affected, in my view.

Wisniewski: Do you see a *continuum* of changes, with cases in a preclinical stage? Our data, and those from Boston (Evans et al 1988), and actually your own data, all point to this continuum.

Wischik: We have demonstrated a continuum during ageing in terms of the loss of normal soluble tau protein, but *not* a continuum with normal ageing in the accumulation of paired helical filaments, in the AD cases we've studied. Normal ageing does not appear to be associated with PHF accumulation. I presume that the movement into successively more severe grades of cognitive deterioration will be associated with the progressive accumulation of paired helical filaments. In one case from Jim Edwardson's series of dialysis patients (see Table 1, p 286), there were simply a few dystrophic neurites (tau-reactive neurons). We are now trying to do a prospective study, to test the specific hypothesis that the ratio of PHF-tau to soluble tau directly correlates with the degree of cognitive deterioration.

Wisniewski: What helps us determine the extent of neurofibrillary pathology is the number of PHF positive and negative plaques. If there are many PHF or tau positive plaques, it tells us that the neurofibrillary pathology is extensive. What do you feel is the relationship between β-amyloidosis and neurofibrillary pathology? As I indicated earlier, they do not always occur together in the same brain. In my opinion, however, in Alzheimer's disease the neurofibrillary pathology is driven by β-amyloidosis.

Wischik: In this connexion, I find the results of John Hardy persuasive (Goate et al 1991). At least in the very rare cases where it occurs, his findings establish a link between the phenomenon of APP dysmetabolism and that of PHF accumulation. Of course, I would like to get frozen brain tissue from his cases, and measure the PHF content, but let's assume that this experiment, if done, would give the result that PHF accumulation is extensive. We would then need to look very carefully for some specific change in the processing of APP that links up with the incorporation of tau into the core of the PHF. We have shown a relationship between PHF-tau and certain portions of the APP molecule. This could provide a link which is activated, presumably, in the genetically determined AD cases, but not in the normal ageing population, necessarily.

Wisniewski: What Dr Hardy showed, we already knew in a way, from studying Down's syndrome. In DS, large numbers of plaques precede the formation of large numbers of NFT. By and large in DS, large numbers of plaques are seen after 30 years and NFT-bearing neurons are seen after 50 years; also, in persons with DS we see deterioration in daily life activities after the age of 50 years. As in AD, in Down's syndrome the numbers of NFT correlate better with cognitive performance than do the β-amyloid plaques. However, the β-amyloidosis seems to produce an environment where many NFT are formed.

Wischik: Yes; it's the point I was trying to make about the necessary and sufficient cause. It appears now that a change in the processing of APP is a necessary precondition for entry into the PHF-generating process, but it doesn't predict entry; something else is required. The two cogs have to engaged in some way for one cog (APP dysmetabolism) to begin to turn the next cog (PHF assembly). And it's the second cog that I am most interested in as a clinician, because it drives the patient into cognitive deterioration.

Blair: You were kind enough to supply us with samples of brain from Alzheimer's disease patients. This disease is often thought of as something superimposed on the normal processes of ageing. As I suggested earlier (p 266), in this disease we are (in most cases) looking at a deficiency in the production of the coenzyme tetrahydrobiopterin (BH_4) (Anderson et al 1987). We find a weak decline in BH_4 with age in the normal brain, in its synthesis in the temporal cortex and in the salvage pathway (dihydropteridine reductase). The data I discussed on the reduction of BH_4 in senile dementia were therefore corrected for age. So we have a declining pathway of a key co-enzyme essential for cognitive functioning.

Wischik: What goes in parallel with that decline is the age-related loss of soluble tau, which is unrelated to PHF accumulation. Also, there are Jim Edwardson's data on the loss of choline acetyltransferase activity with ageing. I am sure that we shall find a whole range of biochemical components that change with ageing. What I wanted to show is the clear distinction between age-related changes and disease-related ones, sufficient to cause clinical phenomena.

Strong: We don't argue with the data, but maybe with the interpretation! At the crux of the issue is the difference between localizationists and non-localizationists in neurology. What you described in Alzheimer's disease, most neurologists would consider as being a cortical dementia, and whether one subscribes to the difference between cortical and subcortical dementias as being clinically relevant is another debate. But if you are proposing a pathophysiology that would affect long association tracts, one might expect to see a subcortical type of dementia, as opposed to a multi-focal type of cortical dementia often observed early in Alzheimer's disease.

As a second point, in hippocampal neurons in culture there is a failure of the compartmentalization of tau; it's the only cytoskeletal protein that's been looked at in dissociated neural culture which does not compartmentalize. I wonder if that is telling us something?

Wischik: Your first point permits me to say that Alzheimer's disease was discovered by a psychiatrist and not a neurologist, and was first presented at a meeting of psychiatrists! As to how one localizes the particular changes in the brain, we would like to be able to relate PHF accumulation in the nucleus basalis of Meynert, and other brainstem nuclei, with losses of functional neurotransmitter systems. We predict a relationship between loss of subcortical neurotransmitter systems and PHF accumulation.

On the tau compartmentalization, K. S. Kosik has done very nice work on tau/MAP-2 compartmentalization (Kosik & Finch 1987). Initially, neurons in culture did not show compartmentalization of MAP-2 and tau; both were uniformly expressed in the cell. Then, as the neurons matured, there was a separation of tau and MAP-2 into neurites that look axonal, by morphological criteria, and those that remain somato-dendritic. His studies addressed the question of how tau is added to microtubules; how does the changeover in microtubule-associated protein occur as tau becomes localized to the axon? Kosik postulated that tau in a phosphorylated state is transported to the tip of the growing neurite and the final exchange occurs at the distal end of the neurite, not the proximal end; his hypothesis is that tau has to be in a phosphorylated state to be guarded from polymerization until it reached its site of action. If so, the whole abnormal phosphorylation scenario would evolve as a consequence, not a cause, of neurofibrillary degeneration. Tau would simply be accumulating in the wrong compartment for some other reason; it then becomes abnormally phosphorylated, because this is part of what happens in the somato-dendritic compartment.

Williams: Is tau moved down the neurite in a vesicular system? I thought not.

Wischik: This is not known. His data were based on immunocytochemical staining of cells in tissue culture. He showed that the change in microtubule-associated protein takes place at the distal end of the growing neurite, so tau has to be in an inactive state, a non-binding state, to reach that point.

McLachlan: Is the phosphorylation site in tau known with reasonable certainty?

Wischik: Nobody knows that. Several phosphorylation sites have been described in tau. None of them so far is in the tandem repeat region. We have used mass spectroscopy to look at this region from the PHF and it is not abnormally phosphorylated. There are sites in the C-terminal tail of tau that may or may not be abnormally phosphorylated, but our data suggest that the tau protein which ends up in the PHF has lost its C-terminal tail, at least the tau molecules that are deep within the core of the PHF. That's also the case for the tau which is incorporated into granulo-vacuolar degeneration. So the relevance of the C-terminal phosphorylation site may be that after cleavage, a regulatory arm on tau has been lost, and this could explain the abnormal compartmentalization. But I don't think there's any good evidence that abnormal phosphorylation of tau *causes* the PHF to assemble.

Perl: In my paper I propose an intracellular 'tanning' type of phenomenon by Al, participating in the cross-linking of the PHFs and rendering them insoluble. The phenomenon of PHF formation is an unusual one; we have an intracellular structure that's so heavily 'cross-linked' that it is virtually resistant to protease digestion and disassembly. What are your thoughts about what might cause this?

Wischik: It is not necessary to invoke 'tanning' to explain protease resistance, whether by alum or anything else. There are several peptides which become protease resistant after polymerization. We have found a 20 amino acid (2 kDa) peptide in PHF core preparations that polymerizes into structures that are left-handed helical ribbons, but with half the diameter and half the periodicity of PHFs (C.B. Caputo et al, unpublished). These structures are protease resistant. This is a synthetic peptide; nothing has happened to it in the way of post-translational modification. Al may be present, because Al is everywhere; but if the postulated Al phenomenon (polymerization of PHFs) is happening at a level as low as 10^{-10} M, this becomes very difficult to deal with experimentally. But I can say that our 2 kDa peptide does polymerize into a protease-resistant structure. Likewise, $\beta/A4$, which is cleaved out of APP, polymerizes into a protease-resistant structure. The very tight packing of a very short peptide into a lattice is sufficient to explain proteolytic resistance.

Wisniewski: The neurofibrillary pathology has a very clear-cut topography. In the cerebellum and the spinal cord, by and large, we don't find the NFT type of pathology. Has anyone looked at tau in the cerebellum? Is there any difference in tau species between cerebrum and cerebellum?

Wischik: We used soluble tau and PHF-tau assays in cerebellum as our negative controls. There is an age-related loss of soluble tau in cerebellum, as there is elsewhere in brain, but there is no PHF accumulation there, as measured by our assay. I should add that the PHFs that we measure are in some cases invisible histologically, and it could have been that in the cerebellum in demented subjects there are many PHFs, but pathologists are not seeing them. The importance of our biochemical assay is that we are detecting PHFs in AD irrespective of whether they are histologically visible; and they are absent from cerebellum.

Martin: You give a very nice description of a 1:1 correspondence between the redistribution of tau and Alzheimer's disease. You suggest that this redistribution constitutes a 'cause' of Alzheimer's disease, rather than simply an indicator for it. Why use such a strong word?

Wischik: 'Causality' is a metaphysical concept but, as a scientist, I suggest that, say, a change in DNA that leads to the end result is a 'cause' in a sense that all life phenomena begin in DNA; but as a clinician I am looking for a molecular change which is adequate to explain the clinical phenomenology.

Martin: So it's an indicator?

Wischik: You could say that, but it is also a *cause* in the sense that it is an adequate molecular change to bring about the clinical phenomenology.

Kerr: So it is not an aetiology, but part of the pathogenesis?

Wischik: Yes; or it's the last molecular change before you get to clinical symptoms. In other words, we are narrowing the gap in our understanding from two ends: from the clinical end, we come to PHF as a very strong candidate for the proximal cause; with the genetic mutation in APP, we come from the genetic end. The two have to meet. It is the aim of current research to show how those two sides of the tunnel connect up, if you like.

Wisniewski: We have yet to prove that β-amyloidosis *leads* to the redistribution of tau in the brain.

McLachlan: In a way, this is like the scrapie story. Several laboratories have shown that the mRNA pool size for tau remains unaltered in AD, compared to controls. Tau transcription is not down-regulated as is the case for tubulin or NF-1. The stability of the tau messenger is not altered. So why is the translation of the tau message not down-regulated? Is there a post-translational conformational change in tau that prevents transcriptional suppression? Could PHF formation be a compensation to handle excess tau in the presence of the deregulation of transcription? The situation in AD is analogous to prion processing. Do you have any evidence in terms of the translation of this particular message?

Wischik: As you say, several laboratories have reported on that. Michel Goedert in our group finds no decline in the amount of mRNA for tau (Goedert et al 1989). Several other groups have confirmed this. In hippocampus there may be slightly more mRNA for tau. What we have shown is a massive loss of the final product in hippocampus, so the increased tau mRNA in the hippocampus may be an attempt by the affected cell to compensate for the extraordinarily and uniformly low levels of tau production across all brain regions. It is as if all the tau that is being synthesized is being withdrawn out of the system immediately, into PHFs.

McLachlan: So there may be an analogy with scrapie, where the prion precursor protein is incompletely degraded to PRP[27-30]? Prusiner has postulated a conformational change in prions that must occur after translation and be responsible for infectivity.

Wischik: I went to his lab., to compare his work on scrapie fibrils with ours on PHFs. I always feel we are following two years behind him! They have not found a post-translational modification, such as abnormal glycosylation. I don't know how it will turn out for the scrapie case, but I suspect the whole story will work itself out in terms of proteolytic cleavages; cuts are produced in otherwise normal proteins that thereby become amyloidogenic.

Blair: One key group in this argument consists of the Down's syndrome cases, because they uniquely proceed to at least the neuropathology of Alzheimer's disease, but not overt clinical dementia. I pointed out (p 183) earlier that Down's syndrome patients have very low binding to transferrin of added metal species, and circulating levels of low molecular weight forms of Al and iron. An alternative approach is to measure transferrin saturation in Down's patients. It is very high, about 80%—a level which would release low M_r species into the plasma and therefore provide enhanced concentrations of aluminium to the brain. This brings things back to the Al story, or a low molecular weight circulating 'mobile' species.

Wischik: We have studied four cases of Down's syndrome with neurofibrillary pathology and one case without. The difference is similar to the difference between controls and Alzheimer's patients.

Williams: Is there any way of linking the tubulin (microtubule) system and the phosphorylation reactions, through GTP, to the changes in tau? I know that in the formation of tubulin microtubules you need a GTP–tau reaction. I believe too that there is a very tightly bound magnesium system in tubulins. Could that connect with the observed differences in the forms of tau (the soluble or non-soluble form), through phosphorylation?

Wischik: One possibility is that phosphorylation makes tau susceptible to proteolytic cleavage. The mechanism might be as follows. Perhaps there is abnormal cytoskeletal transport and Al, perhaps, as Professor Martin says, damages tubulin function in a subtle way, at a very low level. You therefore have a primary abnormality in cytoskeletal protein transport. That means that in the sorting out the neuron has to do to, to determine whether tau will go down the axon, or will remain and enter the somato-dendritic zone, with two different cytoskeletal 'streams' that have to be partitioned, around the nucleus, if the cell's tubulin system is damaged by Al, then tau remains longer in the somato-dendritic compartment and undergoes abnormally prolonged exposure to kinases there. In the non-microtubule-bound state, tau could become a substrate for proteases at unusual sites of cleavage (tau is normally cleaved at the end of the axon and the cell has a very efficient mechanism for degrading it to dipeptides for re-use of the amino acids). Then we could have abnormal cleavage of tau, to produce an amyloidogenic fragment (or a co-amyloidogenic fragment with some other species).

Williams: I am asking this because in the picture that we have now, there is nothing that Al does at all. Could there be a special tight Mg site which becomes a special Al site?

Wischik: That would be one scenario bringing in Al. But there is a counter-argument to that: the argument from Jim Edwardson's laboratory that dialysis subjects who have been exposed to high Al levels for some time don't end up with histological evidence of neurofibrillary pathology. This situation may change when the biochemical analyses are complete.

Edwardson: Derangements of cytoskeletal function may also be relevant to the processing of amyloid precursor protein. Recent evidence indicates that a high proportion of brain APP is bound to microtubules, and it has been suggested that altered cytoskeletal binding of APP might result in abnormal transport and aberrant proteolysis, perhaps with the generation of amyloidogenic fragments (Refolo et al 1991).

Wischik: Then the linkage would be that the C-terminal domain of APP would lead, in some way, to co-polymerization with tau? This is an exciting possibility, and we are currently investigating it (Wischik et al 1991).

Edwardson: In addition to the extracellular processing of APP, there is an intracellular pathway in which the protein is successively shortened from the N-terminal end to produce various C-terminal fragments which contain the amyloidogenic sequence. This alternative pathway may also be important for amyloidogenesis.

Wischik: Japanese papers (Ishiura et al 1989, 1990a,b) show that the enzymes that are capable of cleaving the C-terminal or N-terminal portions of APP are all intracellular rather than extracellular proteases. A body of data now suggests that the processing of APP is intracellular. We've shown in trisomic mice a co-localization in intracellular granular inclusions of PHF-like tau and a portion of APP, within pyramidal cells. So there is an intracellular processing pathway (Richards et al 1991).

Williams: Are those intracellular proteases calcium dependent? Most intracellular proteases *are* switched on by high calcium levels.

Wischik: I don't know about the Ca dependence. There is evidence of calpain proteolytic activity being elevated in Alzheimer's disease.

Perl: You mentioned another form of tau accumulation associated with Alzheimer's disease, namely granulo-vacuolar degeneration (GVD), which as far as I know is not in a PHF configuration and seems to be amorphous ultrastructurally. It is seen in Alzheimer's and in Down's syndrome; it occurs prominently in the ALS/parkinsonism-dementia cases on Guam. Both in AD and in the Guam patients, very high concentrations of Al are detectable by LAMMA in the GVD.

Wischik: Yes. This may provide an important link.

Petersen: As a general point about the use of the word 'cause', I don't really see how the word can be used here. I don't yet see any links between the molecular defect that has been described and the cognitive defect, because the latter must be a signalling problem.

Williams: The world at large believes that scientists investigate 'cause', whether we think we do or not! The public are interested if you use the word 'cause', and they will think that you have a direct line between what you observe in the clinic or laboratory and what happens to, say, aluminium, and that you can show each step on the way. This is the normal linguistic use of the word. But at the level where you move from the clinic or laboratory to the discussion of the brain (and are only really *correlating* events in the brain with clinical observations) and then you move forward to an attempted correlation with Al as an administered chemical (e.g. Al exposure), you can also postulate causal links. Whether you can get the complete set of causal links is another matter. I wouldn't worry too much about using the word 'cause' when you are trying to make a match between closely linked steps, so long as we all know the looseness in the use of the word. I can perfectly well say that the cause of the control of an enzyme lies with cyclic AMP, from detailed knowledge of association and effects, but we do not have such knowledge for Al and Alzheimer's disease because we do not even know how the brain works.

Petersen: There is a difference between an association and a cause—a big difference. Take cystic fibrosis. You can now very clearly show that the identified genetic defect causes a decrease in membrane chloride permeability (the gene product is a chloride channel); this explains reduced fluid secretion in a number of epithelia including the lung, in turn explaining blocked ducts, infection and so on—that is to say, the clinical syndrome of cystic fibrosis. That's what I would call a cause. It is not the case here!

Williams: Dr Wischik may want to take a position on the middle ground of this discussion. He doesn't know the causal connection. We have not shown in this meeting a definitive causal connection between Alzheimer's disease and Al, at any stage. I would agree with that. But, we have discussed 'cause' in terms of factors which correlated. If there is a very close connection between steps of proteolysis, aluminium and other contributing factors, and these are the only things I can correlate with the disease, I begin to think we are narrowing it down very near to a cause. The proof that smoking causes lung cancer remains only a majority verdict, yet we act on it.

Wischik: This is a cultural difference! Professor Petersen and Professor Williams feel comfortable with causal arrows in the molecular domain. As a physician, I feel comfortable with causal arrows between molecules and people, and a good deal of medicine is concerned with those kinds of links. I have demonstrated major changes in associative brain pathways which are known to underlie particular categories of cognitive function. That is adequate for consideration as a cause of the cognitive symptoms.

Wisniewski: It is possible that a shift in tau may lead to cognitive pathology. However, the brains of AD victims that we are studying come from cases with many years of duration of the disease. At the end stage of AD, many neurons are lost. Therefore, until I am able to study cases in the early stages of AD, I shall not be able to determine whether abnormal tau phosphorylation, or neuronal loss, or any of the long list of other pathological changes found in the brains of AD victims, is responsible for dementia. As I said before, in my opinion, dementia in AD is an end result of molecular and structural pathology initiated by β-protein amyloidosis.

Wischik: That's why we need prospective clinico-pathological correlation studies.

References

Anderson J, Blair JA, Armstrong RA 1987 The effect of age on tetrahydrobiopterin metabolism in the human brain. J Neurol Neurosurg Psychiatry 50:231

Evans DA, Funkenstein H, Albert MS et al 1989 Prevalence of Alzheimer's disease in a community population of older persons. Higher than previously reported. JAMA (J Am Med Assoc) 262:2551–2556

Goate A, Chartier-Harlin M-C, Mullan M et al 1991 Segregation of a missense mutation in the amyloid precursor protein gene with familial Alzheimer's disease. Nature (Lond) 349:704–706

Goedert M, Spillantini MG, Jakes R, Rutherford D, Crowther RA 1989 Multiple isoforms of human microtuble-associated protein tau: sequences and localization in neurofibrillary tangles of Alzheimer's disease. Neuron 3:519–526

Ishiura S, Tsukahara T, Tabira T, Sugita H 1989 Putative N-terminal splitting enzyme of amyloid A4 peptides is the multicatalytic proteinase, ingensin, which is widely distributed in mammalian cells. FEBS (Fed Eur Biochem Soc) Lett 257:388–392

Ishiura S, Nishikawa T, Tsukahara T et al 1990a Distribution of Alzheimer's disease amyloid A4-generating enzymes in rat brain tissue. Neurosci Lett 115:329–334

Ishiura S, Tsukahara T, Shimizu T, Arahata K, Sugita H 1990b Identification of a putative amyloid A4-generating enzyme as prolyl endopeptidase. FEBS (Fed Eur Biochem Soc) Lett 260:131–134

Kosik KS, Finch EA 1987 MAP2 and tau segregate into dendritic and axonal domains after the elaboration of morphologically distinct neurites: an immunocytochemical study of cultured rat cerebrum. J Neurosci 7:3142–3153

Refolo LM, Wittenberg IS, Robakis NK 1991 The cytoskeletal association of the Alzheimer's amyloid precursor protein (APP) and its possible functional significance. Soc Neurosci Abstr 17:364.2

Richards SJ, Waters JJ, Beyreuther K et al 1991 Transplants of mouse trisomy 16 hippocampus provide a model of Alzheimer's disease neuropathology. EMBO (Eur Mol Biol Organ) J 10:297–303

Wischik CM, Novak M, Jakes R, Edwards P, Harrington CR 1991 Tau/APP association in Alzheimer neurofibrillary pathology. Kumamoto Med J 42:S15–S16

Summing-up

R. J. P. Williams

Inorganic Chemistry Laboratory, University of Oxford, South Parks Road, Oxford OX1 3QR, UK

What is so good about Ciba Foundation symposia is the interdisciplinary form of the meetings. We have included the extremes of the chemist and the physician in this particular symposium. It is interesting to see what the role of each of the various groups of people has been in relation to the topic.

What the *chemist* has been trying to show is how you are limited by the chemicals in the system, if you want to think about what aluminium can do. The chemist looks at the biological components and says that these are the only ways in which he could see Al coming together within chemical species. He tries to work out some background against which all of us have to stay. At the same time, he introduces the chemicals that could be added so as to interfere with any direct effect of Al— say, silicon. This is another purely chemical question— whether or not various free ion or molecule concentration levels are present and which *must* interact with one another, from known data. I don't care whether the interaction is kinetically or thermodynamically controlled. The given statements are that certain substances would interact with aluminium, given that they were found in the same place. In this symposium I felt that the chemistry of aluminium was coming along quite nicely, and we were seeing new ways of looking at this chemistry with confidence. There are also features we need to go back to and look at, where Al could interact with the biological system, for example with the G proteins. This chemistry has nothing to say yet, of course, about the 'cause' of Alzheimer's disease.

An entirely different way of looking at the chemical correlations in biology is that of the *epidemiologist*, who must ask what chemicals, to which we are exposed, are related to the disease of interest. I found the problem of describing the environment very interesting. What levels of Al are there in our environmental supplies, water and food? This description led in effect to a complete shift to the other end of the story, very near to where the clinician is, because the (environmental) Al is now targeted on people, not chemicals. Again, we developed this aspect very nicely, and we came out with an epidemiological risk factor connecting Al with Alzheimer's disease of only about 1.5. When you arrive at a small figure like that, it means that more refined data are needed if we are to believe that it is a valid figure, and we are on dangerous ground when we connect the correlation with cause. (The risk factor relating lung cancer and smoking is near 40.) Therefore, the more epidemiological information we can get to convince us that there is something in the connection, the better. Only in this way do we act advisedly, with or without understanding.

After that, we turned to how aluminium moves into the body and how it is transported around in it. This is only partly a chemical problem. It became the *physiologist's* problem when we had to look at the various routes and types of movement as Al goes through the intestine, is dealt with by the kidney, and so on. I felt that we were often in considerable trouble here to explain exactly what was going on. The interactions at the barrier of epithelial cells, between the outside and the inside, how Al crossed cell barriers, and all such dynamics, as well as the movement of Al within the body fluids, proved difficult to analyse. We didn't really establish a firm basis here. We are still very weak generally concerning the surfaces of cells; we seem to be better informed inside or outside the cell than on the surface, in this as in other areas of biology.

Then we asked where Al is, if it does get into the brain cell. Here we went into the *analyst's* problems of looking at Al in local cellular detail and only subsequently did we turn to correlating the disposition and associations of Al with chemicals (proteins) in Alzheimer's disease. Sometimes we just had to examine critically series of analytical results done by particular methods. We then looked at the Al associated with the production of plaques, and the formation of neurofibrillary tangles in Alzheimer and other patients. We also looked more specifically at a small protein piece, namely the tau protein. Whenever we tried to leap from one type of analytical observation or another to Alzheimer's disease there remained objections. This was also true when we looked at cell metabolism and Al effects; for example, the interaction with $InsP_3$ or the GTP-dependent signalling systems. Such considerations led us suddenly out into a whole realm of interactive systems, possibly stemming from Al interactions. But there remained at the end (and this is where I think we have to conclude the discussion at the cell level) the nagging doubt as to whether Al is responsible for Alzheimer's disease, or does Al just read to us, like a stain on the system, what was in fact a prior happening?

It seems to me that we haven't convincingly proved or disproved either one of these viewpoints. One of the problems is undoubtedly the level of analytical accuracy that is now being required to describe the distribution of aluminium. One must analyse Al down to at least 10^{-4} M (3 p.p.m.). The concentration of Al ions in the affected system is perhaps only this. So we shall have to think again very carefully about the analytical procedures in cells in any of the post mortem situations. Dr Zatta suggested that we should organize better analytical cohorts and exchange information between the laboratories that are able to analyse at this low level. An interested party should organize this, to establish his credibility as well as that of others.

Against the background of information from the *chemist* as to Al speciation, from the *epidemiologist* as to exposure to Al and correlation with disease, from the *physiologist* as to the uptake routes of Al, and from the *analyst* as to the amounts and the location of Al in the brains of patients, we have had intensive discussion, of course, in which the informed *medical physicians* present have given their views on the likelihood of aluminium being a causative agent in Alzheimer's disease. My firm impression is that although we have not achieved a final solution in these discussions as in the others, we have circumscribed the problems, so that each one of us can go away to use new data and viewpoints in a further attack whence, in a few years, we should be able to place the uses of aluminium and its compounds in an appropriately known perspective and opposite the risk of Alzheimer's disease in the population. In this way we can advise on appropriate action and relieve people from unnecessary worries.

On that concluding point, let us thank the Ciba Foundation for bringing us together in this informal atmosphere to get an insight into one another's problems, as we go back to our laboratories and once more attempt to understand the interaction between aluminium and biological systems.

Index of contributors

Indexes compiled by Liza Weinkove

Subject index